GEOSPATIAL INFORMATION TECHNOLOGY
FOR EMERGENCY RESPONSE

T0186364

International Society for Photogrammetry and Remote Sensing (ISPRS) Book Series

Book Series Editor

Paul Aplin
School of Geography
University of Nottingham
Nottingham, UK

information from imagery

Geospatial Information Technology for Emergency Response

Editors

Sisi Zlatanova
Delft University of Technology, Delft, The Netherlands

Jonathan Li
University of Waterloo, Waterloo, Ontario, Canada

CRC Press
Taylor & Francis Group
Boca Raton London New York

CRC Press is an imprint of the
Taylor & Francis Group, an **informa** business

CRC Press
Taylor & Francis Group
6000 Broken Sound Parkway NW, Suite 300
Boca Raton, FL 33487-2742

First issued in paperback 2019

Typeset by Charon Tec Ltd (A Macmillan Company), Chennai, India

ISBN-13: 978-0-415-42247-5 (hbk)
ISBN-13: 978-0-367-38779-2 (pbk)

British Library Cataloguing in Publication Data
A catalogue record for this book is available from the British Library

Library of Congress Cataloging-in-Publication Data

Geospatial information technology for emergency response / editors, Sisi Zlatanova, Jonathan Li.
    p. cm. - - (International Society for Photogrammetry and remote Sensing book series ; v 6)
    Includes index.
    ISBN 978-0-415-42247-5 (hardback : alk. paper) - - ISBN 9780-0-203-92881-3 (e-book) 1. Emergency management–Geographic information systems. 2. Emergency management–Computer network resources. I. Zlatanova, Sisi. II. Li, Jonathan.

HV551.2.G46 2008
363.34'80285–dc22

2007042702

Visit the Taylor & Francis Web site at
http://www.taylorandfrancis.com

and the CRC Press Web site at
http://www.crcpress.com

Table of Contents

Acknowledgements

The editors of this volume would like to acknowledge the authors and reviewers for giving their valuable time generously to produce a state-of-the-art analysis and evaluation of the applications of geospatial information technology in emergency response. Special thanks go to the review panel members for the evaluation of the manuscripts published in this volume of the ISPRS Book Series: Peggy Agouris (George Mason University, USA), Gennady Andrienko (Fraunhofer Institute Intelligent Analysis and Information Systems, Germany), Natalia Andrienko (Fraunhofer Institute Intelligent Analysis and Information Systems, Germany), Roland Billen (University of Liège, Belgium), Thomas Bittner (The State University of New York at Buffalo, USA), Paul P. Burns (CEO Symbol Seeker Ltd, UK), Budhendra Bhaduri (Oak Ridge National Laboratory, USA), Volker Coors (Stuttgart University of Applied Sciences, Germany), Cherie Ding (Ryerson University, Canada), Suzana Dragicevic (Simon Fraser University, Canada), Matt Duckham (University of Melbourne, Australia), Janet Edwards (Swedish Rescue Services Agency, Sweden), Karen Fabbri (ICT for the Environment, European Commission), Georg Gartner (Technical University of Vienna, Austria), Michael Gruber (Vexcel Imaging Austria), Jörg Haist (Fraunhofer Institute for Computer Graphics, Germany), Christian Heipke (University of Hannover, Germany), Nanna Suryana Herman (Fraunhofer Institute for Autonomous Intelligent Systems, Germany), Shunfu Hu (Southern Illinois University at Edwardsville, USA), Bo Huang (Chinese University of Hong Kong, China), Himmet Karaman (Istanbul Technical University, Turkey), Rob Lemmens (ITC, The Netherlands), Darka Mioc (University of New Brunswick, Canada), Mir Abolfazl Mostafavi (Laval University, Canada), Shailesh Nayak (Indian Space Research Organization, India), David Prosperi (Florida Atlantic University, USA), Alias Abdul Rahman (Universiti Teknologi Malaysia), Brengt Rystert (University of Gävle, Sweden), Gunter Schreier (DLR, Germany), Jie Shan (Purdue University, USA), Ingo Simonis (University of Muenster, Germany), Manolis Stratakis (FORTHnet SA, Greece), Stefan Voigt (DLR, Germany), Monica Wachowicz (Wageningen University, The Netherlands), Bartel van de Walle (Tilburg University, The Netherlands), Stephan Winter (University of Melbourne, Australia), Bisheng Yang (University of Zurich, Switzerland), Xiaofang Zhou (University of Queensland, Australia), and Alexander Zipf (Mainz University of Applied Sciences, Germany). Special thanks go to David Prosperi (University of Florida, USA) for his friendship and support. We are grateful to Wei Xu (Delft University of Technology) for his help in the layout of the book. The advice and counsel of Paul Aplin, ISPRS Book Series editor (2004–2008) have been extremely valuable to improve the quality of this book.

Contributors

Lars Bodum, Centre for 3D GeoInformation, Aalborg University, Fibigerstraede 11, DK-9220 Aalborg, Denmark, Tel: +45-9635-9797, Fax: +45-9815-2444, E-mail: lbo@3dgi.dk

Michael A. Chapman, Department of Civil Engineering, Ryerson University, 350 Victoria Street, Toronto, Ontario M5B 2E3, Canada, Tel: +1-416-979-5000, ext. 6461, Fax: +1-416-979-5122, E-mail: mchapman@ryerson.ca

Michel J.M. Grothe, Department of Geo-Information and ICT, Ministry of Transport, Public Works and Water Management, Derde Werelddreef 1, 2622 HA Delft, The Netherlands, Tel: +31152757063, Fax: +31-152757576, E-mail: michel.grothe@rws.nl

Gerhard Gröger, Institute of Geodesy and Geoinformation, Department of Geoinformation, University of Bonn, Meckenheimer Allee 172, 53115 Bonn, Germany, Tel: +49-228-731764, Fax: +49-228-737153, E-mail: groeger@ikg.uni-bonn.de

Stephan Heuel, Ernst Basler + Partner AG, Zollikerstr. 65, 8702 Zollikon, Switzerland Tel.: +41 44 395 11 95, Fax: +41 44 395 1234, E-mail: stephan.heuel@ebp.ch

Mikael Jern, Department of Science and Technology, Linkoping University, 601 74 Norrköping, Sweden, Tel: +46-11363104, E-mail: mikje@itn.liu.se

Russ Johnson, ESRI, 380 New York Street, Redlands, CA 92373, USA, Tel: +1-909-793-2853, ext. 1836, E-mail: russ_johnson@esri.com

Norman Kerle, Department of Earth System Analysis, International Institute for Geo-Information Science and Earth Observation (ITC), P.O. Box 6, 7500 AA Enschede, The Netherlands, Tel: +31-53-4874476, Fax: +31-53-4874335, E-mail: kerle@itc.nl

Michael Kevany, PlanGraphics Inc., 615 Bennington Lane, Silver Spring, Maryland 20910, USA, Cell: +1-301-466-4335, Fax: +1-301-588-5979, E-mail: mkevany@plangraphics.com

Erik Kjems, VR Media Lab, Aalborg University, Fibigerstraede 11, DK-9220 Aalborg, Denmark, Tel: +45-9635-8079, E-mail: kjems@vrmedialab.dk

Thomas H. Kolbe, Institute for Geodesy and Geoinformation Science, Technical University Berlin, Straße des 17. Juni 135, 10623 Berlin, Germany, Tel: +49-30-3142-3274, Fax: +49-30-3142-1973, E-mail: kolbe@igg.tu-berlin.de

Kris Kolodziej, IndoorLBS.com, 71 Walter Street, San Francisco, CA 94114, USA, Tel: +1-415-621-6672, Fax: +1-415-704-4099, E-mail: kkolodziej@indoorlbs.com

Harry C. Landa, Geo-Information and ICT, Ministry of Transport, Public Works and Water Management, Postbus 5023, 2600 GA Delft, Tel: +31-15-2757147, Cell: +31-6-51398439, E-mail: harry.landa@rws.nl

Jiyeong Lee, Department of Geoinformatics, The University of Seoul, 90 Jeonnong-dong, Dongdaemun-gu, Seoul 130-743, South Korea, Tel: +82-2-2210-5750, Fax +82-2-2246-0186, E-mail: jlee@uos.ac.kr

Jonathan Li, Department of Geography, Faculty of Environmental Studies, University of Waterloo, 200 University Avenue West, Waterloo, Ontario N2L 3G1 Canada, Tel: +1-519-888-4567, ext. 34504, Fax: +1-519-746-0658, E-mail: junli@fes.uwaterloo.ca

Yuan Li, School of Architecture and Civil Engineering, Xiamen University, Xiamen, P. R. China 361005, Tel: +1-86-1306-309-5838, E-mail: liyuan79@xmu.edu.cn

Robert MacFarlane, Training & Doctrine, Emergency Planning College, Cabinet Office Civil Contingencies Secretariat, The Hawkhills, Easingwold, York, YO61 3EG, United Kingdom, Tel: +44-1347-825017, Fax: +44-1347-822575, E-mail: robert.macfarlane@cabinet-office.x.gsi.gov.uk

Guillaume Noel, LIRIS: Research Center for Images and Information Systems, INSA-Lyon, Bât. Blaise Pascal, 20 avenue Einstein, 69621 Villeurbanne Cedex, France, E-mail: noel.guillaume@insa-lyon.fr

Chris Parker, Research Labs, Room C530, Ordnance Survey, Romsey Road, Southampton SO16 4GU, United Kingdom, Tel: +44-23-8079-2878, Fax: +44-23-8030-5072, E-mail: chris.parker@ordnancesurvey.co.uk

Norbert Pfeifer, Institute of Photogrammetry and Remote Sensing, Vienna University of Technology, Gußhausstraße 27–29, 1040, Vienna, Austria, Tel: +43-1-58801-12201, Fax: +43-1-58801-12299, E-mail: np@ipf.tuwien.ac.at

Chris Phillips, Research Labs, Room C530, Ordnance Survey, Romsey Road, Southampton SO16 4GU, United Kingdom, Tel: +44-23-8030-5725, Fax: +44-23-8030-5072, E-mail: Chris.Phillips@ordnancesurvey.co.uk

Lutz Plümer, Institute for Geodesy and Geoinformation, Department of Geoinformation, University of Bonn, Meckenheimer Allee 172, 53115 Bonn, Germany, Tel: +49-228-731750, Fax: +49-228-737153, E-mail: pluemer@ikg.uni-bonn.de

Hardy Pundt, Faculty of Automatisation and Computer Science Geoinformatics, Database Systems, University of Applied Studies and Research, Friedrichstr. 57–59, 38855 Wernigerode, Germany, Tel: +49-3943-659-336, Fax: +49-3943-659-399, E-mail: hpundt@hs-harz.de

Sylvie Servigne, LIRIS: Research Centre for Images and Information, Systems, INSA-Lyon, Bât. Blaise Pascal, 20 avenue Einstein, 69621 Villeurbanne Cedex, France, Tel:+33-4 72-4384 83, Fax: +33-472-438713, E-mail: sylvie.servigne@insa-lyon.fr

John G.M. Steenbruggen, Department of Geo-Information and ICT, Ministry of Transport, Public Works and Water Management, Derde Werelddreef 1, 2622 HA Delft, The Netherlands, Tel: +31-15-275-7301, Fax: +31-15-275-7576, E-mail: john.steenbruggen@rws.nl

Yam Khoon Tor, School of Civil & Environmental Engineering, Nanyang Technological University, Block N1, #01a-06, Nanyang Avenue, 639798, Singapore, Tel: +65-6790-4743, Fax: +65-6791-0676

Kasper M. van Zuilekom, Centre for Transport Studies, University of Twente, P.O. Box 217, 7500 AE Enschede, The Netherlands, Tel: +31-53-489-1098, Fax: +31-53-489-4040, E-mail: K.M.vanZuilekom@utwente.nl

Frans G. von der Dunk, Space Low, University Nebraska, USA; Director Black Holes (http://www.black-holes.eu), Witte Singel 85, 2311 BP Leiden, Tel: +06-48460318, E-mail: f.g.vonderdunk@law.leidenuniv.nl

Yun Zhang, Department of Geodesy and Geomatics Engineering, University of New Brunswick, P.O. Box 4400, Fredericton, New Brunswick E3B 5A3, Canada, Tel: +1-506-453-5140, Fax: +1-506-453-4943, E-mail: yunzhang@unb.ca

Qing Zhu, State Key Laboratory of Information Engineering in Surveying Mapping and Remote Sensing, Wuhan University, P.O. Box C310, 129 LuoYu Road, Wuhan, Hubei 430079, China, Tel: +86-27-6877-8322, Fax: +86-27-6877-8969, E-mail: zhuqing@lmars.whu.edu.cn

Sisi Zlatanova, OTB, Section GIS-technology, Delft University of Technology, Jaffalaan 9, 2628 BX, Delft, P.O. Box 5030, 2600 GA Delft, The Netherlands, Tel: +31-15-278-2714, Fax: +31-15-278-4422, E-mail: s.zlatanova@tudelft.nl

Mark H.P. Zuidgeest, International Institute for Geo-Information Science and Earth Observation, P.O. Box 6, 7500 AA Enschede, The Netherlands, Tel: +31-53-487-4444, Fax: +31-53-487-4400, E-mail: zuidgeest@itc.nl

Geospatial Information Technology for Emergency Response – Zlatanova & Li (eds)
© 2008 Taylor & Francis Group, London, ISBN 978-0-415-42247-5

Introduction

Sisi Zlatanova
Delft University of Technology, Delft, The Netherlands

Jonathan Li
University of Waterloo, Waterloo, Ontario, Canada

Disaster management is generally understood to consist of four phases: Mitigation, Preparedness, Response and Recovery. Mitigation describes activities aimed at reducing the occurrence of emergency situations (e.g., construction specifications for buildings to be able to resist earthquakes, dikes to prevent flooding, etc.). Preparedness focuses on active preparation among rescue forces (e.g., police, ambulance, fire) for emergency situations. Response is the acute phase occurring after the event. Recovery includes all arrangements to remove detriments and a long-term inventory of supplies to deal with irreversible damage. While all phases are interrelated and important, the *response* and *recovery* phrases are often viewed as the most critical in terms of saving lives. Time constraints, stress, equipment with limited capacities (power, display, etc.), and the involvement of numerous organizations in post-disaster operations are only a few of the factors complicating response and recovery. The timely provision of geospatial information can greatly help in the decision-making process, save lives and aid citizens.

The aim of this volume is to share exciting technological advances that allow wider, faster and better utilization of geospatial information in emergency *response* situations. "Fast," "context-aware," and "data integration" are key attributes of emergency response models and decision making frameworks. The chapters in this book describe current accomplishments and challenges in providing geospatial information with these attributes.

The book is organized in six parts. The first part describes practice and legislation, and focuses on the utilization of geospatial information in recent disaster events, as well as resulting legislative attempts to share and access data. The second part focuses on data collection and data products. The third part describes data management and routing in 3D. The fourth part focuses on emerging technologies, including positioning, virtual reality and simulation models. The fifth part focuses on the integration of heterogeneous data. The final part reports on various applications and solutions.

Part 1 Practice and Legislation

In the first chapter, Kevany describes the involvement of an experienced Geographic Information System (GIS) specialist as a participant in response efforts associated with several well known emergencies, such as the 2001 terrorist attack in the USA, the 2004 tsunami in South Asia and the 2005 Hurricane Katrina in the USA. The chapter outlines the "lessons learned" and the corresponding geospatial developments in the USA since 9/11. In chapter 2, von der Dunk provides some practical context to these lessons, concentrating on issues of copyrights, access to remote sensing data, responsibilities and liabilities of data providers, and security and dual-use issues. This chapter also describes the background of charters, directives and resolutions established by international organizations and the United Nations.

Part 2 Data collection and products
The second part includes three chapters that focus on airborne, satellite, and terrestrial sensors and techniques appropriate for acquiring data during an emergency response.

Kerle et al. provide an extended overview of airborne sensors, but also provide important definitions and constraints that frame the other two chapters in this section. Zhang and Kerle focus on satellite technology, including ongoing international remote sensing initiatives that emphasize rapid data collection. Li and Chapman provide an overview of the operational principles and the state-of-the-art development of terrestrial mobile mapping technology. As innovations in this field are relatively new, this chapter also outlines future research needs and describes necessary developments in mobile mapping.

Part 3 3D data management
While significant progress has been made in processing two-dimensional data, and numerous solutions for rapid 3D visualization are also available, the management and analysis of 3D data remains a major challenge. Simply put, the variety of data models, their resolution, details, and methods of representation (B-reps, voxel, Constructive Solid Geometry), etc., are larger, and there are few generally accepted or commercial systems available. The focus in this section is on innovative ideas in the areas of indexing, 3D management and analysis (and more specifically of 3D routing algorithms and visualization), which are emerging technologies in disaster management.

Servigne et al. present two approaches for fast structuring of sensor network measurements used for the monitoring of seismic activities. Arguing that real-time processing of such measurements is most efficient if organized in the main memory of the computer, two indexing schemas have been developed and tested for spatio-temporal data collected from fixed and agile sensors, which can handle updates and perform spatio-temporal analysis by giving preference to the last collected measurements. Lee and Zlatanova argue that reliance on only one data model (representation) might be insufficient for emergency systems. A hybrid model is proposed that combines the benefits of recently investigated and implemented 3D models. Focusing on geometry, topology and the network, they argue that formally described CAD models can be semantically simplified and integrated in emergency response models. Finally, Zhu et al. focus on emergency routing for vehicles. They are confident that 3D navigation is much more efficient and less ambiguous compared to the commonly applied two-dimensional approach. The authors present their innovative 3D algorithm and extensively discuss the construction of 3D dynamic road networks, indicating that this has to be done automatically from both existing and dynamic data.

Part 4 Emerging technologies
Emerging technologies such as 3D indoor positioning, virtual reality technology for training, collaboration and command control, and advanced visuals for decision makers are described, discussed and evaluated in three chapters.

Kolodziej explores the most promising technologies for indoor and combined indoor-outdoor positioning, and the infrastructures needed to support them. TV-GPS positioning technology is featured in this chapter, which also describes the design and implementation of several positioning systems and real-world applications, and shows how these tools are being used to solve problems that can be related to emergency responses. Kjems et al. elaborate on the use of Virtual Reality (VR) techniques for training of first responders in emergency response situations. In addition to an introduction to VR and several examples, the chapter includes a discussion of immersive VR technologies (those that cannot be obtained by ordinary computer monitors). Finally, Jern addresses the emerging field of Visual Analytics (VA), the science of analytical reasoning facilitated by interactive visual interfaces and creative visualization. VA tools help the user detect both expected and unexpected events, provide timely, defensible, and understandable assessments, and finally communicate risk effectively for action. VA tools are illustrated via a specific forecasting tool, "FloodViewer." Perhaps most importantly from the point of view of "context-awareness," the tool provides collaborative visualization tools enabling end users to view and discuss the forecasting results in real time across the network before finally interacting with the media, the police, other officials and the public.

Part 5 Integration of heterogeneous data

Part 5 focuses on semantic interoperability and access to data. Pundt provides an excellent overview of formal ontology, including languages to define ontology and efforts of geospatial communities for establishing ontology standards. Kolbe et al. pay special attention to the exchange of 3D models. CityGML is a semantic model (a kind of ontology) for representing city objects that can be extended easily to incorporate underground objects such as geological formations, underground utilities and construction. CityGML has the potential to contribute to faster employment and dissemination of 3D models during emergency response.

Parker et al. discuss the philosophy of the Ordnance Survey (one of the largest data providers in managing and supplying data for emergency preparation, response and recovery in the United Kingdom). The Ordnance Survey has adopted an elaborate framework to provide data in appropriate forms at any time when requested during emergencies. This chapter focuses on experiences with informational needs at different command and control levels.

Part 6 Applications and solutions

The final part contains three chapters that describe how geospatial information technology is used in different disaster scenarios, illustrated in case studies of transport accidents, floods and fires.

Grothe et al. present the activities of the Ministry of Transport, Public Works and Water Management in The Netherlands. This chapter outlines the use of geospatial information and geoservices in disaster management within the field of national transport and water management. The adopted concepts and corresponding developments are discussed through a crisis scenario expressed by a number of scenes. Zuilekom et al. present a decision support system for preventive evacuation of a population from a dyke-ring area. A framework for modelling evacuations is presented initially, and several methods are presented to assist practitioners in designing evacuation plans. A static quick-scan is developed as an alternative for time-consuming dynamic model runs. The case study focuses on the dike-ring area of Flevoland, The Netherlands. Finally, Johnson examines how GIS helps the fire service to meet the needs of the community.

The chapters in this book are aimed at researchers, practitioners, and students who work in a variety of disciplines with geospatial information technologies for emergency response, and they represent the very best of current thinking from a number of pioneering studies over the past four years. The origins of this book can be traced to the work titled "Knowledge-based technology for improving Urban and Urgent Risk Management (U2RM)" in October 2003, and subsequently to a series of conferences including: the 1st International Symposium on Geoinformation for Disaster Management (Gi4DM2005) held in Delft, The Netherlands, March 21–23, 2005; the 2nd Symposium (Gi4DM2006) held in Goa, India, September 25–26, 2006; and the 3rd Symposium (Gi4DM2007) held in Toronto, Canada, May 23–25, 2007. In addition, several other events provided inspiration and tentative results, including: the Bentley Research Seminar held in May 2005 in Baltimore, USA; the workshop on "Technology for Emergency Response" and the workshop on "Tools for Emergencies and Disaster Management" held in September 2005; "GIS Planet 2005" held in July 2005 in Estoril, Portugal; and the Vespucci Summer School held in July 2005 in Fiesole, Italy. Finally, the efforts of our own ISPRS WG IV/8 over the past two years to continue research and development in the area of spatial data integration for effective emergency services and disaster management have provided guidance and inspiration. The book itself is the result of a collaborative effort involving 33 researchers located in ten countries.

About the editors

Dr. Sisi Zlatanova

Associate Professor at the GIS Technology Section, Delft University of Technology, and currently leading a theme group on 'Geo-information for Crisis Response'. She holds a Ph.D degree in 3D GIS for urban modelling from the Graz University of Technology, Graz, Austria. Prior to joining Delft University, she has worked as software programmer at the Central Cadastre in Sofia, Bulgaria, as Assistant-Professor at UACG, Sofia, Bulgaria, and as researcher at the International Institute for Geo-Information Science and Earth Observation (ITC), Enschede, The Netherlands. Her research interests include the use of spatial technologies in emergency response, in particular where special attention is given to the third dimension: 3D object reconstruction, 3D data structures and geo-databases, 3D spatial relationships (topology) and 3D visualization (VR and AR). Sisi Zlatanova is author of numerous publications on 3D modelling and technology for emergency response, and is currently serving as chair of ISPRS WG IV/8 "Spatial data integration for emergency services" (2004–2008). She is co-editor of the books 'Geo-information for Disaster Management" (Springer Verlag, 2005) and "Geomatics Solutions for Disaster Management" (Springer Verlag, 2007).

Dr. Jonathan Li

Associate Professor and head of the Remote Sensing and Geospatial Technology Lab at the Department of Geography, University of Waterloo, Canada. He holds a PhD degree in Geomatics Engineering from the University of Cape Town, South Africa. Prior to joining the University of Waterloo, he was Associate Professor and Director of the GeoVELab at Ryerson University in Toronto, Canada. His research ranges from urban remote sensing to distributed geospatial information services. His current research interests include 3D city modeling, feature extraction from image and LiDAR data, object-oriented land use classification and change detection, and spatial sensor web for disaster monitoring. Jonathan Li is currently serving as Co-Chair of ISPRS WG IV/8 "Spatial data integration for emergency services" (2004–2008). Jonathan Li was Conference Co-Chair of the Joint CIG/ISPRS Conference on Geomatics for Disaster and Risk Management, May 23–25, 2007, Toronto, Canada and co-editor of the book "Geomatics Solutions for Disaster Management" (Springer Verlag, 2007).

Part 1
Practice and legislation

Geospatial Information Technology for Emergency Response – Zlatanova & Li (eds)
© 2008 Taylor & Francis Group, London, ISBN 978-0-415-42247-5

Improving geospatial information in disaster management through action on lessons learned from major events

M.J. Kevany
PlanGraphics, Inc., Silver Spring, Maryland, U.S.A.

ABSTRACT: Lessons are an important source of information for the development or improvement of GI solutions that support disaster management. Because disasters are so unique, their activities and impacts vary so widely, and the conditions created in a disaster are so chaotic, they do not lend themselves easily to conventional investigation, modeling, and analysis. Therefore, lessons derived from the experiences of disasters can provide very useful guides for development efforts. This chapter presents a series of lessons drawn initially from the response to the attack on the World Trade Center and augmented with lessons from more recent events. A review of these lessons, first promulgated much earlier, and the progress toward adoption are presented to indicate how much is yet to be done. The chapter also presents a review of recent GI developments that contribute to improvements in the areas identified in the lessons. The chapter closes with an identification of significant issues still confronting the effective use of GI in disaster management.

1 INTRODUCTION AND BACKGROUND

This chapter is written, not as the findings of an academic research effort but as the pragmatic views of a practitioner in the field of geospatial information (GI) for disaster management. Rather than reporting on research, it is a call to perform necessary research from a person who experiences the lack of knowledge and of effective techniques in the field of GI for emergency management.

GI is increasingly used in support of all aspects of disaster management but in the main without benefit of sound knowledge-based or careful research. I call for the academic community to fill this knowledge gap and to conduct research into best practices for GI use, effective IT architectures for application to DI, mechanisms for overcoming shortcomings in GI technology as applied to disaster management, and shortcomings such as capabilities for acquiring and managing real-time data.

To assist in targeting the research that I call for, I provide a series of lessons learned from experiences with GI in disaster management, lessons learned in the 'heat of battle' so to speak, in the highly charged environment of actual emergency response operations, lessons learned in integrating GI into the operating environment and with the disaster management systems, lessons learned in exchanging critical disaster information between organizations, overcoming the challenges of interoperability between organizations and integrating GI with enterprise technologies developed to overcome interoperability limitations, and dealing with the wide range of technically sophisticated and technically resistant personnel involved.

The lessons addressed in this chapter are derived from actual experience with GI in the field of emergency – some from actual emergency situation use, in particular, the experiences of the World Trade Center attack of September 11, 2001. Some are also from emergency exercises, and some are from development experiences, such as integration with other technology areas.

Following the attack on the World Trade Center (WTC) on September 11, 2001 ('9/11', as it came to be known), numerous critiques, reports, and papers were produced about the experiences and lessons learned, especially in the area of the use of geospatial information (GI) technologies and geographic information systems (GIS). GIS was used extensively in that response. The experiences of that event raised awareness of the value of GI to emergency response and raised issues from the GI perspective of how best to provide GI value to emergency management. GI support to emergency operations can assist in reducing loss of life, minimize the damage caused, facilitate recovery, and mitigate negative environmental damage caused by emergency events. Many lessons were identified from the WTC experience in various areas from the technology itself to organization and procedures. One of the articles written on the subject was 'Geospatial information for disaster management: Lessons from 9/11' (Kevany 2005). In that article, this author identified and classified a series of lessons that emerged from his experiences in the New York City Emergency Operating Center during the response to the WTC attack.

The following section of this chapter reemphasizes the importance of improving GI use in emergencies by learning from the lessons of prior events and incorporating them, into best practices for future events. To that end, this chapter will review the lessons that were defined in the previous article from the current perspective to indicate the current state of the lessons, to determine what, if anything, has been done to benefit from them, and to renew interest in incorporating the lessons. The chapter also includes a review of more recent events and identifies lessons learned from them, many of which are the same or similar to the prior lessons that are yet to be implemented. While there is much to be gained from the lessons of actual practice in events, the GI and related technologies are evolving rapidly and offer many new opportunities to either implement the lessons or to add new response capabilities. The chapter includes a review of developments in GI technology as applied to emergency management since 9/11/01.

The lessons are organized into the original eight classes:

- General lessons
- Operational lessons
- Organizational lessons
- Data lessons
- Technical lessons
- Customer lessons
- Logistical lessons
- Special challenges.

The events of 9/11/2001 had a tremendous impact on emergency management and response in the United States and all over the world. Funding for emergency management has been increased dramatically since then; a large number of technologies and advances have emerged to be applied to this field; and considerable research is under way to find new solutions and applications of technology to emergency activities. While funds have been applied to equipment, training, and other aspects, technology has been a particular focus of the increases in funding and development.

Unfortunately, the design and development of these technologies has too often had inadequate participation by emergency experts so adoption of technology for active, beneficial use in emergency response has not been as universal as it might have been with more sensitivity to emergency needs and particularly the emergency operating environment.

The focus on technology has occurred without a parallel re-engineering of emergency operating practices and procedures that would maximize the benefit of the technologies. Many of the lessons described in the prior article have not been incorporated into an improved GI in emergency situations today.

This chapter will begin with a review of the classes and individual lessons proposed in the earlier article and will comment on the progress and current situation for each. Next, the chapter will move to major events since 9/11/2001 and lessons learned from them. The chapter will then explore specific areas of GI development relevant to emergency management. A series of recommendations related to implementing the lessons learned for future emergency response will be proposed, and the chapter will close with a summary and identification of the most significant challenges still facing the field of GI in emergency management.

1.1 *Initial observations*

The first thing to be noted is that that 9/11 event led to profound changes in many aspects of our world and more specifically in the use of technology for emergency management. A great deal of attention has been focused on emergency management since that event. The focus has generally been on where the money has been made available and that has generally been on technology. Not nearly enough attention has been given to the effective deployment and use of GI in emergencies. GI has not yet been widely incorporated into emergency preparedness. Few emergency operating plans have been modified to incorporate effective GI use. The lessons of prior emergencies have not been learned or carried out.

Tremendous funding has been made available for emergency technology, and large numbers of technology companies have developed or adapted technology for emergency applications. In fact, it appears that the amount of technology suddenly made available has in some ways overwhelmed the emergency community's ability to assimilate it in areas where few on no technologies were previously available and where the technologies that were available were rudimentary. There are now too many choices available. Technology has moved well ahead of the emergency community's ability to assimilate it.

A very important positive development is that GI has become recognized as a high priority aspect of emergency management for the future. As GI capabilities and products have become available in emergencies, professionals in the field have recognize the value and are ranking GI as a high priority in requirements analyses and requests for funding. As an example, the National Capital Region Interoperability Project, a major 17-jurisdiction effort around the Washington, DC area, determined in the initial phase of its effort that GIS is one of the two top-priority areas for which data exchange is essential. While this is just one example of a specific situation, there is growing recognition of GI importance for all phases of emergency management across the emergency management community.

A critical and persistent vulnerability of GI technology relative to support of emergency management and operations is the temporal aspect. A special challenge of emergency response is the critical nature of time. Seconds and minutes have an impact on emergency response beyond that of days and weeks in conventional GIS use. As stated by a Washington, DC, GIS liaison to the Emergency Management Agency 'If it's not available within 3 minutes, it is of no value.' And yet, as configured and operated for normal day-to-day activities, GI is not optimized to deliver within this time demand. And GI technologies available today are only beginning to be applied to real-time demands such as that of emergency management. This demand is not only for rapid, real-time delivery but also acquisition and management of data in real time and providing the proper sequencing of data in real time, which are so important to emergency management.

A major issue emerging from recent emergencies of major proportions is information exchange and interoperable communications and data exchange. The magnitude of the WTC and major recent natural disasters elevated the requirements from internal, within an individual jurisdiction, to collaboration among multiple organizations. Collaboration among multiple organizations is a necessary aspect of large-scale emergency operations, and such collaboration can benefit greatly from the availability of GI. This poses special challenges to the GI field. While the National Spatial Data Infrastructure program (NSDI) had been initiated in the United States and elsewhere with the goal of developing comprehensive GI data stores that would be valuable in emergencies, as well as normal operations, resources for actual implementation are limited, and much has yet to be accomplished. Recent technology developments offer the opportunity to shift from a massive central data resource to a federated network of existing sources that can be easily accessed through enterprise and service-oriented technology architectures.

A significant aspect of this approach and response to major emergencies in general is interoperability of data exchange. This involves the capability to overcome differences in systems and data structures among organizations involved in an emergency response. Important projects are under way to develop and test technologies and techniques for interoperable data exchange during emergencies.

Perhaps the most significant issue still facing the GI community in regard to emergency management is effective assimilation of the technology into emergency operating procedures. All of this technology is of little value unless the emergency community understands and uses it effectively, and that has not yet been achieved. Emergency personnel should be educated in GI, and GI procedures should be developed to be incorporated into the standard operating procedures of emergency management and response organizations. The education should be structured to address the specific roles of the individuals, addressing the realistic requirements of each job since GI technology is complex and comprehensive and emergency personnel need to understand only specific aspects applied to their assignments. GI applications that can be easily used by emergency personnel to directly support their activities are also needed.

The difference between GI technology and the use of the technology is significant. Having this powerful technology available is essential, but its value comes from its use. The greater the usefulness of the technology and the recognition of that usefulness, the greater the value can be achieved. So the challenge to the GI community is not to just make GI technology available to the emergency community, but also to design and develop the technology in a manner that is truly useful, and to educate the emergency community to that usefulness and how to apply the technology usefully.

2 REVIEW OF PRIOR LESSONS

2.1 *Overview and framework for review of lessons*

The following section provides a review of the progress made and current situation for each of the lessons identified in the prior article. Each of the earlier lessons is listed along with comments on the progress since 9/11/2001. The organizational structure and classification from the original article are replicated here. The comments on progress and the current situation are based on the author's experience and investigations of activities in the field in selected jurisdictions and the national government in the United States. It should be noted that the reporting is based on experiences and observations of practitioners in the field and not on rigorous or exhaustive investigation. Such rigorous investigation is absent at this time and represents a significant gap in knowledge that is essential to improvement of the situation in the future. Little information is available on best practices to those in the field of emergency GI at this time.

2.2 *General lessons*

2.2.1 *Be prepared*
Preparedness continues to be recognized as a most significant component of emergency management. However, great progress has not been made in the area of GI preparedness for emergency use. GI capabilities have not been widely integrated into emergency operations. Few organizations have developed an emergency-specific database of risk conditions, resources, critical facilities, evacuation routes, etc.

Also, lessons from actual emergencies and exercises have not adequately been incorporated in emergency operations as evidenced in the United States by the many failures in the Katrina Hurricane response, such as the absence of a common base of GI data across the multiple jurisdictions impacted, and the recent evaluation of progress in implementation of the recommendations of the 9/11. (The original report, *The 9/11 Commission report: Final Report of the National Commission on Terrorist Attacks Upon the United States*, is available at http://www.911commission.gov/report/911Report.pdf) (The 9/11 Commission 2004).

2.2.2 Recognize the life and death nature of the emergency
It is very difficult for GI experts to truly understand the life and death nature of emergency and its significance to their work until they have been exposed to actual emergency operations and, while the numbers are growing, there are still too few GI experts with hands-on emergency response experience. And few emergency experts have developed skills with the use of GI technologies.

2.2.3 Incorporate GIS and geospatial information technologies into emergency plans
Unfortunately this is a lesson for which very little progress has been made. Where GI is employed actively in emergencies, the expedient solution has typically been the posting of a GI expert in the Emergency Operation centre (EOC) to provide support on an ad-hoc, as needed basis rather than incorporating GI into the organization's operating procedures. Few current emergency plans have defined GI-specific actions.

2.2.4 Maintain off-site backup of data and resources
Emergency organizations are very aware of the need for back-up capacities in their operations, thus those employing GI generally do establish back-up sources for data, though GI system back-up is rarely available. The Internet has opened the potential for access to system capabilities at remote locations, and some emergency industry firms are offering remote GI services through the Internet. Recent deployment of Enterprise Service Bus technology provides opportunities for related organizations to provide backup for each other.

2.2.5 Recognize the impact of heightened security on disaster operations
Considerable investment is being made in security for emergency response facilities and systems in the United States since 9/11. This has had both positive and negative effects. While secure facilities and access to information are beneficial to all types of emergencies, some security limitations on data availability are having an adverse impact on normal day-to-day operations of governmental entities and on the availability of information to the public.

2.2.6 Plan disaster operations to be as close to normal procedures as practical
There is a general recognition of the relationship between day-to-day and emergency activities. Since most current use of GI in emergencies is made on an ad hoc basis, the operations are based almost entirely on normal procedures. However, these procedures are often too limiting and lack effectiveness in the emergency environment.

2.3 Operational lessons

2.3.1 Paper maps are in demand and cause production bottlenecks
Recent emergency events have exhibited a continued dependence on the production of paper maps. Emergency personnel have not yet adopted the direct use of GI, and operations generally involve production of paper maps that are provided to emergency personnel. High-volume production during peak emergency operations is still a limiting factor in the employment of GI in emergencies. The time urgency of emergency response does not allow for spreading map production over an extended time, as is the typical case in normal operations.

2.3.2 Pre-define standard products
Emergency maps integrate data from numerous sources each of which has its own style for display and symbology, and considerable overlap exists in the use of symbols with different definitions across these standards. A national program, the FGDC Homeland Security Working Group, is developing a standard symbology set to be applied to emergency mapping for the emergency management and first responder communities. Information on this symbology is available at: http://www.fgdc.gov/HSWG/index.html (Federal 2005).

Some progress has been made in the definition of at least a few of the most critical emergency maps. In the United States, there has been activity to define the Minimum Essential Data Set

(MEDS), the most necessary data for support of emergency response. There is encouragement for the development of MEDS by local governments, though as yet the actual development is limited.

A national Homeland Security Infrastructure Program (HSIP), has been working on the development of a standard set of GI data on critical infrastructure such as transportation and utilities, though actual development lags in this area as well (Homeland 2004).

Almost all maps used in an emergency are produced on an ad hoc basis with little preplanning or design. There is great dependence, therefore, on the format and content of data maintained for non-emergency purposes.

There continues to be a gap in knowledge of maps and cartography on the part of emergency personnel. Emergency responders continue to be dependent on the GI expert for determination of cartographic format during emergencies, and often the maps are not as effective as they might be if emergency personnel were exposed to design factors or educated in the potential of maps.

2.3.3 *Logistical support and management are important*

Improvements are being made in the development, deployment, and use of mobile computing devices for delivery of GI data and capabilities to those operating in the field. There is still a considerable need for improvement both in terms of the capabilities of the technology and the diffusion of the technology.

GI technology is dependent on development of general communications technology for wireless transmission of data. The deployment of GI to mobile devices is currently dependent on the availability of the devices for emergency operational purposes other than for GI exclusively.

The capability to manage the large volume of GI data, and particularly the ability to display map data in a readable and effective format, often exceeds the requirements of general emergency operations, and this poses special challenges for mobile systems to support GI.

2.3.4 *Anticipate future needs as the event evolves*

Given the very limited advances in the sophisticated use of GI technology, virtually no attention has been given to this aspect. With the limited capability to capture, manage, and display information of the evolving emergency in real time during an emergency, GI technology has not yet been extended to the projection of future conditions.

A recently emerging area is the development of 'alert' capabilities. Employing sophisticated database management, search engine, and business intelligence software, capabilities are being developed to recognize emergency conditions or significant changes as they arise. Alerts can be triggered and provided to emergency managers for action. These capabilities have not yet been widely deployed.

2.4 *Organizational lessons*

2.4.1 *Provide strong geospatial information leadership during a disaster*

This situation varies based on the individual organization involved in the emergency. Little has been done to develop emergency GI leadership through training programs or other mechanisms, therefore, strong leadership is still dependent on the skills and initiative of individual GI experts. Lacking emergency training and with little opportunity to gain experience, GI persons are generally at a disadvantage in terms of exercising leadership relative to emergency managers and responders. GI is not identified as a specific emergency response function in most organizations, therefore, no management or leadership has been assigned formally for GI or the GI role in emergency response.

2.4.2 *Assign geospatial information expert persons or units for 24/7 emergency operations*

Some organizations have recognized the round-the-clock characteristic of emergency response and have formally or informally identified multiple persons to provide staffing for emergency response on a 24/7 basis. Augmentation capabilities from unaffected jurisdictions, other levels of government, or from volunteers has emerged as a solution in recent significant disasters. The GIS Corps was founded in 2003 to provide a formal mechanism for arranging volunteer GI support where disasters overwhelm the capabilities of the local GIS organizations. Its mission is to coordinate short-term, volunteer-based GIS services to underprivileged communities. The GIS Corps has provided assistance in numerous disasters around the world. More information is available from: giscorps.org.

2.4.3 *Provide disaster managers and personnel with appropriate training in the use of GIS*
This is an area for which little progress has been made. A significant gap in the knowledge of, or familiarity with, GI technology or its products by emergency persons still exists limiting the usefulness of GI capabilities and products. The growth in use of GI in emergencies is still primarily limited to production of maps and complete dependence on the GI expert to support emergency operations. Fully effective usefulness of GI will not be achieved until the emergency personnel understand the technology and, particularly, its products, at a level that will allow them to employ GI as a primary tool in their activities. To achieve this, an effective emergency GI training program must be developed and executed. Even as emergency personnel develop an acceptable level of skill with the use of GI, there will be a continuing need for the availability of GI experts to handle the complex or sophisticated GI operations and to develop appropriate emergency GI applications.

2.4.4 *Recognize distributed operations*
GI, where it is employed in emergencies, is limited almost exclusively to map production in or near the EOC, and the distribution of these maps is primarily limited to those operating in the EOC or command centers. Most jurisdictions have established no specific capabilities for distribution of GI to the various locations at which it would be useful. Mechanisms for distribution to those operating in the field have not generally been established, and, where delivery to the field is made, it is generally on an ad hoc basis through mechanisms determined on the spot. A few organizations have acquired mobile command or other vehicles that include GI capabilities for operation on-site or in the vicinity of emergency field operations. These provide direct GI at the scene of the emergency.

2.5 *Data lessons*

2.5.1 *Prepare a comprehensive disaster support database of general-purpose and emergency-specific data*
Most jurisdictions in the United States, Europe, and numerous other areas now have comprehensive GI databases that can be, and in many cases are, used for emergency purposes. Few organizations have created emergency-specific GI databases. Those using GI tend to use the corporate GI data and adapt the data on an ad hoc basis to meet the needs of the emergency. In the United States, the Department of Homeland Security has encouraged development of what are known as 'critical infrastructure' databases of a wide range of transportation, utilities, medical, and other facilities of significance either for response or at risk to emergencies. Even where data are generated during planning and mitigation activities, they are not typically incorporated into the EOC data repository for use during emergencies.

A few pilot projects have been carried out by a combination of federal government agencies in which data from multiple jurisdictions in a region were brought together and integrated into a comprehensive emergency GI database. Valuable experience was achieved that included identification of essential emergency GI data, development and refinement of data models, and definition of procedures for such compilation.

A major effort in the recent Hurricane Katrina was the need to compile a comprehensive GI database for the impacted region since no GI database was available for the numerous organizations involved in the response. Compilation of the database from numerous sources under the emergency conditions required tremendous effort and was extremely expensive, proving that such databases should be developed and available prior to emergency impacts.

2.5.2 Develop a robust data model designed for emergency purposes

Few organizations have developed an emergency GI data model. There is yet no generally available template or guide for an emergency GI data model, as there should be. Agencies of the federal government have been developing MEDS and compiling information on a common emergency data model, but that is not yet published for general use. There are emerging standards for general GI data exchange, including the Minimum Essential Data Set and GML, a special case of XML used for data exchange. Version 3 of GML, soon to be released, will be capable of expressing most of the concepts embodied in the OGC specifications, including coordinate reference systems and all the associated components. The GML schema is horizontal and does not depend on any application domain providing greatly enhanced flexibility in definition of data. GML is being incorporated with justice and emergency management requirements in the U.S. National Information Exchange Model (NIEM) currently in development by the Department of Homeland Security.

2.5.3 Establish strong data management and QA procedures to protect the integrity of data

Few organizations have specific data management or QA procedures to be employed in emergency operations. In general, GI operations in support of emergency management are handled on an ad hoc basis relying on standard, normal practices. Operations remain vulnerable to the real-time volume, intense demand, and mix of sources and reliability of information from those sources under the conditions that prevail during an emergency. Rarely are special QA procedures designed to handle emergency operations employed as of yet.

2.5.4 Provide a flexible capability to update and add data

As with database management, most organizations have not developed emergency-specific update processes or applications. Again, data are generally handled on an ad hoc basis, depending on normal procedures not designed for the real-time demands of emergency operations. Few organizations have developed a catalog of potentially useful information sources available through the Internet or intranets to be employed during an emergency. Discovery of data needed is generally based on informal knowledge or ad-hoc research.

2.5.5 Design the data model to accommodate multiple levels of detail

Since few special emergency databases or data models are used, no specific differentiation of cartographic detail exists for emergency purposes. Corporate data is used in the scale and detail available, and standard tools are used for any generalization needed. In the United States, NSDI has established a scale/detail classification that is beginning to be used in some emergency situations.

2.6 Technical lessons

2.6.1 New technologies must be deployed prior to a disaster

There has been considerable development of new and advanced technologies for emergency management since 9/11. Only a relatively small proportion of the new technologies have actually been deployed, but those that have, with the exception of the Pacific tsunami and Katrina emergencies, have generally been deployed while no significant emergency is under way. As described elsewhere, there is still very limited redefinition of emergency processes to incorporate or exploit the value of these technologies.

2.6.2 Mobile wireless and location technologies offer potential but are vulnerable

Mobile computing is being deployed widely in local governmental organizations for several functions, including emergency response. Most deployments to date do not employ wireless communications, though that capability is growing. Dramatic advances have been made in wireless technologies that are providing opportunities for improved use in emergencies. Flexible remote networks were deployed in the area devastated by Hurricane Katrina that allowed a rapid re-establishment of wireless communication in spite of the damage to the permanent facilities.

2.6.3 *Airborne/Helicopter and other digital imagery have proven to be useful*
With the wide availability of digital cameras and helicopters and the perspective offered by aerial photography, the use of these devices in emergency response has proliferated. Little capability has been developed, however, to support accurate determination of location, elevation, and angle of the photo. These characteristics would be helpful in positioning the images in the correct location relative to maps and other data for use with other sources and to provide the correct perspective to the user.

2.6.4 *Thermal imagery can be useful in disasters involving fires, but scale and timing are critical*
Thermal imagery is increasingly used in forest and wildfire response, though not in urban, plant, or non-fire emergencies. The available practical scale is generally useful for decision-making for large area fires only.

2.7 *Customer lessons*

2.7.1 *Range of customer groups*
The range of customers is perhaps even wider than during the 9/11 events as GI use in emergency management has grown and received broader attention. The actual use, however, continues to be very limited, primarily to those operating in the EOC and support functions that employ GI technology for their normal operations. Where distribution logistics support it, GI products are useful to both management and response personnel in virtually all aspects of emergency response. Reports from emergencies have indicated GI use for a wide range of activities from those managing and combating the emergency, to search and rescue, to support operations in transportation, medical care, evacuation and shelter, security, and recovery.

2.7.2 *Range of customers' technical skills and needs*
Slowly, the GI technical skills of those involved in emergency management are growing, though the majority of persons involved are still quite limited in their GI knowledge and skills. Most GI operation continues to be performed by GI experts who generate products for emergency personnel. Hard copy maps continue to be the primary GI products, and the understanding of their effective use is still limited to general knowledge of maps with little special emergency GI training available.

2.7.3 *Accurate and timely public information is important*
A few organizations are providing GI-based information for public use either directly through an emergency Web site or page or by delivery to public media services. Little has been accomplished in terms of development of emergency-specific design for public information, although emergency-specific Web sites are available in New York City and Washington, D.C.

2.7.4 *Disaster response customers require maps with easily identifiable ground locations and orientation*
Few organizations have developed special maps designed for use in the field during emergencies or that provide special landmarks that may still be available for orientation following the destruction of a disaster. Most maps used in emergencies are still the basic maps of the organization with minor modification for use in emergencies.

2.8 *Logistical lessons*

The logistical lessons are all closely related and are addressed here in summary form. Distribution of GI information remains a serious challenge. Advances have been made in communications networks and capabilities that are connecting more nodes in the emergency network with digital and GI access to information. Paper maps continue to be the primary GI product, however, and the delivery of paper maps is still a logistical challenge. The responders operating in the field pose the most serious challenge to the delivery of information, and only the most advanced organizations are able to provide digital communication to mobile and field units.

As noted above, improved communications are extending electronic capabilities to more office locations, but the field locations generally remain without electronic or effective logistical support for the delivery of GI products.

3   RECENT SIGNIFICANT EVENTS AND EXPERIENCES

GI lessons continue to be learned, validated, or challenged as GI experience is gained in the series of major and minor events since 9/11 and as development of GI technology and applications grows. The major tsunami that devastated large areas of Asia and Hurricane Katrina that hit the Gulf Coast of the United States have again had a dramatic impact, increasing recognition of GI value, exposing shortcomings in GI support, and furthering the body of valuable GI experience. They have confirmed past lessons and generated new lessons.

The first lesson identified earlier, and a basic concept of emergency management – 'be prepared' – was one of the most glaring shortcomings in the response to Hurricane Katrina. That overwhelming event exposed major gaps in preparation at all levels of government. Most organizations were not well prepared for effective response. The roles and coordination between levels of government and participating organizations were not executed successfully. An event of this magnitude requires a highly coordinated effort among all levels of government and the private sector (Egge 2005). That lesson is apparently yet to be adopted.

The tsunami, hurricanes, and earthquakes raised concern for events that cause massive damage. The tsunami and Katrina overwhelmed normal emergency capacity and actually destroyed most of the local response capacity. Local responders need a triage approach to adjust the response strategy to realistically align available resources with the actual demands of the event. A GI-based situation assessment tool is needed to quickly recognize thresholds of impact magnitude so that the appropriate response strategy that recognizes and matches the response capacity with the situation requirements can be applied.

The magnitude and geographic extent of these events created an additional logistical support dimension that called for response from external sources well beyond that required for most emergencies.

In events of this magnitude, a massive amount of resources must be employed from external sources. Some of those must be moved to the site of the impact, and others must operate from remote locations. This was the case with the GI/photogrammetric interpretation that was performed for the tsunami response at unaffected remote locations, the results of which were then provided to responders on site.

Locating available resources, including GI systems, data, and personnel, and arranging for their delivery to the impact area and managing the volume and organizational diversity were serious challenges to those in charge of emergency response.

A strategy is needed to cope with major destruction that seriously impacts and reduces the capacity to respond. The strategy should include arrangements with external organizations for back-up resources that might be from higher-level (state, region, national) government organizations, or similar jurisdictions in other regions not subject to the same events or impacts. Higher-level governmental organizations should develop a flexible, mobile capacity to provide rapid replacement for lost local response resources.

3.1   *Observations on Hurricane Katrina*

GI technology was widely used in the Katrina response by a large number of public and private organizations. GI use in Katrina provides an indication of the extent of emergency functions that can benefit from GI support such as the following:

- Forecasting and tracking incident progress
- Dispatching search and rescue teams

- Assessing damage
- Logistical management of the distribution of materials and supplies
- Demographic analysis of the population at risk and requirements for evacuation and shelters
- Tracking the status of available hospitals and shelter capacity
- Routing emergency vehicles and evacuation
- Planning for evacuation of people at risk
- Managing police and security force deployment
- Planning for debris removal and disposal
- Analysis of utility outage and infrastructure damage
- Visualization of event situation, e.g., extent of flooding.

It was necessary to create a portal for discovery, accessing, and integrating a wide range of data from many sources as part of the response effort.

In a personal communication with this writer, Jack Dangermond, president of ESRI, has offered valuable observations on the GI experiences of Hurricane Katrina (Dangermond 2005). Among the observations identified by Dangermond are:

- Little Geospatial preparedness existed prior to the event, and success required considerable heroic efforts, though because of lack of GI preparation, many efforts were repetitive, inconsistent, and performed with considerable cost.
- Data sharing was challenging in both technical and policy areas, consuming considerable time and resources and confirming that data sharing needs to be part of the emergency response plan.
- During the response, it was necessary to develop a multi-purpose database. To accomplish that, a new interoperability process was used integrating and disseminating local, state, and federal data. Data was converted, transformed, and integrated into a uniform data model using spatial ETL procedures that changed formats, data models, schema, and projections/datum.

Mr. Dangermond observed that a national system in place prior to Katrina would have:

- Provided immediate preparedness
- Facilitated multi-agency collaboration
- Improved response
  - Saved lives, property, and time
  - Made better use of people
  - Saved considerable money.

3.2 *Observations from development activities*

The U.S. Department of Homeland Security, recognizing the importance of GI to its areas of responsibility, funded multiple GI initiatives in its ITEP Program. As a part of that program, Towson University and the Maryland Emergency Management Agency conducted a pilot development and test of MEGIN (Maryland 2005). That project involved the development and testing of a Web-based tool for search, discovery, and access to GI data from a wide variety of sources. MEGIN also includes a security component that limits access to data to those who are authorized by the data owner.

A related ITEP project for the National Capital Region, CapStat (Washington 2006), provided interoperable access to GI data among participating jurisdictions in the region, including the MEGIN host in Maryland. These projects, along with the prior EMMA GI interface tool, identified valuable lessons in emergency management requirements for GI capabilities and data and mechanisms for accessing GI, the mechanisms for delivering GI, the usefulness of ESB technology, and the significance of organizational challenges to successful GI deployment. As an example, CapStat encountered the legal constraints to data sharing caused by differences in legislation and policy among participating jurisdictions where Maryland control and limits the distribution of GI data and DC publishes its data on an open public Web site.

A key concept of the CapStat project involves achieving interoperable access to data using an ESB to overcome incompatibilities rather than requiring the establishment and maintenance of a central repository or universal use of a common software. In supporting this concept, CapStat surfaced several lessons on GI interoperability, including:

- Unique qualities of GI Web services pose special requirements for deployment to an ESB.
- Differences between Web feature service and Web map service are significant to the satisfaction of emergency requirements.
- ESB can provide multiple transformations to support exchanges between organizations such as transformation of symbology from jurisdiction symbology to the emergency standard, transformation between software versions, and transformations between different attribute tables.

Another lesson from development and exercise experience was defined in DC where the integration of GI with a new Incident Management System (IMS) indicated the need to geocode data in the earliest stage of data acquisition by the IMS to enable GI applications to use real-time event data from the IMS in support of emergency operations.

4 SIGNIFICANT TECHNOLOGY DEVELOPMENTS

As introduced earlier, several significant GI technology developments are being applied to emergency management. The GI technology situation can be viewed from two perspectives – the essential requirements for technology support tools noted in the lessons above, and the current state of technology development. Unfortunately, as noted often in this chapter, some development of technology appears to be occurring independently of the actual requirements of emergency managers and responders. From the perspective of the current state of technology development, the following section provides an overview of a select set of key developments.

4.1 *Portal*

During an emergency, a very wide range of data may be useful, including basic IMS data, GI data and numerous other sources such as traffic surveillance cameras, hazardous materials guidance, flooding or plume models, and others. Virtually all of these data may be considered GI data since all are related to a geographic location, and that location is significant to emergency personnel. Various data types need to be integrated onto the GI map layers on-the-fly to generate and display a common operational picture for participants. Access to these sources must be streamlined to be useful in the urgency of an emergency situation. Streamlining access can best be accomplished through the employment of portal technology.

A portal, available to all in the EOC and other authorized personnel operating elsewhere, allows emergency managers and personnel to easily select and visualize useful information in a common window on their PC workstation. The portal allows easy selection of a range of information from the internal database and from numerous network sources for display by the click of a button. Navigation to these sources can be achieved through a set of carefully crafted software tools with data discovery provided by a store of metadata that describe the source, location, and navigation path to the source. The software component will also include tools (such as the SOA described in the Interoperability section of this paper) to perform necessary processing to deliver the data to be rendered in the portal display to suit the operator's requirements.

The key focus of the portal is to provide easily operated, timely access to all data necessary to make informed decisions when responding to an actual emergency or for training and simulation purposes.

4.2 GI in emergency Incident Management Systems

As the importance of GI has gained recognition in the emergency field, significant improvements have been developed in the ability of IMS to employ GI data and capabilities. Several commercial and public IMS are available to support emergency management activities. Since most were developed outside the realm of GI technology, the approach has generally been to add to the IMS access to GI functions. GI enhancements can be viewed at three levels of IMS-GI interface or integration:

1. Few are GI-centric with GI functions serving as the core around which the event reporting, resource management, and other functions are integrated.
2. Some provide GI capabilities as a module of the IMS, using standard GIS software or specially developed GI capabilities.
3. Most provide a module to support an interface to the organization's general purpose or a separate GIS, through which emergency applications are supported.

Based on the approach to GI integration, the solution may provide external, Internet-available GI data or may allow access to the organization's internal GI database. Organizations such as the Maryland Emergency Management Agency, with the GI centric approach to integration, are finding GI technology particularly valuable.

4.3 Rapid data acquisition

Data acquisition to support emergency response and recovery has been an area of considerable progress since 9/11. Recent major emergencies such as the terrifying tsunami in the Pacific and Hurricane Katrina in the United States have benefited from acquisition of post-impact data from aerial photography, satellite imagery, and other remote sensing technologies. Improvements have been made in the areas of acquisition and delivery technologies and administrative procedures. The capabilities of the general photogrammetric and satellite remote sensing technologies have been improving dramatically over the past decade and are capable of providing valuable data for emergency response. Rapid capture for a specific area and delivery following an emergency, however, has posed the primary constraints to usefulness in emergencies. The first hurdle has been the administrative or legal one – the procedures that make it possible for an emergency organization to obtain authorization to acquire data immediately following an event. Pre-established contracts with private photogrammetric firms and pre-authorization of release of government-controlled satellite imagery are being established to remove administrative constraints. Processes and software tools are also being developed to support the rapid processing and delivery of digital imagery for use by emergency managers.

Other approaches to rapid post-impact data acquisition being developed and enhanced include use of streaming video or digital cameras to capture on-the-scene images that can be transmitted and processed by GI and IMS software. These devices are used either on the ground or from a helicopter or fixed wing aircraft.

The images, while useful in their raw form, can be made more useful by developing procedures and software that will position the images relative to GI maps so that interpretation by users can be enhanced.

4.4 Mobile/field-accessible GI

A major lesson that was addressed regarded the delivery of GI to managers and responders in the field both for field use of available GI data and applications and for collecting information in the field for delivery to the EOC. Two solutions exist – mobile command vehicles and individual mobile devices. New York; Washington, DC; and other major jurisdictions have acquired a bus or van outfitted with GIS equipment and sophisticated communications systems. These special vehicles operate in field locations in proximity to the event. They provide GI capabilities to field managers on site. Some include complete GI systems and databases, as well as wireless communications.

The second level is mobile computing on a range of devices from cell phones and PDAs to tablets and laptop computers. These provide varying levels of capabilities from simple displays to interactive operation and wireless communication.

4.5 *Actionable intelligence and alerts*

Time is critical in emergency response. The sooner emergency managers and responders become aware of emergency situations, the more rapidly they can take action, the less damaging the impact, and the more effective the action taken.

Some types of emergencies emerge over time from an accumulation of minor or worsening events rather than by a sudden flash. Most emergencies evolve over time, growing in intensity and waning and/or moving along a geographic path. Also, during an emergency, many incidents are occurring, and it is often difficult to recognize patterns or indications of conditions that require immediate response. Early detection of these conditions is critical.

Technology is now available to monitor real-time data flows, recognize conditions, convert data to actionable intelligence for emergency managers, and raise alerts for responsible parties.

A combination of technology tools can be integrated to recognize alert conditions. These include:

- Web services to acquire data from selected sources available either from emergency reporting or sites on the Internet such as the NOAA weather services
- A search engine to monitor the flow of data from sources
- A DBMS to manage acquired data
- Business intelligence software to apply alert determination rules to data.

Alerts call the manager's attention to a situation but leave the decision-making or determination of action to be taken to the emergency manager. Alerts can be raised from monitoring emergency event reports, from tracking emergency event data, or from monitoring selected conditions during normal times and applying predefined rules to determine when an alert condition exists. Alerts can be routed to the appropriate consumers through messaging software such as Enterprise Service Bus (ESB).

4.6 *Interoperability*

Interoperability is a complex issue. The purpose of interoperability is for organizations to exchange information without constraints caused by differences in systems or data structures. The ability to exchange in most cases, however, is challenged by differences in systems, data structures, data models, data definitions, and other technical and procedural issues among organizations requiring interoperability.

Interoperability can be achieved in various ways using various combinations of procedures and technologies. At the simplest level, interoperability can be achieved by all organizations wishing to exchange data employing a central data repository and common system. This approach has been attempted for years with very limited success. It requires complete agreement of the makeup of the central repository and system and tremendous resources to establish and maintain the data in particular.

Another relatively simple approach is for all organizations involved to use the same version of the same software, the same data model, and the same data definitions. This level of commonality is generally not feasible in reality, especially where already existing systems are being used by individual organizations.

In that real world context, it is necessary to employ an architecture, software, standards, and procedures to overcome the inherent incompatibilities to achieve effective interoperability. The challenge of interoperability extends beyond the field of GI, and sophisticated architectures and software have been developed in the general field of information technology that facilitate the development of interoperable exchange mechanisms. Service Oriented Architecture (SOA), employing Web services, ESB software, and standard XML schemas, is a recently developed approach to interoperability and enterprise integration that is now being applied to GI implementations and emergency functions with success.

5 CURRENT CHALLENGES AND RECOMMENDATIONS

Based on the review of lessons and on the current state of GI for emergency management, the following challenges and recommendations have been identified with the hope that they will provide a focus for an increased effort on the part of the GI field to improve the use of this valuable resource by emergency management. To facilitate focus, the issues are presented as the author's view of six significant issues that are outstanding today.

5.1 *Six significant GI-emergency challenges today*

5.1.1 *GI education*

For GI to be used effectively, emergency personnel must understand and value its use. While emergency personnel are increasingly aware of the usefulness of GI, they lack an understanding of the full capability and value of this technology beyond merely map production. There is therefore a challenge for the GI community to provide a comprehensive education for emergency personnel in the technology and its uses. The comprehensive program should include multiple facets.

It should include the publication of articles in the media of the emergency field that describe the use and value of GI, success stories, guidance on acquisition and use, and other topics that provide general understanding and encouragement for the use of GI.

The program should include more specific publications that provide detailed guidance on best practices for use and development of GI applied to emergency functions.

GI training sessions should be provided independently and in conjunction with emergency meetings and conferences.

GI information and training should also be provided through Internet channels that are recognized by the emergency community.

Another important aspect of the GI emergency education program is training and familiarization with emergency operations and conditions for the GI personnel who will develop and support emergency activities. This training should address the potential uses of GI, database issues, and the unique high intensity environment of emergency operations.

5.1.2 *Best Practices*

There is an absence of information on Best Practices in the application of GI technologies to emergency functions. While GI use in support of emergency management is not yet mature, many lessons have been learned, and knowledge of what will constitute best practices in several areas already exists. There is a need, however, to capture that knowledge and formulate it into descriptions of best practices in the various aspects of GI in emergency. Best practices should be defined in the areas of GI applications and operations, data acquisition and management, deployment of GI in the emergency environment, and GI processes.

5.1.3 *Templates*

Effective development and use of GI in emergencies can benefit greatly from the availability of models and templates of effective approaches that can be deployed or adapted by emergency organizations. The models and templates define specific components for implementation of best practices. Deployment of GI capabilities for emergency management can be expedited by the availability of templates for everything from individual applications to data models to operating procedures.

5.1.4 *Mobile GI*

There is a critical need for software, communications, and logistical support to deliver GI to all emergency operation locations, including emergency responders operating in the field. Where GI is being used, it is generally limited to the emergency command or operating center. The potential value of GI can be greatly enhanced by extending its availability to persons operating in the field, at the scene of the emergency impact, as well as those operating in support locations of participating organizations. Mobile communications supporting IT are now available or becoming available to most emergency organizations. In parallel, mobile GI applications are now available and maturing in numerous functional areas. So the development of mobile emergency GI applications is a practical reality today and one of the highest priority opportunity areas in the emergency GI field.

5.1.5 *Data acquisition and filtering*

Information is the lifeblood of emergency management, and the range of potentially useful information for any specific emergency is almost limitless. There is a challenge, therefore, to provide the tools for the search, discovery, access, integration, consumption, and rendering of relevant emergency GI or location-related data from the multiple sources that are now available both among the internal databases of the organization and over the Internet.

To avoid the potential for 'information overload' from so many sources in so many forms, there is a corollary requirement for tools to target queries to relevant sources and to filter the large volumes of potential data to reduce them to the 'actionable information' that will effectively support emergency decision-making and actions.

5.1.6 *GI preparedness*

The area of GI emergency preparedness poses challenges in several areas. The primary challenge is adoption of GI use by the emergency management community. This challenge may be met by the education program posed in the first challenge, GI education.

Once accepted for use, GI technology and its use must be incorporated into the standard operating procedures of emergency organizations. This requires an analysis of the current procedures, identification of steps in the procedures where GI and its products can be useful, and modification of the current procedures to steps for GI use. The development of effective GI operating procedures can benefit from the availability of templates addressed in the prior challenge number 3.

6 SUMMARY AND CLOSING

While there has been a great deal of activity and recognition of GI as a critical resource, much improvement is still required if emergency organizations are to be able to take full benefit from GI technologies and from the lessons learned from 9/11 and other experiences with significant emergencies. While a wide range of new and improved technologies has been made available, not all are being used effectively or exploited fully. Little improvement has been made in GI preparedness, incorporating GI in the emergency planning processes, GI education for emergency personnel, and other non-technology aspects of GI support for emergency management. There has been, and continues to be, too little involvement of emergency management experts in the design and development of technology for emergency purposes, and, thus, the value and level of deployment are constrained.

There is a significant challenge to move emergency management GI use from dependence on paper maps that are slow to produce, hard to deliver, and quickly out of date in an emergency to more effective location-based query, dynamic situation display, practical spatial analysis, and direct use by emergency personnel.

I call for the academic community to fill the knowledge gap that currently exists – to conduct research into best practices for GI use, effective IT architectures for application to GI and mechanisms for overcoming shortcomings in GI technology as applied to emergency management, and shortcomings such as capabilities for acquiring and managing real-time data.

To assist in targeting the research that I call for, I have provided a series of lessons learned for experiences with GI in disaster management, lessons learned in the 'heat of battle' so to speak, in the highly charged environment of actual emergency response operations – lessons learned in integrating GI into the operating environment and with emergency systems; lessons learned in exchanging critical disaster information between organizations and overcoming the challenges of interoperability between organizations; lessons learned integrating GI with enterprise technologies developed to overcome interoperability limitations; and lessons from dealing with the wide range of technically sophisticated and technically resistant personnel involved.

Lessons are very important but they must be acted upon to be meaningful!

REFERENCES

Dangermond, J., pers.comm. 2005.

Egge, D. 2005. U.S. is given failing grades by 9/11 panel; Bipartisan group faults counterterrorism progress. Washington Post, 6 December 2005, A1.

Federal Geospatial Data Committee, Homeland Security Working Group, Symbology Subgroup, Homeland Security Mapping Standard – Point Symbology for Emergency Management, Draft INCITS 415, ANSI, 2005.

Homeland Security Infrastructure Program Tiger Team Report, USGS/National Geospatial-Intelligence Agency, 2004.

Maryland Emergency Geographic Information Network (MEGIN), Matthew Felton, Towson University, July 2005.

Kevany, M.J. 2005. Geospatial information for disaster management: Lessons from 9/11. In P. van Oosterom, S. Zlatanova, E. Fendel (ed.), Geo-information for disaster management: 443–464. Berlin: Springer, pp 443–464.

The 9/11 Commission Report: Final Report of the National Commission on Terrorist Attacks Upon the United States, August 2004.

Washington, DC, CapStat ITEP Project Final Report, Office of the Chief Technology Officer, Washington, DC, 2006.

Geospatial Information Technology for Emergency Response – Zlatanova & Li (eds)
© 2008 Taylor & Francis Group, London, ISBN 978-0-415-42247-5

Legal aspects of using space-derived geospatial information for emergency response, with particular reference to the Charter on Space and Major Disasters

F.G. von der Dunk[1]

Nebraska University, USA; Black Holes, The Netherlands

ABSTRACT: Increasing attention is being paid today to the potential offered by geospatial information in particular if generated with the help of satellites to contribute to mitigation of major disasters – tsunamis and earthquakes as much as man-made disasters. The current contribution seeks to outline some of the major legal issues involved in the use space-derived data for emergency response, focusing on four topics: copyrights, access to remote sensing data, responsibilities and liabilities, and security and dual use-issues involved. This contribution forms part of the Leiden Faculty of Law research programme 'securing the rule of law in a world of multilevel jurisdiction: coherence, institutional principles and fundamental rights'.

1 INTRODUCTION

Major disasters, man-made as much as natural, seem to be rapidly increasing in both size and frequency over the last years, though this impression may be due partly to the increasing media coverage of such events – the images from the tsunami that hit South and Southeast Asia, then the catastrophic earthquake in Pakistan, are still fresh on everyone's mind. What is beyond doubt, is the increasing attention being paid to the potential offered by geospatial information, in particular if generated with the help of satellites in outer space, to contribute to mitigation measures in the various phases recognised, from preparedness and alert to long-time rehabilitation.

The most visible aspect thereof no doubt concerns the establishment of the Charter on Space and Major Disasters as of 2000 (Charter on Space and Major Disasters, 1999). The Charter, essentially the first rudimentary 'organisation' of activities in the field, was established by a number of leading space agencies with operational remote sensing capabilities, initiated by ESA and the French space agency CNES in 1999 as a follow-up to the Unispace III Conference where the potential of earth observation in the context of major disasters was prominently discussed. The Canadian Space Agency (CSA), the US National Oceanic and Atmospheric Administration (NOAA), the Indian Space Agency ISRO, the Argentine National Commission on Space Activities CONAE, the Japanese Aerospace Exploration Agency (JAXA) and most recently the British National Space Centre (BNSC) joined on behalf of the Disaster Monitoring Constellation (DMC), so that the Charter currently counts eight full-fledged partners (Charter on Space and Major Disasters, 1999). The United States Geological Survey (USGS) has also joined the Charter as part of the US team.

In many respects, however, the establishment of the Charter merely represents the most institutionalised context for using geospatial information for disaster and emergency response purposes: most of the legal issues playing within the context of the Charter are of wider relevance for the field as a whole. Thus, the current contribution constitutes an effort to discuss some of those international

---

[1] The author would like to thank in particular Gunter Schreier for pointing out some interesting additional facts which he took the liberty of incorporating in this contribution. Of course, the current version is exclusively the author's responsibility.

legal issues considered of relevance from such a more general perspective. The main limitation here is that of focusing on satellite-derived geospatial information as opposed to *in situ* or airborne methods of generating geospatial information.

This is, indeed, a substantial limitation: mainly because of the novelty of the issue, there is as of yet *no* law or regulation dedicated to, and tailor-made for, the issue. While this does not result in a total legal vacuum, it does mean that recourse will generally be had to a few more general legal regimes, not at all developed with the prospect of deriving geospatial information from satellites in mind, yet turning out to have some bearing thereon as well. The novelty of the issue, moreover, will also mean there is as of yet insufficient practice to go into any details as to what precise impact those general regimes would have on space-derived geospatial information. Law, after all, only comes alive when persons or legal entities start using it for the purposes of defending their own self-perceived interests.

The current contribution will thus address the following four sets of legal international issues in somewhat greater detail:

1. The application of copyrights to geospatial information products relevant for emergency response as far as resulting from satellite activities.
2. The international regime applicable to access to data, which result from remote sensing, and the application of copyrights.
3. Responsibilities and (in particular) liabilities which may result from satellite-based geospatial information operations and activities.
4. Security and dual-use issues in the context of using geospatial information for emergency response, to the extent that existing international arrangements may have a bearing on the legal context within which certain emergency services or products might be provided.

This certainly does not pretend to offer an *exhaustive* list of relevant legal and/or organisational issues involved. However, such other legal issues as privacy (in view of the current state-of-the-art potential of some satellites to offer very detailed 'pictures') or telecom law (as relevant for various telecom-related aspects of remote sensing satellite operations) would be one step further removed from the core aspect of using satellite-generated data for emergency response. Hence, they will not be considered here.

2 THE APPLICATION OF COPYRIGHTS

2.1 *Copyrights and satellite remote sensing*

Copyrights is the most relevant version of intellectual property rights in the current context since it directly refers to the intellectual ownership over the data and information generated. In view of their importance in stimulating creative activities, the first legal regimes to provide for copyrights – and a certain balance between the creator's interests of protection and the public interests of access – were developed already centuries ago. Obviously, this has been done without very much taking into account the possibility that space-based data and information could also be involved. Still, once satellites started to generate data, subject to more experience with space-derived geospatial information as well as more analysis, such regimes might explicitly or implicitly come to apply to such data as well.

The generation of geospatial information by means of satellite is but a version of satellite remote sensing: the core of the systems providing the data and information to be used for emergency purposes consists of remote sensing satellites. These satellites operate in outer space, which is an area not subject to any territorial sovereignty (Outer Space Treaty, 1967, Art II) As a consequence, freedom of activity is the point of departure and any limits to such freedom have to be derived from existing legal principles or from rules, obligations and rights of other states stemming from international treaties, including the UN Charter, or international customary law (Outer Space Treaty, 1967, Art I, III) For private parties involved, moreover, national regimes may (further) limit the opportunities to make use of the freedom of exploration and use of outer space. This also includes copyrights, much as they did not take space-specific aspects into account. Still, except where specific aspects of satellite operations generating geospatial information would be explicitly or implicitly prohibited or conditioned, such operations are basically allowed.

Historically, copyright regimes have been developed first and foremost at the national level. In general, copyright protection may be obtained for an original work of authorship fixed in a tangible medium. Relatively early on the international ramifications of hugely diverging national copyright regimes having become clear, one of the oldest international treaties provides for a first effort to align those national regimes, which resulted in a measure of 'mutual recognition (Berne Convention, 1886, ATS 1901, No126). This 1886 Berne Convention was followed by a number of other international treaties further harmonising national regimes as to their international effects (Universal Copyright Convention, 1952). However, in spite of these international efforts, such fundamental differences as between a 'first-to-file' regime (to which all European countries adhere) and a 'first-to-invent' regime (to which *inter alia* the United States adheres) continue to exist.

2.2 *The European context*

In the international arena, developments towards harmonisation have thus far stalled essentially at the level of 'mutual recognition', leaving much to be desired especially in terms of substance. For that reason, further to the above it is instructive to take a look at how in Europe specifically the issue of applying copyrights protection to satellite remote sensing has been dealt with, in view of the fundamental involvement of two intergovernmental organisations in space activities: the European Union and the European Space Agency (ESA).

In the beginning, within Europe the topic of copyrights in the context of remote sensing was considered a matter for ESA because of its key role in European space activities, including remote sensing activities. Thus, 'the Agency shall, with regard to the resulting inventions and technical data, secure such rights as may be appropriate for the protection of its interests, of those of the Member States participating in the relevant programme, and of those of persons and bodies under their jurisdiction' (ESA Convention, 1975, Art III(3)). However, it rapidly became clear that ESA's own competencies were too limited for establishment of a more comprehensive legal regime; it could only effectuate relevant protection through, and as far as could be provided by, individual contracts.

On the other hand, when potential applications within Europe became a distinct probability, the European Commission became interested in the issue, in view of the possibility to use intellectual property rights as anti-competitive tools. Individual companies could, for example, use copyrights to sell licenses for exclusive access or usage in specific national territories, thus artificially carving up the Internal Market into nationally separated markets in contravention of relevant EU principles. A study initiated by the Commission has resulted in recommendations to make the then-draft Community Directive on the protection of databases applicable to remote sensing data. This concerned the so-called 'Gaudrat study' of April 1993, which concluded that the best way to effectuate any protection of remote sensing data would be to bring them under the heading of databases, rather than for instance copyrights.

The problem of appropriate legal protection of the data resulting from remote sensing resulted from the way in which the concept of copyrights had been developed historically. One of the main problems with raw data is that it does not satisfy the originality criterion for protection by copyright: there usually is no creative human intervention involved in producing them – especially if they are generated automatically or in a pre-programmed fashion. Collections of raw, corrected or treated data also fail to satisfy the originality criterion if there is no creative human intervention involved in producing collections of such data, read databases. This, of course, equally applies to the specific area of satellite-derived geospatial information data.

Still, for want of better legal tools, most operators in Europe used copyright protection to protect their data resulting from activities in outer space. Of course, in the absence also of any specific Community legislation on the matter, risks abounded that protection could differ between European states due to varying national copyright laws and/or varying interpretations thereof.

In this regard, the resulting Community Directive 96/9 established a *sui generis* right of data base protection (Directive 96/9/EC, 1996). It obliges the member states to include databases, amongst them those containing remote sensing data, in their national intellectual property rights regimes, in conformity with the parameters further provided by the Directive. It applies both to nationals (including companies) from EU member states undertaking such activities, and to such activities if undertaken from the territory of any of the EU member states (Directive 96/9/EC, 1996, Art. 11(1), (2)).

In other words: any satellite activities generating geospatial information conducted either by EU nationals or from the territory of an EU member state could enjoy the protection of Directive 96/9 – for example to limit access to the relevant data. Outside of these situations, that is if neither an EU national (whether person or legal entity) is crucially involved in the generation of date, nor such generation is (at least for a major part) conducted and undertaken from an EU member state, such protection exclusively depends upon the national regime of the state in question, where applicable in conformity with international treaties to which such a state is party.

In terms of substance, the Directive protects creative databases under copyright law and creates a unique protection – the *sui generis* right – for those databases which do not meet the requirement of originality, as long as they are individually accessible and require a substantial investment to be generated. In other words, the *sui generis* right extends protection to databases containing material not protected by copyright. As a result, data derived from activities in outer space and assembled in an original database are protected within the territory of the EU member states.

The protection offered by the Directive basically consists of two sets of rights, defined in Article 7(2) as the 'extraction right' and the 're-utilisation right' respectively, both principally resting with the creator/owner of the database and for him or her to license others to use. The 'extraction right' refers to the right to permanently or temporarily transfer all, or a substantial part, of the contents of a database to another medium by any means or in any form. Likewise, the 're-utilisation right' refers to the right to make available to the public all or a substantial part of the contents of a database by the distribution of copies, by renting, by on-line transmission or any other form of transmission. The first sale of a copy of a database within the European Union by the right holder, or with his consent, exhausts the right to control resale of that copy within the Union. The Directive by now has been transposed into national legislation by all EU member states, as was (of course) required by its terms.

2.3 *Copyrights and remote sensing data for emergency response purposes*

Let us go back to the issue of geospatial information data in support of emergency management. Not just within Europe, but everywhere copyrights will obviously constitute a major parameter determining the scope of usage of satellite geospatial information data being allowed or practically possible in the context of emergency response, since they give the owner of the data a very fundamental legal tool to control such data.

In many cases, the entities generating relevant data will be public in character and legal role, in a perhaps varied but generally large measure. Here the issue would sometimes be whether they can effectively own copyrights in the first place. At the same time, it may be pointed out that such a public entity will have considerably less interests in using copyrights as a tool to limit access to relevant data, certainly as long as not security-sensitive.[2] One main idea behind such constructs as the Charter after all is to provide what may be called public goods and services paid for at least in major part by the tax payers, which should benefit as much of society as possible, copyrights being used as little as possible to obstruct such benefits from being realised.

On the other hand, public investments in space-derived geospatial information should not allow private companies to take a free ride for their own, private and usually commercial purposes, piggy-backing on overly liberal access policies. In such cases, copyright may indeed present a handy tool

---

[2] See further *infra*, Section 5.

for allowing some control over the downstream use of any satellite-generated data, which requires not only independent ownership of copyrights, but also attendant copyright strategies and policies.

The practical effects of such control tools of course ultimately depend on the general effectiveness of law monitoring and enforcement. There is certainly no perfect defence possible against malicious intentions, since in principle every state, organisation or relevant entity can request remote sensing data. However, firstly the organisational structure for data request and delivery acts as a filter in a number of cases.

Secondly, as a consequence of the existence of copyrights, at least legally speaking instruments would be available to (try and) ensure that usage is made of the data downstream exclusively for proper purposes. One could draft an (additional) international protocol requiring any requesting organisation to formally declare usage to be only for specific, well-delineated purposes. More pragmatically, one could also include relevant clauses in copyright licenses which certain data owners might require from users before allowing them access.

As a matter of fact, the Charter already knows of such a process to protect data exchanged under its sway. Raw satellite data is only exchanged between the Charter partners and entities defined by the 'Charter Manager', whereas others, including affected states and aid organisations, will only receive derived geospatial information such as maps, tables and prints. This process, clarified in advance and known to every Charter participant, might have the obvious drawback of impeding the rapid and efficient usage of the Charter's data in a given event, if for example the information-derivers are missing relevant information which the affected states or aid organisations, if they were to analyse the data, would not miss out on, but it prevents at least raw data from being used for unintended, possibly abusive purposes.

Whether such measures would be sufficient 'in the real world' to ban malicious use to a satisfactory extent is of course another matter, but with licenses and relevant clauses on usage one at least would have the legal tools to fight such use and criminalise those who undertake it. That certainly does not apply to Charter-induced data generation only, but to *any* geospatial information data with any real, read in particular potential commercial, value.

Furthermore, in a number of cases relevant data might be generated by (completely or partially) privately-owned and -operated satellite systems, like the French SPOT image, the Canadian Radarsat or various private US Very High Resolution-data satellite systems. Such private operators in principle would use their copyrights to control access to the relevant data, read to make money by allowing such access in individual cases against fees. It is their principled entitlement to decide whether, for example for reasons of public relations and public image, data would be provided when requested for emergency purposes, subject to any further conditions such as referring to usage other than directly emergency-related.

3   ACCESS TO REMOTE SENSING DATA

3.1   *From copyrights to data access rights*

The previous Section dealt with the issue of copyrights, which provides an ad hoc-tool to deal with access to certain data sets – by establishing a specific balance between the rights of the general public to have access to a certain set, and the rights of the copyright owner to limit such access, as subject to applicable legislation. Apart from this issue at the international level there are a few legal parameters relevant for satellite-derived data, approaching the issue as it were from the other end: that of obligations to allow access to remote sensing data – which might, in principle, come into conflict with applicable rights of copyright owners to limit such access. If such data access rights are unequivocally established by comprehensive legal regulation, they would actually override any potential rights to limit access by copyright owners, but the situation is usually not so clear-cut. This makes it difficult at this point to establish more detailed guidance as to what happens in case of such a conflict.

The parameters currently calling for immediate attention would be found in three areas in particular: the international legal regime for access to remote sensing data in general, the specific development of the Charter on Space and Major Disasters referred to earlier and general humanitarian obligations.

3.2 *The international regime for access to remote sensing data*

As referred to before, one of the most fundamental rules of space law is the principle of freedom of space activities (Outer Space Treaty, 1967, Art. I). Consequently, the point of departure under international space law is that the activity of using satellites for remote sensing purposes is allowed. The Outer Space Treaty itself only provides for a few principles to which any space activities should conform, such as international cooperation, mandatory supervision and authorisation of private space activities (for which a state is held responsible without further qualification), and sincere efforts to minimise harmful effects of one's space activities, for example as to the environment (Outer Space Treaty, 1967, Art III, VI, IX).

More in particular, states are also held liable for damage caused by space objects involved in any private activities as long as they would have been involved in the launching of the space object concerned in a sufficiently substantial manner (in addition of course to liability for damage caused by their own space objects) (Outer Space Treaty, 1967, Art. VII). This regime was further elaborated by means of the Liability Convention of 1972 which formally qualifies such involvement as that of a 'launching State' (Liability Convention, 1972, Art I(c), II, III).

The issue of remote sensing specifically, as a sub-set of space activities, at the global level has only been dealt with in any detail by UN General Assembly Resolution 41/65, adopted by consensus on 3 December 1986 (Resolution 41/65, 1986).This adoption by consensus, as well as the general respect accorded to its contents, leads most experts to consider those contents as reflecting customary international law – as Resolution *per se* is not binding. The Resolution acknowledges the freedom of remote sensing activities, as one particular manifestation of the freedom of space activities subject only to international law (Resolution 41/65, 1986, Principle III). Further to this, the Resolution provides some important parameters for remote sensing activities, including those that are geospatial information related, as follows.

At the outset it should be noted that the Resolution applies to remote sensing activities 'for the purpose of improving natural resources management, land use and the protection of the environment' (Resolution 41/65, 1986, Principle I(a)). Since such usage arguably would not require the quality of spatial resolution of better than in the range of 10 metres, Very High Resolution (VHR) data issues might actually fall outside the scope of the Resolution. They certainly were not taken into consideration – or even envisaged – at the time the Resolution was drafted. In view of the fact that much geospatial information data would likely fall within the range of VHR data, this may present a rather important issue in regard of which to further elaborate the law, so as to at least establish the desired clarity.

In other words: the Resolution does not clarify to what extent the individual discretion of states, European Union and international remote sensing operators as to how to deal for example with dissemination and usage issues regarding VHR data would still be intact. Privacy aspects typical of VHR remote sensing data dissemination at the very least have not been considered. Another issue following from this, somewhat narrow, definition of remote sensing for the purposes of the Resolution, is that it might be taken to exclude from its scope any military activities.

Then, Principle II provides that 'Remote sensing activities shall be carried out for the benefit and in the interests of all countries, irrespective of their degree of economic, social or scientific and technological development, and taking into particular consideration the needs of the developing countries'. This Principle reflects the similarly-phrased Article I of the Outer Space Treaty. Obviously, it very much supports the general use of data, and information derived from them, for emergency response purposes, although it also raises some questions as to the extent in which such benefits are to be shared in a mandatory fashion.

Here, the frequently-found and rather general reference to 'the benefit and (…) interest of all countries' with special consideration for the developing countries was developed further in 1996, by means of another UN Resolution (Resolution 51/122, 1996). This Resolution left complete freedom to states 'to determine all aspects' of such cooperation, and furthermore repeatedly referred to the requirement of 'an equitable and mutually acceptable basis' for any activities undertaken in its implementation (Resolution 51/122, 1996, Principles 2, 3).

Principle IV of Resolution 41/65 then deals with the core issue of satellite remote sensing: the dilemma between the freedom of use of outer space, in its particular manifestation of freedom of information-gathering making use of satellites, and the principle of sovereignty of states over their own territory, more in particular over their own wealth and natural resources. These two concepts at the time of drafting the Resolution were considered to collide in particular where the 'sensed state' finds itself in a situation that a 'sensing state' might obtain valuable information, especially in economic terms, with regard to the territory of the 'sensed state' which that state itself does not possess.

A balance of sorts has been established by Resolution 41/65, which in the final analysis tilts towards the freedom of space activities. The principle of full and permanent sovereignty, it is true, is to be respected, consequently legitimate rights and interests of the 'sensed state' shall not be harmed, and also the benefit and interest of all countries shall be taken into account (that is, including those of the 'sensed state') (Resolution 41/65, 1986, Principle IV).

This is no mere theory. In the recent activation of the Charter in the case of Pakistan, VHR data were available – and in some cases already used – to monitor the areas affected by earthquakes. In spite of the clear emergency character of the context in which this took place, however, the United Nations authorities involved then requested all Charter participants not to use VHR data for fear to alienate the government of Pakistan in view of the potential impact on security or other crucial interests of Pakistan. Luckily, it turned out the Pakistani government took a relaxed approach and made it clear that, as far as it was concerned, VHR data could be used for the intended purposes, but it is very well possible that other countries in other circumstances would not be so relaxed about this.

All this, however, does not alter the fact that the 'sensed state' neither has a veto to prevent it from being 'sensed', nor an exclusive, free or even merely preferential right of access to the data – and neither is it entitled automatically to becoming a partner in the relevant remote sensing operations (Resolution 41/65, 1986, Principle XIII). This becomes especially clear when these principles are seen in conjunction with Principle XII, since for the purpose of a particular set of remote sensing data – whether geospatial information related or not – concerning its territory the 'sensed state' is no different from any other state interested in such data.

Principle XII namely provides: 'As soon as the primary data and the processed data concerning the territory under its jurisdiction are produced, the sensed State shall have access to them *on a non-discriminatory basis and on reasonable cost terms*. The sensed State shall also have access to the available analysed information concerning the territory under its jurisdiction in the possession of any State participating in remote sensing activities on the same basis and terms, particular regard being given to the needs and interests of the developing countries.'[3]

In general it has not been considered 'discrimination' when data disseminators – so far especially governments or intergovernmental organisations – apply different prices to scientific and/or non-commercial users on the one hand and commercial users on the other. Consequently, on a national or regional level distinctions are usually made between users from the scientific, educational or other evidently-public sectors (which normally have to pay nothing or only cost-based fees) and commercial users (who have to pay substantially higher, essentially commercial fees). Geospatial information data for emergency management purposes would squarely fall within the former category.

Also, where public authorities co-fund, subsidise or substantively support remote sensing activities, it seems obvious that they would have a right of access distinct from those of others to the resulting data, as this would not be tantamount to 'discrimination' in the real sense of the word. However, due to the vagueness at the level of the principles contained in Resolution 41/65, national and regional implementation of this principle has taken place in many different ways.

The difference between primary and processed data on the one hand and analysed information on the other hand is further noteworthy, in particular as geospatial information data would usually

---

[3] Emphasis added.

refer to either processed data or analysed information (rather than to primary data as such). As to the first, a 'sensed state' will only have access to the data concerning its territory if the 'sensing state' or any entity for whom it is responsible is interested in marketing and selling those data – and then, of course, at the same ('non-discriminatory') price and in conformity with the other relevant conditions. As to the second, the – already inarticulate – right of access ('as soon as' data have been produced) is further diluted; this time no time limit at all is provided for. Moreover, a right of non-discriminatory access for a 'sensed state' exists only with regard to analysed information in the hands of a 'sensing state' – not, therefore, in the hands of any entity for whom it is internationally responsible. At least, that has been the interpretation to date of experts, states and international organisations alike.

Gradually, some practice is becoming clear in this respect. For example, whilst InfoTerra Germany has the commercial distribution rights of TerraSAR-X, it has licensed (only) Japan to receive data, where Japan may opt to programme TerraSAR to acquire data over North Korea. Thus, while the satellite owner in the last resort is Germany (more precisely: the German Space Agency, DLR, on behalf of the government), the commercial rights lie with InfoTerra and the operator (in the relevant region) is Japan – with the sensed state being North-Korea. This raises some legal issues, for example as to who is responsible for any 'violation' of the Principles of Resolution 41/65, or any relevant rule of the Outer Space Treaty?

In terms of further legal parameters to the freedom to distribute remote sensing data, finally two further Principles contained in Resolution 41/65 are of special importance with a view to emergency response.

Firstly, Principle X provides: 'Remote sensing shall promote the protection of the Earth's natural environment. To this end, States participating in remote sensing activities that have identified information in their possession that can be used to avert any phenomenon harmful to the Earth's natural environment shall disclose such information to States concerned.'

The clear moral value of this Principle, coupled with general duties of care, international cooperation and respect for benefit and interest of all countries, makes it rather difficult for states not adhere to it, or even not to make private or other disseminators or users adhere to it. Thus, although directed again at states, and probably even in the absence of explicit obligations on the domestic/private level for disseminators and users, neglecting these provisions in disseminating or using remote sensing data might not be legally excusable on the international plane any longer. This might even mean that if a satellite operator has obtained satellite data that would clearly show global warming to lead to future degradation of the global environment, and such information is not duly transmitted to other states, it would violate Principle X.

Since, under Principle X, remote sensing shall promote the protection of the Earth's natural environment, it may be asked what is included in that term 'natural environment'. The Principle would not apply to 'man-made environments', certainly not according to the letter of the law. Since, however, the drafters of the text of the UN Resolution simply would not have had the possibility of dealing effectively at an international level with disasters in 'man-made environments' such as factory explosions in mind, a development in interpretation could come to include such events. It is interesting to note from this perspective that the Charter does not limit itself so much to the 'natural environment' as the UN Resolution does; thus, the train explosion in North Korea a few years back triggered the Charter into operation just as much as the tsunami did.

Secondly, Principle XI provides in a fashion rather similar to Principle X: 'Remote sensing shall promote the protection of mankind from natural disasters. To this end, States participating in remote sensing activities that have identified processed data and analysed information in their possession that may be useful to States affected by natural disasters, or likely to be affected by impending natural disasters, shall transmit such data and information to States concerned as promptly as possible.' The Charter, from this perspective, constitutes an institutional and structured elaboration of this Principle, and thus represents the next step to actually implementing it and making it work.

Principle XI largely mirrors Principle X; the latter dealing with man-originating threats to the natural environment of the earth, the former with nature's threats against mankind. Consequently, the evaluation of Principle X largely applies here as well; for example, when it comes to the *prima facie*-focus on states possessing data, or as regards the vagueness of terminology, from which no actual conditions for disclosure can readily be distilled. Nevertheless, the obvious moral value of this Principle too would imply close-to-binding effects also upon non-state disseminators or users – at least it results in an international responsibility for relevant states to make sure these entities would adhere to the Principle. Both Principles, in short, clearly support liberal access to geospatial information data for emergency management purposes.

Finally, one noticeable difference with Principle X is that Principle XI explicitly applies to 'processed data' in addition to 'analyzed information', as opposed to mere 'information' as it is contained in the former Principle.

3.3   *The Charter on Space and Major Disasters*

Of major impact in the area of disaster and emergency response, the Charter on Space and Major Disasters was already briefly introduced *supra*. Prior to the Charter's existence, generally speaking there was a lack of awareness on the side of the potential victims as much as of the providers regarding the potential usefulness of such data, coupled with a general attitude on the side of potential data providers of unease: what are my risks in terms of giving away valuable and/or sensitive data? How do others deal with such requests? How should I handle this? More in practical terms finally, there was no general system or format to handle any such requests; with the Charter there is at least such a system, with people and states knowing who has what role, and what they can normally expect when calling for help in this domain.

The Charter focuses directly and exclusively on the mitigation of major disasters and their harmful effects without creating any new international organisation. It may therefore be said that it constitutes, so far, the sole international structured *system* for handling space-derived geospatial information data for emergency management. While there are no obligations to conduct geospatial information emergency management operations through the Charter, as such it clearly represents the most advanced context therefore, justifying extended discussion here. The Charter also represents a specific manifestation of such general principles of space law as pertaining to the benefit of all countries and the requirement to allow free and uninhibited access to data if natural or man-made disasters are at hand, as discussed above in the context of Resolution 41/65 (Outer Space Treaty, 1967, Art I, Resolution 41/65, 1986, Principle X, XI).

The Charter, formally declared operational on 1 November 2000, aims at providing a unified system of space data acquisition and delivery to those affected by natural or man-made disasters. Formally, such information would have to be requested by the affected state, even if in practice it may (have to) end up in the hands of third states and relief organisations supporting the affected state, which might not even have the technical means to work with the satellite information. This would also raise some legal issues worthy of further discussion and investigation.

Each member agency has committed resources to support the provisions of the Charter and thus helps to mitigate the effects of disasters on human life and property: ESA provides data from ERS and Envisat, CNES from the SPOT satellites, CSA from the Radarsat satellites, ISRO from the IRS satellites, NOAA from the POES and GOES satellites and CONAE from SAC-C.

Article 6(1) of the Charter stipulates that requests to adhere to the Charter may be made by any space system operator or space agency with access to space facilities agreeing to contribute to the commitments made by the parties. In other words, it is a *de facto* prerequisite for membership to the Charter to possess capability to operate satellite systems; or at least to start doing so in the near future. Those space facilities are not necessarily limited to earth observation satellites or instruments; 'space systems for observation, meteorology, positioning, telecommunications and TV broadcasting or elements thereof such as on-board instruments, terminals, beacons, receivers, VSAT's and archives' are also included (Charter on Space and Major Disasters, 1999, Art.1). Indeed, GOES and POES for example are meteorological satellites.

Upon request by a 'beneficiary body', the member agencies acquire the data on the area affected by the disaster from their satellites, process them into images, analyse them further if necessary, and distribute the resulting information free of charge to those states affected by the disaster via 'associated bodies'[4]. A state affected by disaster (or one intent upon coming to the rescue) that wishes to access relevant data needs to contact either associated bodies or 'cooperating bodies'[5] acting in partnership with an associated body.

The effective determination of which satellites should provide data for a particular disaster is to be facilitated by *a priori* scenario-writing, although this seems to be largely theory so far. The partners agree to engage themselves in writing a range of scenarios to anticipate which data and information would be useful for different types of crisis. The parties shall together analyse recent crises for which space facilities could have provided or did provide effective assistance to the authorities and rescue services concerned (Charter on Space and Major Disasters, 1999, Art. 4(2)), draw conclusions and prepare sample response plans for future events. The scenarios cover the issue of the type of sensors effective for specific disasters, and even more specifically include selection criteria for a specific satellite. Such scenario analyses save time when decisions are due with respect to provision of the most appropriate data to crisis victims, and hence facilitate rapid assistance.

A number of legal issues with respect to the operation of the Charter remain to be solved. The underlying point of note is that parties to the Charter continue to be obliged to follow all the international agreements they are party to, including those on copyrights, data access and liability as discussed in this contribution. The mere fact of signing a Charter, even if it would be fully legally binding, does not allow such signatories to ignore other international duties which they have to abide by.

In any particular case, one would have to look at which state, party to the Charter, has become party to which agreement, for if a state has not become party to an international treaty it is, basically, legally free to ignore its contents. In the case of the treaties dealt with in the present contribution, these include all or at least most of the Charter partners amongst their parties. If a state would consider that its 'obligations' or interests with respect to the Charter would be interfered with by obligations resting upon it as the result of an international treaty, it could – within the terms of that treaty, e.g. as to a one-year-advance-notice – denounce its membership to that treaty, or announce that certain reservations would henceforth apply.

Services under the Charter are provided on a 'best efforts' basis, implying that Charter members will take all necessary measures in rendering aid but do not guarantee successful results. A specific provision in the Charter waives the liability of satellite operators called upon to provide data under the Charter: 'The parties shall ensure that associated bodies which, at the request of the country or countries affected by disaster, call on the assistance of the parties undertake to: '(…) confirm that no legal action will be taken against the parties in the event of bodily injury, damage or financial loss arising from the execution or non-execution of activities, services, or supplies arising out of the Charter' (Charter on Space and Major Disasters, 1999, Art. 5(4)). So the member agencies would not assume liability arising from services offered under the Charter. Death cases are also subject to the waiver of liability, even though this is not stipulated specifically in the above clause.

This waiver of liability, however, does not comprehensively solve the problem. Firstly, since the Charter is concluded among the partner agencies but (obviously) not with all the potential crisis victims, the waiver of liability is not mutually agreed upon in any comprehensive sense. Therefore, certainly in those cases where the victim of a crisis is not (in) one of the countries to which the Charter partners belong, the unilaterally-declared waiver of liability raises questions as to its validity.

Furthermore, the Charter provides for a waiver of liability only concerning cases arising between the affected country and the Charter partners. It does not mention, for instance, cases arising from

---

[4] An 'associated body' is 'an institution or service responsible for rescue and civil protection, defence and security under the authority of a State whose jurisdiction covers an agency or operator that is a party to the Charter'; Art. 5(2), Charter on Space and Major Disasters.

[5] Cooperating bodies includes the European Union, the UN Bureau for the Coordination of Humanitarian Affairs and other recognized national or international organizations with which the parties may have cause to cooperate in pursuance of the Charter. A 'cooperating body' does not operate a space system but acts in partnership with an associated body which does; see Art. 3(5), Charter on Space and Major Disasters.

potential liability of intermediate actors with respect to Charter partners or countries affected by a disaster. The Charter does not stipulate whether such a state can assert a legal claim against intermediate actors directly, in case these are somehow involved in the damage being caused.

The above finally raises issues regarding the so-called 'Good Samaritan' principle, a principle known in various national jurisdictions. This principle essentially means that a person who injures another in imminent danger while attempting to aid him (as long as not under an obligation to do so), is not to be charged with contributory negligence unless the rescue attempt is an unreasonable one or the rescuer acts unreasonably in performing the attempted rescue (http://pa.essortment.com/goodsamaritanl_redg.htm). Its purpose is to prevent people from being unduly reluctant to help a stranger in need, for fear of legal repercussions should they make some mistake in doing so.

The 'Good Samaritan' doctrine has been used widely in different jurisdictions throughout the world. In Canada and the United States it is incorporated by means of specific acts. The principle is also reflected in different national laws of European states. If the rescuer has actually worsened the condition of the imperilled person many techniques are available to assess the rescuer's conduct: from mitigation of damages in Dutch law to the presumption of a low standard of care in French and English law. Since the 'Good Samaritan' principle is incorporated into domestic law of many states, it is generally considered to reflect customary international law.

What it means in the context of the Charter, however, and whether its main criteria and parameters are overruled by it, remains an issue to be dealt with in further detail. For example, the principle is usually found to apply only when there is no specific (legal) obligation resting upon someone to come to the rescue. Are states or governmental agencies in the possession of relevant knowledge, alternatively of technological means to easily acquire such knowledge, however, not obliged to share such information? In other words, do the Charter partners qualify as 'Good Samaritans' so as to be able to invoke this principle in their defence?

3.4   *General humanitarian obligations*

As already indicated, both the international space law-rules pertinent for remote sensing and the Charter on Space and Major Disasters are representations of a broader duty under general international law for states to assist other states and their peoples in cases of larger humanitarian disasters, whether natural or man-made. This excludes, understandably but of course very unfortunately, those man-made disasters created by wars, persecution and other forms of violence, since in particular those states where events in these categories take place are generally unwilling to have other states come to the rescue on humanitarian grounds.

Suffice it therefore here to make reference briefly to the existence of these underlying general humanitarian principles. Though they would apply also in cases not covered by either the international space law-regime or the Charter (whether *ratione materiae* or *ratione personae*), and as such would have a general bearing on a number of emergency response-related activities, their main disadvantage from a more practical perspective is their very broad and vague content. At every turn, a different set of issues and situations are at stake, making it very difficult to determine what, in any particular case, such general humanitarian duties would amount to in terms of, for example, concrete actions or measures.

For that reason, these obligations should be best perceived as obligations-of-effort, as opposed to obligations-of-result. Their practical reach remains to be determined for each specific instance, and in the last resort they may serve more as guidelines to prefer one course of action over another if, all other things essentially being equal, the one course would be more in tune with such humanitarian obligations.

3.5   *Implications for emergency response purposes*

Mirroring to some extent the copyrights issue, data access represents a major area of legal issues relevant for the present theme. For emergency response purposes especially the general international law regime on access to remote sensing data and the more specific requirements under the Charter resting upon key satellite operators should be taken into account. These regimes would considerably limit the discretion of any such key operator in deciding whether and how to distribute certain data.

Such limitations should essentially ensure that wherever geographical information becomes available that is of value for the purpose of emergency management activities – whether in the context of the Charter or not – of whatever nature (and perhaps, although this goes beyond the scope of the present contribution, also when not generated by satellites), it shall be made available without further ado for such purposes. Operators in the possession of such data, if worried that inappropriate use thereof might result from granting liberal access to their data, would do best to become part of the Charter-structure to the extent possible (if they are not already part thereof): even if also the Charter does not, as of yet, provide for a solid and general measure of protection against abuse, it is the only structure currently available where at least *some* protection can be enjoyed.

4 RESPONSIBILITY AND LIABILITY ISSUES

4.1 *State responsibility and satellite-based information for emergency response*

A further (very general and in first instance abstract) aspect of basic importance concerns that of responsibility and liability under international (space) law. As for responsibility, the general form of international accountability, states are responsible in broad terms for ensuring that activities conducted on their territory or within their jurisdiction do not violate the rights of other states. See for a general abstract treatise of state responsibility and how it works in the context of space activities e.g. Von der Dunk, 1998.

In addition however to such state responsibility principles as they arise under general international law, Article VI of the Outer Space Treaty has caused a specific version of those principles to be applicable to space activities (Outer Space Treaty, 1967, Art. VI). Space activities, or more precisely 'activities in outer space', would certainly include everything from the operation of a ground station controlling (part of) a satellite system to the activities of the latter itself, up to the generation of any geospatial information data (Outer Space Treaty, 1967, Art. VI).

Furthermore, Article VI of the Outer Space Treaty provides that states are internationally responsible for 'national activities in outer space', including cases where these are 'carried on (...) by non-governmental entities'. This responsibility pertains to 'assuring that national activities are carried out in conformity with the provisions set forth in the present Treaty'. States are thus responsible for activities undertaken in outer space in case these activities violate obligations under international space law. Moreover, states are responsible to the same extent for private activities as they are for their own, public activities – even, or perhaps more to the point: precisely – in the context of emergency management, in view of the clear public and humanitarian character of the relevant activities.

Whilst Article VI then begs the question: for which categories of private space activities is which particular state to be held responsible on the international plane, it would be beyond the purpose of the present article to deal with those issues. In any case, the answer to this question would lie in the interpretation of the key term 'national activities'. However, no authoritative definition of the (scope of) 'national activities' of a state for which it is to be held responsible has been provided by the Outer Space Treaty or elsewhere, and consequently no agreement exists as to the interpretation of this term. From the author's perspective, the most effective and sound interpretation of private 'national activities' would make states internationally responsible precisely for those activities over which they can exercise legal control (for further discussion see von der Dunk, 1998). In other words: a state would be held responsible for those private activities that are undertaken from within its jurisdiction.

4.2 *The concept of liability*

When analysing liability in the case of geospatial information products and/or services generated by satellites, one has to realise that again most of existing law and regulation is non-specific. Most of the legal environments further elaborating the concept, consequences and parameters of liability, moreover, are nationally defined, that is: operate within the territory of one particular state (only), even if (for example under space law) international regimes may be superimposed.

Liability therefore itself, as a concept and term used in numerous national as well as a considerable number of international legal regimes, may be differently interpreted, applied and, in particular, further elaborated, in each case. The consequence thereof is that at the international level quite often a large measure of confusion has arisen as to the scope, meaning and consequences in law of that concept.

'Liability' has for example been defined as the 'condition of being responsible for a possible or actual loss, penalty, evil, expense or burden', and as 'the state of being bound or obliged in law or justice to do, pay, or make good something' (Garner, 1999). For the purpose of discussion here, this may be restated as: 'the accountability of a person or legal entity to compensate damage caused to another person or legal entity, in accordance with specified legal principles and rules and based upon specified sources of law'. Thus, liability depends upon a specific legal regime, which itself determines the boundaries of the particular liability regime at issue for example as regards where it applies, which persons or legal entities are involved on both sides of the damage (the causing respectively suffering side), what type of liability is provided for, how compensation is being dealt with, and suchlike.

Furthermore, the fundamental threefold distinction between contractual liability, non-contractual liability and product liability should be noted, leaving aside for the moment the question as to the extent in which each of those types of liability would actually come to be involved in the context of geospatial information supported emergency management activities. The key issue distinguishing the three different types of liability focuses on the legal relationship between the alleged victim of the damage at issue and the alleged responsible therefore – in other words: between the claimant and the defendant.

Contractual liability should be defined as 'the liability which arises from a contract or agreement', and thus deals with liability as between partners to a contract, regarding activities undertaken in relation to respectively damage suffered in the context of that contract and its subject matter. Black's Law Dictionary, 295 and West's Law & Commercial Dictionary in Five Languages, Vol. I, 339, define 'contractual obligation' as 'the obligation which arises from a contract or agreement.' 'Contractual liability' is essentially a term coming from national law, and, by way of common denominator is explicit and formalised by way of the contract, already in existence at the time the relevant accident leading to damage occurs. Hence, for the purpose of analysis here it coincides in a principled sense with inter-party liability as it is often discussed on the public international level, where international treaties between states would essentially take the place of contracts.

From a legal point of view, dealing with contractual or inter-party liability is a matter of the freedom of parties to contract between themselves. This freedom may only – exceptionally – be restricted by an overriding public interest to ensure that contracts are generally fair, if indeed such public interests are expressed through a law or other general statute.

Non-contractual liability, in view of the above definition of 'contractual liability' then logically constitutes liability for damage occurring outside a contractual relationship, most prominently where the person or entity suffering the damage is in no way formally or contractually related to the person or entity causing it (or at least any damage caused would not be covered by any such formal or contractual relationship), and likely neither aware of the possibility of damage occurring nor able to take precautionary measures against it. Thus, it equates at this level of abstraction with the tort liability[6] of national legal systems, respectively the third-party liability known in international law: its common denominator would thus be that the legal relationship is implicit, not formalised and solely based precisely on the fact that one party is the proven cause of the damage sustained by the other party.

---

[6] 'Tort' is defined as 'a private or civil wrong or injury, other than breach of contract, for which the court will provide a remedy in the form of an action for damages' (Garner, 1999, pp. 1334); West's Law & Commercial Dictionary in Five Languages, Vol. II, 660.

As a consequence, in contrast with contractual liability, protecting the interests of third parties through non-contractual liability regimes clearly in itself is a public matter, to be taken care of preferably by legislative means, since by definition bystanders cannot protect their interests themselves by contract or otherwise. Hence, this is also the type of liability which a public legislative document on the international level is most often concerned with. On the national level, this equates with the need for, preferably, a clear written law or statute, or – in common law countries – at least clear jurisprudence and customary law.

Product liability finally is defined as: 'the legal liability of manufacturers and sellers to compensate buyers, users, and even bystanders, for damages or injuries suffered because of defects in goods purchased' (Garner, 1999, p. 1089; West's Law & Commercial Dictionary in Five Languages, Vol. II, 358). Thus, it is of a fundamentally different nature; not imposing liability upon someone for activities undertaken and damage suffered as a consequence, but imposing it upon someone having manufactured and/or sold a product by which, in the course of using it, damage has been caused. In a sense this constitutes an indirect form of liability, as the occurrence which triggers liability claims may take place (long) after the manufacturer or seller – the entity to be held liable – has had any involvement with the cause of the occurrence – the product. The relevant legal relationship here is effectively created through the product concerned.

Also product liability, even if elements thereof may have found their way into contracts (on the sale of the product), in the last resort has usually been considered a matter of general public interests being preserved through the enunciation of explicit laws, statutes or (occasionally, that is: largely in the case of the European Union) international legal documents of a binding nature.

4.3 Contractual liability in the context of emergency response activities

For contractual liability, of course any analysis would only be relevant in as far as in the context of emergency response activities contracts would be required, for instance with satellite data providers. In any case, two main categories of contracts could be at issue: public contracts and private (commercial) contracts. In either case of course potential liability will at the primary level depend on the contract terms negotiated between parties. The claimant will then have to prove that the service or product provider did not comply with its obligations of providing certain data or services.

Further to that, however, a private entity's contractual liability would be limited to the services and products it provides under the relevant contract, whether or not it would itself provide additional data or services downstream or confine itself to the provision of raw data only.

By contrast, some contractual relationships may be of a totally different nature since dealing with safety and security services: the value-added service providers would then (likely) be public entities or entities that benefit from public prerogatives justified by the fact that they are running a public activity. Hence these relationships would be more likely to take the form of public contracts, and be subject to public contract law, whether national or at a European level.

The most important thing to note, however, is that such contractual liability does not deal with any damage caused to those victimised or threatened by the emergency situation at issue, nor at the other end with damage caused to those trying to come to the rescue. Thus, it is an issue perhaps not of primary relevance in the present context.

4.4 Non-contractual liability in the context of emergency response activities

The main element of non-contractual liability to be discussed here, in the context of victims of emergency situations and addressees of emergency response activities, concerns the issue of 'negligence' involved in the provision of relevant data and services. Which activities in the present context could or would qualify as a negligent public act or negligent omission, and if so, what would that mean in terms of liability? The United States' National Oceanic and Atmospheric Administration (NOAA) has already been taken to court for its 'failure' to warn (adequately) against the December 2004-tsunami. Would there be an inherent obligation to provide certain guarantees? Or would it be lawful to waive or disclaim liability for (absence of) provision of relevant guarantees?

States under international law assume a certain responsibility for ensuring that relevant activities conform to rules of international law. States may not be held liable automatically at the international level, unless this has been expressly provided for in a treaty somehow applicable to the matter. Nevertheless, relevant operators or data providers might remain liable for negligence under national law, though one would have to study such relevant national laws in considerable more detail before more substantive conclusions would be feasible.

One specific regime at the international level meriting to be mentioned here concerns that of international space law. As far as direct physical damage caused by space activities is concerned, this is ruled by Article VII of the Outer Space Treaty, further elaborated by the Liability Convention. This regime provides for liability for damage caused by a space object resting upon the 'launching State(s)' of that space object; the concept of 'launching State' being defined in a fourfold fashion. The term 'launching State' means: (i) a state which launches or procures the launching of a space object; (ii) a State from whose territory or facility a space object is launched (Liability Convention, 1972, Art. I(c)). It may be noted once more, that such state liability would apply regardless of whether the actual operation causing the damage was privately conducted or even if the whole satellite venture would be a private one (see also von der Dunk 1998).

This is, however, not the whole story when it comes to liability for satellite activities in the context of emergency response. The international space law regime for liability is only relevant for damages caused by a satellite physically harming another space object or causing terrestrial damage – arguably even restricted to such damage caused by physical impact, that is a crash. In the case of emergency response, while this is not a negligible issue, attention also needs to be paid to the possible damage caused by the user of data or information, for example when that user, wrongfully trusting the data and services provided to him, causes damage which may in turn trigger other liability regimes to become applicable – with the user being held liable for such damage! Such other liability regimes may be both of a very general nature – tort or wrongful act – or of a more specific nature, yet still (arguably) applicable. In this context finally the 'Good Samaritan' principle once more may play a role in determining ultimate liabilities for damage occurring.

4.5 *Product liability in the context of emergency response activities*

Finally a few words on product liability in the current context. The generation and distribution of emergency response data and other products could involve product liability claims against the relevant providers. Two aspects of such activities are actual candidates for product liability suits: the equipment used to generate, transmit or receive data, and the data products themselves. Existing product liability law was, of course, not at all designed for such activities, and considerable *lacunae* and inconsistencies might arise when applying it to them nevertheless. It may, once more, be illustrative to zero in on the European context as established within the European Community legal framework, to illustrate how product liability law might be applied in the context of emergency response.

The Council of the Community adopted Directive 85/374 in June 1985, amending it by means of Directive 1933/34[7] in May 1999, to harmonise the product liability regimes of the member states (Council Directive 85/374, 1985). Directive 85/374 provides in Article 1 that the producer shall be liable for damage caused by a defect in his product.

Further to this general rule, the Directives contain the following main elements: liability without fault of the producer; burden of proof on the claimant as to the damage, the defect and the causal relationship between the two; joint and several liability of all the operators in the production chain, providing a financial guarantee for compensation of the damage; exoneration of the producer when he proves the existence of certain facts explicitly set out in the Directives; liability limited in time by virtue of uniform deadlines; and illegality of clauses limiting or excluding liability towards the victim.

---

[7] Directive 1999/34/EC of the European Parliament and of the Council of 10 May 1999 amending the Council Directive 85/374/EEC on the approximation of laws, regulations and administrative provisions of the Member States concerning liability for defective products, OJ L 141, 04/06/1999, 20.

The Directives allow member states to derogate from the common rules adopted with regard to three issues: (1) to include unprocessed agricultural products in its scope of application; (2) to not exonerate the producer even if he proves that the state of scientific and technical knowledge at the time when he put the product into circulation was not such as to enable the existence of a defect to be discovered; or (3) to fix a financial ceiling of not less than €70 million for damage resulting from death or personal injury and caused by identical items with the same defect.

This is, of course, the general regime for product liability within the European Union. Applying the Directives to products generated in the context of emergency response activities would be subject to a number of criteria being fulfilled. This concerns: (1) whether an emergency response data product will qualify as a product under them; (2) the extent to which the claimant is able to establish a defect in such product; (3) the extent to which the claimant is able to establish the alleged damage and the causal relationship between the defect and the damage; (4) establishment of the fact that a relevant entity is the producer within the meaning of product liability law and the Directives in particular; and (5) whether that producer has any justifiable and recognised defence.

In view of the liability cap in the Directives, and the prescription and liability periods introduced, it is possible, particularly in jurisdictions where contractual liability or general law of negligence offer better opportunity to him, that a claimant would choose alternative avenues for claims. This possibility is left open by the Directives; they shall not affect any rights which an injured person may have according to the rules of the law of contractual or non-contractual liability, or a special liability system.

4.6 Summarising: the liability issue and emergency response

The effect of the liability issue on emergency response activities should not be underestimated. The willingness to undertake such activities would, after all, be considerably lessened if liability claims would be possible at each and every turn. Relevant partners will face a number of non-contractual liabilities where there would be little opportunities to fundamentally deflect or alter such liabilities – and consequently will have to somehow face them and deal with them.

In the field of contractual liabilities, by contrast, relevant operators and information providers have a large discretion to determine the extent of such liabilities. Thus the question from the other end arises to what extent these would be prepared and willing to accept liabilities.

In terms of product liability finally, as was illustrated by the EU example, regimes may exist that have a bearing on emergency response products also, subject to a number of criteria being fulfilled.

It is beyond the scope of this contribution to develop further details on how to handle liability in the case of emergency management using space-based geospatial information data. To start with, more experience needs to be had with relevant operations somehow resulting in damage (rather than mitigating it), and how the resulting, conflicting interests were handled in practice. More importantly, probably, an analysis would then be necessary of the ways in which liabilities, and such more specific concepts as the 'Good Samaritan' principle, are elaborated and implemented within the national jurisdictions at least of the major countries and as representing the major legal systems of the world.

5   SECURITY AND DUAL-USE ISSUES

5.1   The Wassenaar Arrangement

The Wassenaar Arrangement is a global, formally non-binding arrangement on export controls for conventional weapons and sensitive dual-use goods and technologies (Wassenaar Arrangement, 1995). It was designed to promote transparency and greater responsibility in transfers of conventional arms, dual-use goods and dual-use technologies. Participating states commit themselves to ensure through national policies and, where appropriate, regulations that cross-border transfers of these items do not contribute to the development or enhancement of military capabilities of other states (Wassenaar Arrangement, 1995, Art. I(1)).

The decision to allow or deny transfer of any item, however, remains the sole responsibility of each individual state (Wassenaar Arrangement, Art. II(3)). Thus, export controls differ from state to state (in terms of documentation required, license fees, length of time to get a license, and duration of validity of the license).

The participating states only agree to notify transfers and denials, as well as to control all items in the List of Dual-Use Goods and Technologies and the List of Munitions, annexed to the Arrangement (Wassenaar Arrangement, 1995, Art. II(4), III(1), Appendix 5). Controls do not apply to technology or software in the public domain, to basic scientific research or to the minimum necessary information for patent applications. The Lists have two annexes of sensitive items and of very sensitive items respectively, to which different levels of control should be applied, and are reviewed regularly to reflect technological developments.

The participating states finally agree to exchange general information on risks associated with transfers of conventional arms and dual-use goods and technologies in order to consider, where necessary, the scope for coordinating national control policies to combat these risks (Wassenaar Arrangement, 1995, Art. IV(1)).

As to emergency response activities, subject to further analysis but above all experience, some of the products and services envisaged in their context might turn out to be, explicitly but especially implicitly, included in the relevant List. If so, the question arises what could be done about that and about the resulting potential obstacles to distribution of relevant satellite-generated information.

5.2 *Regulation 1334/2000*

The Wassenaar Arrangement as such does not recognise the European Union in any substantive manner. Partially as a result thereof, within Europe the same issue was also dealt with in a more classical, legally binding format by means of Regulation 1334/2000, which sets up a regime for the control of exports of dual-use items and technology for the EU member states (Regulation 1334/2000, 2000). An authorisation is required for export of the dual-use items listed in Annex I (which is essentially similar to the Wassenaar Arrangement's List of Dual-Use Goods and Technologies). If the prospective exporter is aware that an item, even if it is not listed in Annex I, might be used in a way proscribed by the Regulation, it is still bound to apply the applicable provisions (Regulation 1334/2000, 2000, Art. 4). Under the Regulation export is defined to include transmission of software or technology by electronic media, fax or telephone to a destination outside the Union.

As with the Wassenaar Arrangement, under Regulation 1334/2000 the responsibility for deciding on applications for export authorisations lies with the national authorities. Some items on the List of Dual-Use Items and Technology (Annex 1) are not controlled if they accompany the user and are for the user's personal use: Regulation No. 1334/2000 'does not apply to the supply of services or the transmission of technology if that supply or transmission involves cross-border movement of natural persons' (Regulation 1334/2000, 2000, Art. 3(3)).

The Regulation establishes a Community General Export Authorisation (CGEA) as set out in Annex II for certain exports. Annex II, Part 1, specifies that the CGEA is possible for all dual-use items listed in Annex I, except those specified in Annex II, Part 2, dealing with the more security-sensitive items. National export authorities are not automatically obliged to provide a CGEA, however, and, in any event, the exporter must comply with certain reporting requirements, as set out in Annex II, Part 3.

For all other items, authorisation shall be granted by the member state where the exporter is located (Regulation 1334/2000, 2000, Art. 6). This authorisation may be an individual, global or general authorisation. Member states must maintain or introduce in national legislation the possibility of granting a global authorisation to a specific exporter for dual-use items valid for export to one or more specified countries. The competent authorities may refuse to grant an export authorisation and may annul, suspend, modify or revoke an export authorisation which they have already granted (Regulation 1334/2000, 2000, Art. 9). Exporters are required to keep detailed records of their exports.

Once more, with a view to emergency response activities, the Regulation may turn out, under its present status and contents, to unduly obstruct the distribution of some of the relevant products and services. This ultimately depends, of course, upon the extent to which those products and services may, *prima facie*, be seen as dealing with dual-use and/or sensitive software or information.

5.3 *The United Nations system for international security*

For completeness' sake, it would be appropriate to refer here briefly to the general global system for dealing with international security issues, as developed in the context of the United Nations. The United Nations under the UN Charter has been given the major task by the member states to try and establish alternatively preserve international peace and security, using the various competencies allotted to it (UN Charter, 1945). Those competencies in particular rest with the UN General Assembly, which has the possibility to issue (non-binding) Resolutions as well as a role in despatching peace-keeping or peace-making forces, but especially with the Security Council, which has the power to issue binding Resolutions in this regard.

Under this system, the Security Council may, for example, impose boycotts, economic blockades or even authorise full-fledged military actions if it considers international peace and security sufficiently threatened (UN Charter, 1945, Art. 41, 42). Throughout the last decades, these powers have been used in such cases as Yugoslav civil wars *vis-à-vis* Serbia in particular, the Iraqi invasion in Kuwait in 1990, and the military actions against Afghanistan in 2002 and Iraq in 2003.

The main point to keep in mind for emergency response activities is that, should any such measures be imposed by the Security Council in the future, the relevant operators and geospatial information providers would be bound to comply with them as well. It could be imagined in particular that certain data products or services would not be allowed to be delivered to certain parties, or that certain international cooperation ventures with certain parties would have to be suspended or cancelled in cases where the Security Council would determine a threat to international peace and security to exist.

In such an event, a close reading of the actual decision by the Security Council would be requisite, since it will have to draw a very delicate balance between the political needs behind for example the suspension or cancellation of international cooperation and the obvious humanitarian needs resulting from the disaster at issue requiring geospatial information for alleviating the disaster's consequences. A comparison with the 'Food for oil' programme of the United Nations *vis-à-vis* Iraq at the time Iraq was already being internationally isolated in punishment for its refusal to comply with inspections of their purported facilities for weapons of mass destruction is illustrative from this point of view.

5.4 *Summarising: security issues and emergency response*

Discussion of issues of security and dual-use character in the context of emergency response activities is not that farfetched. Data generated by those activities or information based on such data, could very well be found to be subject to the legal regime, summary as it may be, applicable to international trans-frontier movement of security-sensitive information or become involved in international actions trying to preserve international peace and security.

It would therefore sooner or later be necessary to address these issues in more detail: analyse those situations where the issue has, or could have, come up, and then offer further suggestions to ensure that security interests and the humanitarian interests of emergency response are fairly and transparently balanced. For example, in the context of the Wassenaar Arrangement (and for Europe Regulation 1334/2000) exceptions could be drafted here necessary to enhance the clarify of what is, and what is not, appropriate in any given case of geospatial information supported emergency management involving potential security interests.

6 FINAL REMARKS

Maybe the above, first analysis raises more legal questions than it provides answers. This is, however, no doubt due to the novelty of the issue of emergency response on such an international level and with the fundamental involvement of satellite technology. To reiterate, one would need considerably more and deeper analysis, but in particular experience with legal, pre-legal or para-legal discussions and disputes, before any thorough discussion of the many issues specific to geospatial information data derived from space and used to support emergency management operations could be undertaken. It may perhaps come as an unwelcome surprise that such activities, normally undertaken with the best of intentions, might be subjected to legal scrutiny and run into legal obstacles or at least raise legal issues which may make potential rescuers think twice before doing the seemingly obvious.

Nevertheless, major legal issues (and in their wake also organisational ones) can indeed already be seen to arise, as the above has hopefully demonstrated, and precisely in order to ensure that the best intentions are allowed maximum leeway whilst undesirable side-effects are mitigated or even, preferably, ruled out, work should be done to solve those legal issues in the most appropriate way. In a number of respects, moreover, the European situation is particularly relevant and/or illustrative, in view of the fundamental legal developments taking place in the EU context and the extended legislative opportunities to deal with issues relevant for emergency response.

It is all about balance. A proper balance will have to be found for example between justified interests of a copyright owner in protecting his intellectual property regarding certain data products useful for emergency response action and the obvious public interest in allowing such data products to be, in essence, so used. Similarly, interests stemming from security perspectives should not unduly hinder humanitarian efforts to respond effectively and swiftly to disasters or emergencies. The attendant responsibilities and liabilities, which will not of themselves go away by the mere fact of an action being of a humanitarian, emergency-response related character, will have to be distributed appropriately.

In view moreover of the international character, both of many of the major emergencies and disasters and of the use of satellite images to try and deal with them, such a balance should preferably have a strong international component. While there is no denying the relevance of national interests and national sovereignties in today's world in spite of creeping globalisation, and many legal issues cannot but be solved at a national level, a certain international understanding based on sound international – read essentially inter-state – agreements seems to be indispensable.

Actually, the Charter on Space and Major Disasters represents the, so far, furthest step in that direction. Without creating as of yet an institutional structure or even undisputed legal obligations, it has brought into focus the serious and substantial willingness of a number of satellite operators to (allow others to) use geospatial information generated by their satellites for overly humanitarian purposes. The almost weekly growing number of Charter activations moreover show that the practical value of such constructs for many is not at issue anymore. Finally, from the perspective of international law it is very interesting to note the range of states having so far triggered – or at least grudgingly accepted – activation of the Charter since threatened by or suffering from major disasters: developing countries as much as developed ones, from the South to the North, from the East to the West, literally from the United States to North Korea (in which last case it actually was the United Nations Office for Outer Space Affairs that activated the Charter).

All this means that global acceptance of certain duties cannot be far, and that seems to predict a bright future for further efforts to clarify the legal issues that need to be solved – such as those elucidated in the present contribution. It certainly means that using satellite-generated data and information is considered to be a prime example of the 'benefits and interests' of mankind and all countries, which Article I of the Outer Space Treaty posits as one of the legal cornerstones of all usage of outer space.

REFERENCES

Berne Convention, 1886, Berne Convention for the protection of literary and artistic works, available at http://www.wipo.int/treaties/en/ip/berne/trtdocs_wo001.html (last accessed October 2006)

Garner, A.B., 1999, Black's Law Dictionary, Sevent Edition, 823; West's Law & Commercial Dictionary in Five Languages, Vol. II, 47

Charter on Space and Major Disasters, 1999, Charter On Cooperation To Achieve The Coordinated Use Of Space Facilities In The Event Of Natural Or Technological Disasters, available at http://www.disasterscharter.org/main_e.html. (last accesses October 2006)

Council Directive 85/374/EEC, 1985, Council Directive 85/374/EEC on the approximation of the laws, regulations and administrative procedures of the Member States concerning product liability, available at: http://europa.eu.int/eur-lex/lex/LexUriServ/LexUriServ.do?uri=CELEX:31985L0374:EN:HTML (last accessed October, 2006)

Directive 96/9/EC, 1996, Directive of the European Parliament and of the Council on the legal protection of databases, available at: http://europa.eu/scadplus/leg/en/lvb/l26028.htm (last accessed October 2006)

ESA Convention, 1975, Convention for the Establishment of a European Space Agency, available at: http://www.esa.int/convention/ (last accessed October 2006)

Liability Convention, 1972, Convention on International Liability for Damage Caused by Space Objects. Available at: http://www.islandone.org/Treaties/BH595.html (last accessed October 2006)

Outer Space Treaty, 1967, Treaty on Principles Governing the Activities of States in the Exploration and Use of Outer Space, including the Moon and Other Celestial Bodies (hereafter Outer Space Treaty), available at: http://www.state.gov/t/ac/trt/5181.htm (last accesses October 2006)

Regulation 1334/2000, 2000, Council Regulation setting up a Community regime for the control of exports of dual-use items and technology, No. 1334/2000/EC, available at: http://trade.ec.europa.eu/doclib/docs/2004/september/tradoc_118992.pdf (last accessed October, 2006)

Resolution 41/65, 1986, Principles Relating to Remote Sensing of the Earth from Outer Space, available at: http://ioc.unesco.org/oceanteacher/oceanteacher2/01_GlobOcToday/01_SciOc/05_RemSens/UNprinciples.htm (last accessed October 2006)

Resolution 51/122, 1996, Declaration on International Cooperation in the Exploration and Use of Outer Space for the Benefit and in the Interest of all States, available at: http://www.un.org/documents/ga/res/51/a51r122.htm (last accessed October, 2006)

UN Charter, 1945, Charter of the United Nations, available at: http://www.un.org/aboutun/charter/ (last accessed October 2006)

Universal Copyright Convention, 1952, available at: http://www.unesco.org/culture/laws/copyright/html_eng/page1.shtml (last accessed October 2006)

von der Dunk, F.G., 1998, Private Enterprise and Public Interest in the European 'Spacescape' – Towards Harmonized National Space Legislation for Private Space Activities in Europe (1998), Dissertation, xv + 311 + XCII pp.

Wassenaar Arrangement, 1995, Wassenaar Arrangement on Export Controls for Conventional Arms and Dual-Use Goods and Technologies, available at: http://www.wassenaar.org/ (last accessed October, 2006)

Part 2
Data collection and products

Real-time data collection and information generation using airborne sensors

N. Kerle

Department of Earth Systems Analysis, International Institute for Geo-Information Science and Earth Observation (ITC), Enschede, The Netherlands

S. Heuel

Ernst Basler + Partner AG, Zollikon, Switzerland

N. Pfeifer

Institute of Photogrammetry and Remote Sensing, Vienna University of Technology, Vienna, Austria

ABSTRACT: In this chapter we review existing airborne platforms and sensors, as well as the products that can be generated from the data they acquire, and assess their utility in the context of emergency response. Airborne remote sensing has a much longer history then satellite-based Earth observation, which has led to the development of a wide variety of vehicles that can carry different sensors, ranging from tethered balloons to Unmanned Aerial Vehicles (UAVs) that operate at stratospheric heights. We base our evaluation on a number of objective assessment criteria that include technical and commercial aspects, as well as various considerations for emergency response, such as required processing time or use for a given disaster type. We specifically show how different emergencies can lead to a range of observable consequences that call for suitably adapted remote sensing tools, and how more sophisticated products, e.g. 3D or 4D, or integrated in a GIS environment, can lead to improved response capability. We review the use of optical, thermal, laser scanning and radar data, and show how they are best suited for different disaster scenarios. Thus the decision on which data type or product to use in a given emergency situation is a multi-criteria problem that requires good understanding of the available information types, their utility for the disaster in question, but also practical and flexible thinking to identify potential alternatives. We show that airborne remote sensing has great potential to aid in an emergency situation, though also note that the provision of value-added products within the 3 day response window we consider here will not always be possible. We conclude with a table that summarises the relevant characteristics of the main remote sensing data types and products with respect to their suitability for different disaster types, which can serve as guidance for decision makers.

1 INTRODUCTION

Natural or industrial emergencies, with their wide variety of spatio-temporal and damage characteristics, call for an equally diverse arsenal of tools to aid in their response. Remote sensing provides an entire suite of sensors and platforms that allow pertinent data to be collected on any given form of disaster. To that end relevant sensors can be deployed on the ground, in the air, or in space. In this section we give an overview of the utility of airborne remote sensing to provide real-time spatial data in emergency situations, while the following sections address the prospects, as well as limitations, of spaceborne and terrestrial sensors, as well as sensor webs. We provide here an introduction to the currently available platforms and sensors, and review them in terms of their technical and economical characteristics, as well as their actual utility in a crisis situation. We discuss the products that can be generated from those sensor data, and the specific purpose they may serve in crisis response, provide a comparative overview of existing technologies, and discuss relevant current technical developments. The discussion is illustrated with examples and links to further literature.

1.1 *Emergency response and airborne data*

Airborne remote sensing is typically equated with sensors mounted on airplanes. However, the field is far more diverse and comprises a multitude of platforms that range from small low-flying remote controlled model aircraft to Unmanned Aerial Vehicles (UAVs) that fly at stratospheric heights of up to 30 km. Depending on technical constraints such as payload these platforms can be equipped with different sensors. In section 3, we provide a detailed overview of the various available platforms, and in 4, we introduce the sensors and the products their data can provide.

The purpose of this chapter is not to provide cookbook-type guidance on how to respond to a given disaster with remote sensing. The variety of possible platform-sensor combinations alone precludes such exhaustive treatment. This is further compounded by the disaster-type specific utility of remote sensing data. By this we mean that each disaster type, in its spatio-temporal characteristics as well as its physical consequences, demands an appropriate technical solution. For example, flooding is frequently associated with cloud cover, thus requiring either an active radar sensor that can penetrate clouds, or else an optical system flown beneath cloud level. Further complexity is added to the possible range of technical solutions by the different products that can be generated from the raw data, creating a 3-dimensional solution space within which the appropriate response strategy for a given disaster situation can be found.

The approach chosen also has to be supported by the financial and technical means available. Good examples of extreme scenarios of disaster-type dependence are the 2004 Asian tsunami and the 2001 collapse of the World Trade Center in New York, both devastating and sudden events. The former was vast in scale, requiring coastlines in several countries to be surveyed, most of which lacked the required technology. In New York, with unconstrained financial means and political will, as well as small affected area, multi-source data were acquired daily and products made available to ground forces in near real time (Rodarmel et al., 2002; Williamson and Baker, 2002).

Thus in this section we provide an overview that is suitably generic but strongly illustrated with specific case studies. However, it has to be stressed that some basic understanding of remote sensing is necessary to decide on a suitable technology in a crisis situation. For additional information on basic properties of remote sensing technology, as well as image analysis and interpretation concepts see for example Lillesand et al. (2004) or Mather (2004), and McGlone et al. (2004) for relevant photogrammetric concepts.

2 ASSESSMENT CRITERIA

When assessing different techniques for their value in emergency response, clearly defined criteria have to be used. In the context of this book this includes the standard technical parameters of remote sensing systems, their cost and availability on the market, but also their actual utility in terms of their state of development or automation, suitability for specific disaster types, required post-processing, or need for specific operator expertise. These are defined in this section, and relate less to the platforms but more to the sensors they carry, as those provide the actual data. The sensor types we consider in section 4 are (i) uncalibrated and calibrated cameras, (ii) laser scanners, (iii) thermal sensors, and (iv) airborne radar.

2.1 *Technical aspects*

(1) Spatial resolution
Spatial resolution refers to the smallest unit of area measured, and indicates the minimum detail of objects that can be distinguished. Generally speaking an inverse relationship exists between flying

height and spatial resolution. Airborne sensors, therefore, typically have higher spatial resolutions (cm to a few meters) than satellite sensors (meters to km), although modern optical satellites also achieve panchromatic resolutions of close to 0.5 m. For large scale disasters such as flooding a resolution of 100–500 m may be sufficient, while for more localized events, especially in urban areas, sub-meter detail may be necessary.

(2) Spatial coverage
Spatial coverage refers to the total ground area covered by one image. Similar to spatial resolution it is typically inversely related to flying height. While satellites may cover over 100,000 km^2 in one image, a sensor on a low-flying helicopter may only image a fraction of a km^2 at once, albeit at higher spatial resolution. While the latter may be suitable for a localized event, surveying a large area affected by an earthquake with such limited spatial coverage is impractical and costly, and a compromise between spatial resolution and coverage may be needed.

(3) Accuracy and precision
Accuracy is the degree of conformity of a measurement to a value accepted as true. It is a function of the errors affecting the measurement process and relates to the *quality of a measurement result*. Precision relates to the quality and *reproducibility* of an operation used to make a measurement, and the *degree of measurement*. For example, a thermal scanner may record the temperature of a given pixel at a precision of 2 significant decimals (e.g. 15.45°C), yet may be off by 2°C due to incorrect calibration (low accuracy). Therefore, both accuracy and precision are important, yet increases in either are also costly.

(4) Calibrated vs. non-calibrated
A measuring instrument is called calibrated if it allows measurements that are precise and accurate according to a defined task. In the context of airborne sensors we use the term calibration as the determination of geometric parameters: first, the exact interior parameters of a sensor (e.g. the focal length for a camera), and second, the exact exterior parameters, meaning the position and angular attitude of the platform with respect to the actual object (note that traditionally photogrammetrists use the term "calibration" only for interior parameters). Only with well calibrated interior *and* exterior parameters it is possible to obtain accurate geometric information. However, it is sometimes not possible to have access to completely calibrated sensors, especially not in emergency response cases. We, therefore, also include partially or completely uncalibrated sensors, such as digital video and still consumer cameras. For these cases, the missing calibration parameters, e.g. the position and altitude, have to be recovered during a pre-processing step to be able to relate the imagery to the real world.

(5) Analogue or digital
Traditional aerial photography used analogue cameras, which are gradually being replaced with digital systems, a development mirrored by video and consumer still cameras. Analogue photo cameras still tend to have higher spatial resolution, and procedures to process them (e.g. photogrammetrically) are well established. However, digital cameras require less exposure light, record also in the near infrared (up to approximately 1.1 μm), and digital data can be processed and distributed directly and without the need for scanning. Given the time pressure following a disaster, digital systems are thus advantageous. However, in many parts of the world analogue cameras continue to be used, and if rapid deployment and processing protocols are in place, they too can acquire timely data (see 2.1(6)). Other sensors, such as laser scanners, acquire digital data by design.

(6) Deployment time
Satellites are continuously in orbit, and their data collection either has to be simply activated or reprogrammed. The principal limiting factor is orbiting constraints, whereby the earliest image acquisition by a given sensor may be days away. With the exception of HALE UAVs (see 3.5), which may be used for continued observations above a designated area, all airborne sensors have to

be specifically deployed following an event. The required time is a function of the number of available suitable systems, the location of the disaster, and financial considerations. Given the severity of an event (and associated political will), standard aerial cameras can typically be deployed within days. Police or media helicopters that may carry still or video cameras, however, may be deployable within hours. Laser and radar scanners, of which far fewer operational systems exist, may only be available weeks after a given event. For immediate emergency response, deployment time is the most critical parameter.

(7) 2D and 3D topographic mapping capability
The information content, and therefore value, of information tends to increase with higher dimensionality. Relief information added to a multispectral or thermal image allows a more in-depth analysis of a given phenomenon, including a disaster. Some sensors, such as laser scanners, collect height information by design, while others, such as stereo (i.e. overlapping) aerial photographs or video frames allow extraction of 3D information using photogrammetric techniques. Pre- and post-disaster 3D information allow valuable change detection, such as identification of collapsed buildings.

(8) Environmental deployment restrictions
Some sensors are restricted in their operation by environmental factors. Some active sensors (i.e. those emitting a signal and recording the return, such as radar) can operate at night and poor weather conditions, while laser scanners, also active, cannot be used during rain due to its different signal wavelength. Passive optical sensors in particular are limited to daylight conditions, and require cloud and smoke free situations. Such environmental limitations are of particular concern during crisis situations where a wait for more favourable conditions may not be an option.

2.2 *Commercial aspects*

(1) Cost
The cost of data is a function of many factors. It depends on the type and availability of a given sensor, the possible need to transport it in remote locations, and the deployment infrastructure required (e.g. ground processing infrastructure, GPS stations). A sensor can be highly versatile and mobile (e.g. a still or video camera), or dependent on a fixed frame (e.g. a radar sensor or laser scanner integrated in a customized aircraft). If a sensor has to be shipped to another country or continent first, the cost will increase (although in that case data acquisition will most likely also move outside the 3-day window considered here). Lastly, some sensors acquire data that are already of value in their raw format (e.g. digital photographs). Others require sophisticated (and costly) post-processing (e.g. laser and radar data) to be of value to disaster response forces.

(2) Availability
As indicated above, some sensors are widely available and thus readily deployed. Of others, in particular radar sensors, only a limited number of operational systems exist. Despite their use, for that reason we only provide a limited discussion on radar sensors.

2.3 *Utility aspects*

(1) State of technological development
Given the long history of airborne remote sensing, a range of methods, such as photogrammetry, are well established and implemented in standard software packages. Other approaches, e.g. laser scanning, are well developed in their actual data acquisition, while the data processing is still subject to extensive research. This is also true for the processing of non-calibrated data (e.g. photographs taken from consumer cameras). There are also platforms that are still in the development or testing phase, such as HALE UAVs or powered kites. In our evaluations below we distinguish between research ideas, research or commercial prototype, commercially used, and well established methods.

(2) Required expertise for operational use

Sophistication and maturity of a data acquisition method not only affects cost and availability, but also the need for additional expertise. Some products, such as videos or photographs can be used to some extent without any special training or assistance, while additional basic processing, such as the generation of Digital Elevation Models (DEMs) or change detection is possible for individuals with limited training. On the other hand, the operation of an UAV or a radar sensor requires higher levels of expertise, and is typically not directly possible for disaster response forces. Thus in a crisis situation not only the suitable data type has to be considered, but also the need for such extra expertise, and whether that is feasible in time and in the disaster zone.

(3) Required (post-)processing time

As stated in 2.2(1) some methods require no processing beyond the actual data/image acquisition, while others only collect data that are not useful in their raw form. Here we also need to distinguish between data, i.e. what is collected by a sensor, and information. For a photograph there may be little difference, since relevant information on the extent or consequences of a disaster may be readily extractable from the photograph (the raw data), even without georeferencing. However, 3D topography, radar-based flooding maps, change detection etc. all require additional processing, leading to higher costs and delays in information availability.

(4) Suitability for different disaster types

Even though different sensor types can be straightforwardly grouped into different technical categories, their utility in emergency situations is strongly disaster-type dependent. For example, laser and radar sensors are active instruments, yet to map the extent of flooding the most useful are radar and infrared optical sensors, since radar waves penetrate clouds and water appears black in infrared bands (yet only in cloud-free conditions). Hence in the comparison below we rate the utility of a given sensor/product for the following principal disaster types: earthquake (EQ), forest fire (FF) and urban fire (UF), flooding (FL), landslides (LS), volcanic eruptions (VE), and surface lifeline damage (e.g. power cables, pipelines; SLD). Note that for some disaster types, such as flooding or landslides, the physical expressions to be detected are of one principal type (water and landslide scar, respectively), allowing for a single remote sensing solution to be identified. Others, however, such as volcanic activity can occur in different ways or combination thereof, thus requiring several different sensors. A thermal sensor would record changes in surface temperature, a multispectral sensor detect ash or gas emissions, while a radar instrument could detect changes in surface morphology. Other disaster types that defy easy analysis are earthquakes, which can lead to building collapse, lifeline destruction, fires or other secondary damage, or industrial accidents.

2.4 Time vs. information

The previous section has illustrated that a range of criteria can be used to assess the utility of a given platform or sensor for a given disaster type. To analyse such a multivariate situation requires in-depth knowledge of remote sensing. Thus we also provide more generalized guidance for emergency managers who want to assess the potential of incorporating Earth observation technology in their response activities. To that end we first have to introduce the time vs. information concept, and emphasise that different emergency response activities, and thus different stakeholders, require different types of information at different times. For example, the international media is interested in rapid but more general vital statistics of an event for its reporting. Work to prevent post-disaster disease outbreaks, on the other hand, requires far more detailed and location-specific information, both socio-economic and environmental, at a later stage.

The time vs. information concept is illustrated in Figure 1. Following an unheralded event, initially only knowledge of the disaster type and its (approximate) location exists. The aforementioned disaster-type dependency becomes clear here. For an industrial accident, for example a fire in a refinery, the type of the event (explosion, fire) and its location are immediately clear, although not necessarily its cause. For an earthquake in a remote area the seismic signals recorded by different

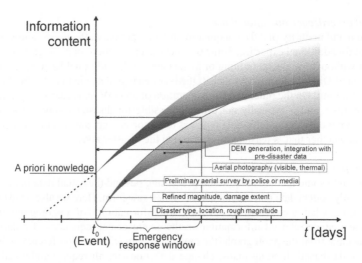

Figure 1.   The time vs. information concept. The principal focus of this book, and thus the information that can realistically be derived from remote sensing data, is on the first 3 days after a disaster.

stations only reveal the rough location and magnitude, and are subsequently refined. Depending on the availability of a suitable sensor and the environmental conditions, initial surveys by the police or news media may be carried out, to be followed by more structured and sophisticated surveys in the following days or weeks. The times at which those activities are carried out depend on the criteria discussed in 2, in particular on the commercial and utility aspects, but also deployment time and environmental conditions. Hence the amount of derived information at a given time t_0 (the event) $+ x$ is variable. The situation also changes profoundly with availability of a priori data, in particular in form of advance warning, but also with the availability of emergency response protocols and pre-disaster spatial data (preparedness). The shape of the information content areas in Figure 1 is thus highly variable and situation-dependent. Nevertheless, it is possible to make approximate statements on the realistic availability of data and derived products at different times after the event.

In the following we will provide an overview of existing airborne platforms, followed by the available sensors and products that can be derived.

3   AIRBORNE PLATFORMS

In principle a sensor can be attached to anything that flies – and it has. The first aerial photograph was taken from a manned tethered balloon in 1858. Later cameras were strapped to kites and carrier pigeons, and subsequently to powered vehicles ranging from rockets to remote-controlled model aircraft to Unmanned Aerial Vehicles (UAVs). The type of platform influences (i) which sensor(s) can be deployed, (ii) the principle acquisition parameters (spatial and temporal resolution, ground coverage), (iii) the cost of data acquisition, (iv) the environmental conditions within which can be operated, and (v) the level of control, accuracy, and convenience. Existing platforms can be categorised in a number of ways, such as low-vs. high-flying, manned vs. unmanned, experimental vs. established, or low vs. high cost. We structure our review according to flying height, although overlap exists (see Figure 3). As regards passive sensing, all platforms discussed here in principle allow sensor deployment for acquisition of vertical, low- or high oblique images (Figure 2). Oblique images are suitable for panoramic views of large areas, but are difficult to integrate with existing spatial data due to geometric distortions. In addition they lead to extensive occlusion, making image processing in particular in urban areas more difficult. Vertical imagery covers smaller areas but is

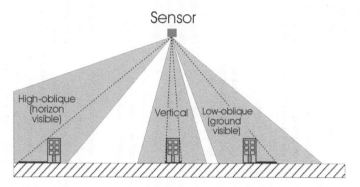

Figure 2. Vertical vs. high- and low-oblique imaging. Note the increasing occlusion in oblique imagery. Black bars indicate areas that cannot be imaged for the given camera orientation.

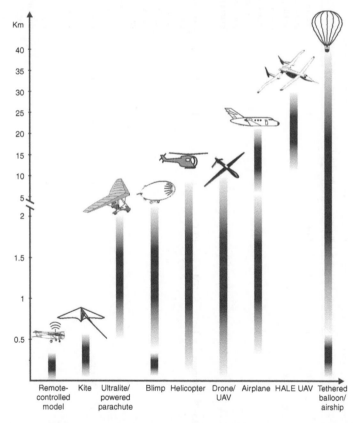

Figure 3. Overview of approximate minimum and maximum operating altitudes of different airborne plat-forms (shaded bars). Note that there are two different kinds of blimps – low-flying unmanned and those piloted at higher altitudes. Similarly we can distinguish between (i) airplanes acquiring stereo-photography that operate at altitudes of 1–5 km and aircraft flying at greater heights, and (ii) relatively simple and inexpensive balloons that may reach heights of approximately 1.5 km, and sophisticated and expensive polyurethane ones that reach stratospheric altitudes. See Table 1 for details.

Table 1. Overview of the basic characteristics, capabilities and limitations of available airborne platforms. Some values, in particular those related to capabilities, are approximations due to the range of models available.

System	Altitude range (km)	Pay-load (kg)	Deployment time	Ground coverage[1]	Manned/ Unmanned	Development status[2]	Availability[1]	Cost[1] Acquisition	Cost[1] Operation	Advantages	Limitations	Examples systems and references
Blimp	0.1–0.4	6[3]	Minutes	L	UM	PT-C	L	L-M	L	– inexpensive to acquire and operate – stable platform – easy to transport and maintain – also images from very low flying heights – manoeuvrable	– helium has to be stored/ transported – large volume relative to payload – danger of power lines, trees, towers, etc. – high weather dependence – flying height difficult to keep constant	(Ries and Marzolff, 2003)
Airship	3	>500	Hours	M	M							
Kite	0.05–0.6	0.5–2[4]	Minutes	L	UM	PT-C	L	L-M	L	– can carry several cameras – can be fitted with an attitude regulator for controlled imaging – can be used under a wide range of weather conditions – easy to transport and maintain – does not fall under airspace regulation	– small payload – risk of blurred images caused by sudden line movements – tethered, and thus stationary – relatively high risk to deployed sensors	(Boike and Yoshikawa, 2003; Oh and Green, 2003)
Remote controlled model aircraft	0.01–0.3	3–11[5]	Minutes	L	UM	PT	L	L-M	L	– live-camera downlink possible for accurate imaging – relatively stable and controlled – flexible navigation – low cost – does not fall under airspace regulation	– requires constant line of sight of operator – small area coverage – sophisticated systems require monitoring software, communication links, video link for monitoring the image overlapping skilled operator needed	(Eisenbeiss, 2004; Jang et al., 2004); Hirobo and Eagle 90

50

Platform									Advantages	Disadvantages	References
Balloon	1.5 5–40	0.5– >500[6]	Minutes Hours	L M-H	UM	C-E	L-M	L-E	L-M	– repetition easily possible – low cost – high payload for larger balloons	– helium has to be stored/transported – risk of blurred images caused by sudden line movement – use in calm wind conditions only – tethered, and thus stationary	EASOE-Campaign on MIPAS-B (Kramer, 2002)
Ultralite/ powered parachute	0.5–1.5	200	Minutes	M	M	PT-C	L	M	L	– stable platform and simple operation – take off distance as low as 15 m – can be used without additional infrastructure support – moderate construction and low running cost – slow speed and low flying height for high resolution data	– not suitable for mapping large areas – geometric quality of acquired imagery is variable – limited to flying in good weather (wind less than 20 km/h) – software must be custom designed (commercial off the shelf software is not widely available)	(Dare, 2005) (Spatial Scientific)
Helicopter	0.1–9[7]	750	Hours	H	M	E	M-H	H	H	– flexible – hovering possible – high payload and operating range	– less mobile to deploy – pilot and possible sensor operator needed – support infrastructure needed – high operating costs	Heliscat on DASA/MBB Helicopter, BO-105 (Kramer, 2002)
Airplane	0.3 up to 22[8]	Up to 1500	Hours	H	M	E	M-H	H	H	– large area coverage – stable platform – good altitude and attitude control – large payload and space for auxiliary equipment	– pilot and possible sensor operator needed – takeoff and landing strip/field needed – high operating cost	DO 228-212 (Tufte et al., 2004); ER-2, M-55, INGARA, LARSEN, LFS (Kramer, 2002)

(Continued)

Table 1. (Continued)

System	Altitude range (km)	Pay-load (kg)	Deployment time	Ground coverage[1]	Manned/ Unmanned[1]	Development status[2]	Availability[1]	Cost[1] Acquisition	Cost[1] Operation	Advantages	Limitations	Examples systems and references
UAV (drone) (non-military)	0.1–4.5	5–20	Minutes – hours	L-M	UM	C	L-M	L-H	L-H	– slow and stable flight – real time image transmission – different sensors can be mounted – pictures are taken via remote control or preset plan – high resolution images at the desired temporal resolution – autonomous flight possible for advanced models	– skilled teleoperators necessary – maximum wind speed of ca. 20 km/h for smaller models – more sophisticated and autonomous models are expensive and require extensive avionics systems and ground support – falls under airspace regulation (if >100 m)	(Ambrosia et al., 2003)
UAV (HALE)	12–30		Not applicable (long-term operation)[9]	H	UM	PT	L	H	M	– environmentally friendly – flies above air traffic control zone – flight duration up to several months (thus high temporal frequency imaging of area of interest) – low operating cost	– skilled teleoperators necessary – needs proper ground infrastructure, flight planning, data downlink, flight control	Global Observer (2004 to present); Helios, Pegasus (VITO)

[1] L – low, M – moderate, H – high.
[2] PT – prototype, C – commercially used, E – well established.
[3] 10 m length, 6 m diameter (Ries and Marzolff, 2003).
[4] For 3 m wingspan and flown at 16 km h⁻¹ wind (Oh and Green, 2003); historic maximum of 20 kg over San Francisco in 1906.
[5] Bell 222.
[6] Floatograph Technologies.
[7] Eurocopter landing on Mount Everest in 2005; regular helicopters have lower operating altitude ranges.
[8] M-55 Stratospheric Aircraft.
[9] Note that there are also plans to build several Pegasus systems that could also be shipped to a location where they are needed. In that case the deployment time would be days to weeks.

more suitable for topographic mapping and execution of pre-planned flight paths. Radar imaging is by design oblique (side-looking).

3.1 Balloons, kites and blimps

Of all platforms used in remote sensing, balloons and kites have the longest active history, with kites being the first platform used for the assessment of a large disaster. In 1906, following the San Francisco earthquake, a 20 kg camera was flown on a kite-train over the destroyed city, with the topmost kite reaching a height of almost 600 m. Even though these platforms appear cumbersome and ungainly, they score in terms of flexibility, easy of use, and low cost. All three platform types are thus actively used and advanced. Although still a niche application, Kite Aerial Photography (KAP) is now a recognised field, and technical advances have led to more controlled image acquisition (Oh and Green, 2003; Tielkes, 2003). For single kites, the typical payload ranges between 0.5–2 kg, and operating altitude around 150–200 m. Platform cost is determined by the size and material of the kite, and increases if radio control for camera orientation or shutter control is added. Complete sets for image acquisition are commercially available (Haefner, 2004). While kites can operate in moderate to strong wind, balloons (typically filled with helium) require calm conditions. With approximately 1 l of helium per 1 g payload (i.e. 1 m^3 for 1 kg), relatively large balloons are needed to carry a sensor. A further disadvantage is that the helium needs to be transported to the deployment site. Given their connection to the ground, neither kites nor balloon are suitable to survey larger areas. However, very low-cost image acquisition has been demonstrated (Lindholm, 2004). Blimps or airships, either filled with helium or hot air, are powered and thus manoeuvrable. First used in 1901 in France they still serve in research and surveillance, with payloads of up to 6 kg (Ries and Marzolff, 2003). They are also more stable than kites and balloons, can operate at altitudes of approximately 400 m, and in windy conditions of up to 18 km/h.

3.2 Remote controlled model aircraft

Remote controlled aircraft offer another controlled means to collecting imagery over small areas. They are operated in real-time via radio control, and require line of sight between aircraft and operator. Depending on the model size, payloads can be as high as 11 kg, allowing more sophisticated or several instruments to be deployed. Unlike UAVs (section 3.5), model aircraft typically lack autonomy and are thus restricted to an area defined by the line-of sight radius. Their operating altitude is typically less than 300 m. Model aircraft are not subject to regulations and can thus be flown also in urban areas. Both remotely controlled helicopters and airplanes are being used for detailed urban mapping/monitoring (e.g. Jang et al., 2004; Thome and Thome, 2000).

3.3 Ultralights and powered parachutes

Ultralight aircraft and powered parachutes represent a more advanced and powerful platform, allowing higher payloads, larger ground coverage and flying height. While the cost of acquisition is moderate, operating costs are low. Dare (2005) reported on a powered parachute system that costs approximately 20,000 USD in acquisition, but only 25 USD per hour to operate. In addition to higher cost, piloting skills are required, and in many countries these vehicles fall under aviation regulations, and are thus subject to licensing and limited operability.

3.4 Airplanes and helicopters

Airplanes have been the dominant platform in aerial remote sensing since shortly after the invention of powered flight. Flying typically at altitudes of 1000–6000 m (though in extreme cases also at stratospheric altitudes), they facilitate planned surveying of large areas for topographic mapping. Depending on their size they can carry large payloads exceeding 1000 kg, allowing for several instruments to be operated jointly. Within the ER context planes are a suitable platform for large-scale disasters, but also to carry relatively heavy instruments such as laser or radar sensors. Helicopters are more limited in their maximum operating altitude, but can carry payloads comparable with airplanes, and do not require runways for takeoff and landing, and are able to hover over a scene. Operating costs are higher for helicopters than for small airplanes.

3.5 *Unmanned Aerial Vehicles (UAVs)*

The emerging platform of choice is Unmanned Aerial Vehicles (UAVs), also called drones, largely fuelled by their suitability for military applications. UAVs range from the very small (weighing only grams) to full-scale aircraft, and date back to the early 1930. Initially not autonomous at all and thus similar to the remote controlled aircraft discussed above, UAVs have since developed into an independent aeronautical engineering field (see for example http://www.vectorsite.net/twuav.html). UAVs offer substantial potential for emergency response applications, and we distinguish here between tropospheric (i.e. flying at altitudes below approx. 12 km), and semi-autonomous, and High Altitude Long Endurance (HALE) UAVs.

Low flying drones are primarily being used by the military, for surveillance or equipped with weapons, and are either radio-controlled or employ guidance systems. Even small UAVs weighing only several kg can carry imaging payloads and employ built-in navigation capabilities, such as waypoint tracking or obstacle avoidance based on a DEM. UAVs are quickly deployed, some not requiring takeoff or landing facilities, and are particularly suitable for hazardous environments, such as forest fire sites (Ambrosia et al., 2003), as well as monotonous or routine mapping, for example of pipelines (Hausamann et al., 2005). Small and inexpensive versions can thus facilitate rapid scene assessments, while large version provide a stable platform for multi-instrument surveys.

HALE UAVs differ from conventional drones in their mission duration and altitude. They are designed to operate on solar power, thus allowing autonomous missions of up to several months duration, and fly above the troposphere in altitudes of 12–30 km. Their principal limitations are power-supply and limited payload. Once deployed, however, they can circle over an area of interest and provide continuous imaging. Previous systems were NASA's Pathfinder and Helios missions. Helios reached a record elevation of 29.5 km on solar power alone, but crashed in 2003. Currently Pegasus (*Policy support for European Governments by Acquisition of information from Satellite and UAV-borne Sensors*) system is being built by VITO (Belgium, Biesemans et al., 2005), designed to fly for up to 9 months and carry an initial payload of 2 kg (a multi-spectral scanner), to be extended later to accommodate also laser, thermal and radar sensors.

HALE UAVs have the added advantage of flying above regulated airspace. On the other hand only drones flying below 100 m can be deployed freely, while the remaining systems suffer from lack of standard and regulations and are thus deployable only under certain, country-specific conditions.

In this section we provided a comprehensive overview of existing airborne platforms, emphasising how different technologies may be most appropriate for a given disaster type, operating environment or budget. However, it should be noted that currently airplanes and helicopters are by far the most common platforms used today in emergency response.

4 SENSORS AND PRODUCT

4.1 *Calibrated and non-calibrated cameras*

Traditionally, airborne imaging has been the primary technique for large-scale surveying, using high quality cameras to capture the terrain. Typically airplanes are deployed at a flying height between 1,000 and 10,000 meters, but other platforms have also been used, especially helicopters or UAVs. The resulting images give an immediate overview of the scenery and can be directly used for visual interpretation. The integration of the images into existing maps, however, is not straight-forward. Due to perspective distortions some pre-processing steps are required to generate georeferenced data. The two most important pre-processing steps are: (i) recovery of the geometric relationship

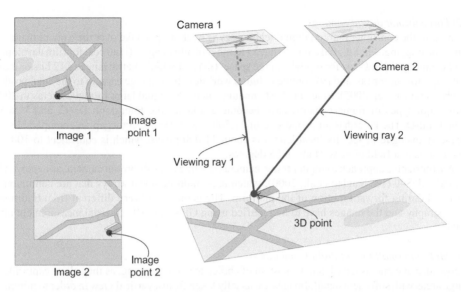

Figure 4. Illustration of image triangulation: On the left, two images of the same area are shown, together with the 3D situation (right). If two image points of the same object are identified, one can reconstruct the 3D point by intersecting the viewing rays that are induced by the image points.

between the (digital or analogue) image and the physical camera ("interior orientation"), and (ii) determination of the camera position and viewing direction, i.e. attitude, at the time of the exposure ("exterior orientation").

To obtain 3D measurements from airborne imagery, more than one image is required. In overlapping (stereo) images taken at different positions homologous points are identified, the resulting viewing rays of which are intersected ("triangulation", see Figure 4). Airborne imaging is central to the field of photogrammetry, i.e. the science of measuring and interpreting photographic images (see McGlone et al., 2004; Mikhail et al., 2001). For uncalibrated cameras, extensive research has been done particularly in the field of Computer Vision (see e.g. Hartley and Zisserman, 2000).

4.1.1 *Sensors*
Cameras in airborne imaging capture light onto a sensitive medium during an exposure time and, therefore, are passive sensors as opposed to laser scanner sensors, which actively measure distances, see 4.2. There is a great variety of different cameras: they may be (i) analogue or digital, (ii) large, medium or small format, (iii) frame or slit cameras, (iv) calibrated or non-calibrated and (v) still or video.

(1) Film cameras
As mentioned above, traditional film cameras used for aerial imaging have highly calibrated interior geometry and very low lens distortions (less than 3 μm), and typically a large image format of 23 by 23 cm. They are well established in the industry, although very expensive at up to €500,000 for a full set including devices such as forward motion compensation and gyro-stabilized platforms. The main commercial systems have a focal length of either 150 mm or 300 mm with a field of view between 55 to 90 degrees. The cameras weigh about 150–250 kg.

Nowadays, almost all image processing is done digitally, whereby the developed film has to be scanned to a digital image using photogrammetric scanners, which range between €40,000 and €150,000. A typical scanning resolution of 20 μm results in 132 megapixel, which needs about 400 MB on the hard disk for an uncompressed colour image at 8 bit per channel.

(2) Large format digital cameras
Similar to the consumer market, digital technology is beginning to take over the camera market in airborne imaging, since digital cameras have a number of advantages: (i) no cost for film developing and scanning, (ii) low cost for digital storage space, (iii) better radiometric range of 12 bits or more, (iv) superior sharpness, and (v) complete digital workflow from image capture to the final product (Leberl and Gruber, 2005). Since the first announcements of digital large format cameras in 2000, two major types of frame cameras have become available, the Intergraph DMC, and Vexcel's UltraCam-D. The latter has an image size of $11,500 \times 7,500 = 86.25$ megapixel, comparable to a scanned analogue image. Its focal length is about 11,000 pixels, which is equivalent to 100 mm and results in a field of view of about 55 degrees.

An alternative approach to digital frame cameras is a line or push-broom camera, as deployed by Leica's ADS40 (see Sandau et al., 2000), which uses multiple linear arrays that are continuously exposed during forward motion of the platform. The geometry is very different from traditional photography and the images have to be rectified using GPS and IMU (inertial measurement unit) data.

(3) Mid- and small format digital cameras
Large format cameras have been the system of choice in aerial imaging, as they allow capturing of large areas with sufficient detail. Images are usually taken from a vertical view in order to minimize distortions and to have a view that is close to orthographic map-projections.

However, there are also medium and small format digital cameras which may be used for aerial imaging, they are less suitable for top-notch photogrammetric mapping and interpretation of large areas of terrain, but also less expensive and easier to employ. Medium format cameras start at around $4,000 \times 4,000$ pixels, while lower resolution cameras are considered to be small format. Digital SLR cameras are now available at about €3,000 and are suitable for small terrain or oblique views. Note that these cameras usually do not have fixed interior parameters, and that the focal length may vary due to zooming.

(4) Video
So far we have concentrated on still cameras, but in recent years, the use of video for mapping has been increasing, see Kerle et al. (2005) or Kumar (2001). Video imagery has significant advantages, such as low cost, ease of use, and the ability to capture dynamic events, but there are also some challenges to overcome. The cameras deployed are usually not calibrated and have a much lower resolution (typically close to TV imagery), resulting either in very limited terrain coverage or low detail. Part of the low-resolution problems may be addressed by the high redundancy of observations of 25 frames per second, cf. super-resolution techniques, see Capel and Zisserman (2003). As with medium and small format cameras, video imagery is suited best for oblique views to cover larger areas.

4.1.2 *Interior and exterior orientation*
For all the camera sensors discussed above, the most important prerequisite for mapping is the recovery of the exterior orientation, i.e. the position and attitude of the camera at the time of the image capture, also called georeferencing. Again, there are several approaches that mostly depend on the additional knowledge and hardware available.

The simplest way for the user is directly to record position and attitude of the camera by using an additionally attached GPS and IMU, a process called direct georeferencing. If no GPS/IMU hardware is available, one has to rely on image measurements for interior and exterior calibration: corresponding image points have to be identified manually or automatically. To establish the relation between the images and the real world, one also needs some ground control information, i.e. 3D point coordinates ("ground control points") of some image points. The estimation procedures for calibration depend heavily on the knowledge of the camera and the terrain, and there is no unique solution. In the end all solutions finish with an optimal final estimation, the so-called bundle adjustment, which simultaneously takes all observations into account. In case of a large number of aerial images, the orientation reconstruction process is called aerial triangulation. Even with available GPS/IMU data, it is advisable to perform an aerial triangulation.

The interior and exterior orientation process is more stable for calibrated cameras, but also for uncalibrated ones this process can be performed (Hartley and Zisserman, 2000). There are some commercial systems, which can perform the orientation, including expensive digital photogrammetric workstations, simple photogrammetric software (PhotoModeler, ImageModeler), and software specifically designed for video reconstruction (MatchMover, Boujou).

4.1.3 *Assessment*

Nowadays the spatial resolution of digital and analogue large format cameras is comparable: Assuming a flying height of 1,500 m, a focal length of 150 mm results in an image scale of about 1:10,000. After a successful aerial triangulation the horizontal coordinates can be calculated with a horizontal and vertical precision of 5–10 cm and ca. 15 cm, respectively. These results assume calibrated cameras and some ground control points in the terrain. If the camera is not calibrated, the results will be less accurate. Additionally, the focal length is usually less than 150 mm for uncalibrated cameras, which means even less accuracy. Other influencing factors are the resolution of the image and the radiometric and geometric quality of the images.

Assuming a film camera is to be used to image a large area in a crisis response situation, camera deployment and generation of data products, including a successful aerial orientation, is unlikely. However, using digital cameras the time window of 3 days is more realistic. This is especially true if GPS/IMU data are available along with the images.

The world-wide number of digital large format cameras was expected to be around 200 by the year 2006 (Leberl and Gruber, 2005), compared to the current number of large format cameras of about 500 to 600. In contrast to large format cameras, the availability of small-format cameras and video is much more likely. The major drawbacks here are that it is not possible to achieve the same precision as with large aerial cameras, and the spatial coverage is only limited, or else the spatial resolution is low in case of high-flying platforms. Additionally, the maturity of software tools for processing uncalibrated images has not reached a level to be used without sufficient user knowledge or severe restrictions on the problem.

4.1.4 *Products*

The most typical products derived from airborne images are:

- The *image* itself, which gives an immediate overview and can be the basis for a qualitative survey or historic tracking of events. No processing is needed here, but the link to maps or other geographic data is missing, although one may tag or pinpoint the image to a digital map.
- A *stereo image pair* provides a qualitative 3D impression if the pictures were taken with approximately parallel viewing directions. Usually special glasses (digital data) or a stereoscope (analogue images) are necessary for visualization, as well as a sufficient overlap of the images. The images then need to be arranged such that a 3D impression is possible, which can be done by hand; for quantitative measurements one needs to compute the correct orientation of the cameras.
- Multiple overlapping images can be stitched together to a *mosaic* or a *panorama*. This can be done with or without georeferencing, with vertical or nadir images, and also with uncalibrated cameras. However, distortions will be introduced, depending on the perspectivity. Mosaics are most useful for small format imagery, especially video.
- An *orthophoto* is one of the major results of a photogrammetric process: the original image is rectified to conform to the orthographic view of a map. A prerequisite for the accurate extraction of orthophotos is the availability of a 3D terrain model to remove distortions due to perspectivity. One can also merge images taken from different viewpoints to create a *true orthophoto* to achieve a more accurate result taking into account occlusions, see Figure 5.

These products are all image-based. However, from the original images, it is also possible directly to extract 3D information such as *3D points*, *3D contour lines*, etc. Especially noteworthy is

Figure 5.   Top and bottom left: Cut-outs from aerial photos taken from different viewpoints. Bottom right: true orthophoto derived by merging the photos and compensating for occlusions.

Figure 6.   Partial 3D reconstruction of a disaster scene from oblique video data (from Kerle et al., 2005). Such information can aid visual damage assessment or quantitative change detection when a pre-disaster surface model is available. Original colour image is reproduced in greyscale here.

the automatic generation of *digital surface models* (DSM, see 4.2). After interior and exterior calibration, automatic techniques can generate 3D point clouds by extraction and matching of corresponding points in the images. Note that the creation of an accurate DSM within 3 days after the incident is only possible if we assume the availability of experienced staff and both hardware and software resources for processing the original images. Some approximations using non-calibrated images are possible, especially based on video footage (Figure 6), but commercial offerings for these tasks are relatively new and not directly usable for emergency response.

4.1.5 *Assessment of the products*

The obvious advantages of airborne images and their derivative products are the opportunity to interpret the content directly. Only if the visibility of the affected area might not be given, the application of image sensors is not possible (as in fires or volcanic eruptions (FF, UF, VE)). Qualitative and quantitative survey of earthquake damages or flooding is possible (EQ, Kerle et al., 2005).

With standard procedures, airborne images from calibrated cameras are easy to integrate into maps using mosaics or approximate orthophotos. The generation of accurate and true orthophotos and other products is challenging in the ER context, but may not be necessary in most cases. Utilizing the reference of the images to map coordinates, damage assessment can be related to existing geographic data.

Although non-calibrated imagery has the advantage that it can be acquired easily, the process of recovering the interior and exterior orientation can not be done in a standard way and changes as the case arises. They are also only suited for local disasters due to their limited spatial coverage.

All products that are derived from airborne images need to be processed by a trained operator who can react to unexpected problems that might arise.

4.2 *Laser scanners*

Airborne laser scanning (ALS) is an active technique for measuring directly and efficiently 3D coordinates (x, y, z) of points on surfaces, including the terrain but also buildings, vegetation (leaves) and above-ground utility lines (ISPRS Journal, 1999). It is also known as (scanning) LIDAR (LIght Detection And Ranging), LADAR (LAser Detection And Ranging), and ALSM (Airborne Lidar Swath Mapping). Data are acquired from a helicopter or an airplane referenced to a global coordinate system. Other information, such as object reflectivity, can also be recorded for each point. ALS delivers directly a so-called point-cloud (Figure 7b), i.e. directly without laborious manual processing steps. In the following the principle of ALS, as well as typical products will be introduced and discussed in the context of emergency relief.

4.2.1 *Airborne laser scanning – the sensor*

While ALS does not produce photographic images, it has other advantages, such as (i) penetration of vegetation and thus recording of the ground surface also in wooded areas, (ii) the high degree of automation, ranging from data acquisition to digital terrain model (DTM) generation, (iii) a high point density (1 point/15 m^2 to 15 points/m^2) that allows a very detailed terrain description, (iv) a vertical accuracy of about ± 10 cm, and (v) as an active system allowing measurements at night or over areas without texture (e.g. snow). Penetration of haze, smoke, water vapour clouds, rain, and snow-fall is only possible for short ranges or limited atmospheric disturbance by these effects (Jelalian, 1992). Thus data acquisition during rain or fog or through clouds is not possible.

An ALS is a multi-sensor measurement system that can be broken down into several components (see also Figure 7):

- The laser scanner, consisting of the laser range finder (LRF), measuring the distance r from the sensor on the airborne platform to a reflecting surface, and the scanner, i.e. a beam deflection device, that periodically deflects the laser sideways.
- The satellite positioning system, GPS (alternatively Glonass or Galileo), which is used to determine the position (x, y, z) of the sensor platform.
- The inertial measurement unit (IMU), used to determine the angular attitude of the platform (e.g. roll, pitch, and heading).

The laser range finder operates by measuring the two-way travel time required for a pulse of laser light to travel to the target, i.e. the location of reflection, and back to the detector of the LRF. The distance r can be computed from the travel time Δt by the known speed of light c: $r = c\Delta t/2$. Current systems are capable of operating between 1,000 m and 8,000 m above ground level. Not all systems can be operated at low altitudes because of eye-safety consideration. Range measurements are rapid and can be done multiple times per second, allowing measurement frequencies of up to

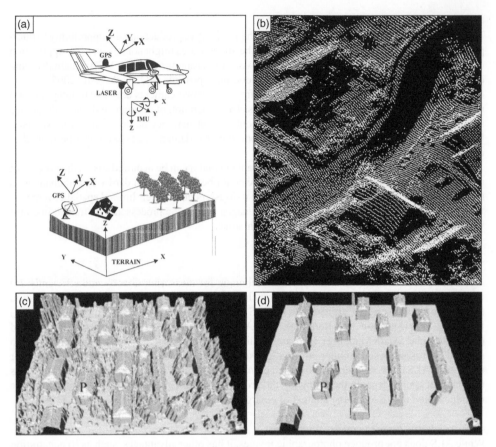

Figure 7. Overview of airborne laser scanning (ALS). (a) components of an ALS system, (b) point cloud (raw data, courtesy of George Vosselman), (c) unfiltered and (d) filtered dataset (Dash et al., 2004).

250 kHz (250,000 point measurements per second) for the fastest commercial systems available in 2007. Additionally the reflection strength of the surface can be recorded (Höfle et al., 2007). Some surfaces, such as snow, scatter back nearly all of the incident energy, while others, such as tar used for street paving, absorb a large portion of the energy and scatter back only a small portion. Measurements are also difficult for materials that reflect the light in a mirror-like fashion, e.g. wet roof tiles.

The laser beam has a certain diameter, typically 50 cm to 1 m, the so-called footprint, and multiple reflecting surfaces may be found within one footprint. In those cases more than one "echo" of the emitted signal can be recorded, e.g. one from the top of the vegetation, and a second from the ground. Therefore, with one laser shot two or even more points can be measured.

The scanning mechanism is used to deflect the laser beam across the flight direction. The opening angle of a laser scanner is typically between 10° and 50° to either side. While the airborne platform is moving forward (typically 75 m/s for an airplane, lower speeds for helicopter deployment), terrain and object points are measured below it, looking sideways to the left and right. This leads to a strip-wise data collection, or a so-called swath. For mapping of extended areas multiple strips have to be acquired.

The flight path of the platform is measured with GPS and an IMU. Satellite navigation is available worldwide, but it has to be used in a differential mode for achieving the required accuracy (Schwarz and El-Sheimy, 2004). A base station, i.e. an operating receiver, is necessary at a known reference point during the entire data acquisition mission, which should not be more than 30 km away from the mission area, although in highly developed areas existing networks of base stations can be used. The data recorded on the platform and at the base station are used for computing the flight path first and then the point coordinates. The position and attitude of the sensor can be determined with a precision of a few centimetres and an angular precision of at least 0.01°.

4.2.2 Airborne laser scanning – technology assessment

Summing the errors of the laser scanner and the precision of the flight path leads to a typical vertical and horizontal point accuracy of ± 10 cm and ± 0.6 m, respectively, for a flying height of 1200 m. Precision is higher than accuracy, reaching up to 2 cm for vertical and 10 cm for the horizontal direction, and especially horizontal precision deteriorates with increasing flying height. In a typical scenario the flying height may be 1200 m, resulting in a swath width of 1000 m, and within 17 seconds of flying time the area of 1 km² is covered with a spatial resolution of about 1 point per m². Lower flying heights allow higher point densities at the cost of reduced area coverage. Data acquisition is inherently in 3D and completely digital, and data delivery, i.e. the point cloud or a DSM (see below), can be performed within 24 hours. Such fast turnaround times were demonstrated for the World Trade Center disaster (Rodarmel et al., 2002; Williamson and Baker, 2002). Flying height was much lower in this case, the area itself was well known and a GPS base station could be set up quickly. In other scenarios data delivery may not be as fast, especially if the laser scanner has to be mounted in the airplane first.

Some 130 laser scanner systems were sold worldwide until 2004 (LID, 2004; Toth, 2004), but airborne laser scanning is still a relatively new and fast developing technology, and older instruments do not have the capabilities of current systems, in terms of precision, point density, and ease of use. For operational use a suitable aerial platform, a pilot, a laser scanner technician, and a person for processing the data are required. Because of the large data amount gathered by laser scanning, standard hardware is typically not sufficient in terms of memory and computing power required.

4.2.3 Airborne laser scanning – products

(1) Digital Surface Model

By triangulating the point cloud or computing a regular raster with interpolation algorithms a Digital Surface Model (DSM) can be created (Fig. 7c). As the laser beam is reflected either by the ground, buildings or vegetation, this model often represents solid surfaces (the ground or the roofs of buildings). In vegetated areas typically the ground and the tree tops, or intermediary stories are represented. It is useful for a first impression of the acquired data, and manual measurements (e.g. building height, width of a street) can be performed manually in such a model. The models can either be viewed as colour coded height or shaded relief maps or in 3D perspective with virtual walk- or fly-through.

(2) Change detection

A DSM can also be compared to an earlier acquired (e.g. pre-disaster) model of the scene. Performing such a comparison automatically, and detecting relevant changes while neglecting changes smaller than the measurement accuracy and non-stationary objects such as cars, is called change detection. The a priori model can have been also acquired by airborne laser scanning, but also by another means. It is important to note here that a DSM contains no explicit information on the features making up the surface, i.e. whether a given group of pixels represents a house, the terrain itself, or other objects. For example for the automatic detection of collapsed buildings this means that the a priori model has to contain this information (Dash et al., 2004; Tuong Thuy et al., 2004; Rehor, 2007).

(3) DEM filtering

A DSM is not suitable for hydrological modelling, such as water runoff calculations, because it does not only show the ground but also spikes in the terrain corresponding to vegetation and buildings. The process of determining the terrain itself and removing the off-terrain points is called filtering. Its result is a Digital Terrain Model (DTM, Fig. 7d, shown with the houses added again). Different algorithms have been developed and compared (Sithole and Vosselman, 2004). In different and even complex scenarios these algorithms typically classify 90% or more of the points correctly, although especially at steep slopes misclassifications can occur. While these algorithms require only a few parameters to be set manually at the start by an experienced user, some post processing is required for checking the results and correcting classification errors. Single misclassified points often do not play an important role, whereas linear misclassifications, e.g. total removal of small dikes or generation of holes in dikes, can require hours (up to days) of work for larger areas.

(4) Draping of other information
A DSM or DEM can be used for adding the 3rd dimension to 2D information from calibrated (optical or thermal) images. By draping these images over the "laser surface" the height information corresponding to the colour can be visualized, leading to textured 3D models presented with simple computer graphics or advanced virtual reality tools. These models are easy to understand and interpret, because their appearance is similar to our human viewing experience. The draping algorithms run automatically, but problems occur if the view point of the sensors is different, leading to wrong height/colour association.

4.2.4 *Assessment of airborne laser scanning products*
The DSM is an unprocessed product which, by virtue of its 3D display of terrain objects, is easy to interpret for laymen and engineers. In the context of urban fires, landslides, volcanic eruptions and surface lifeline damage, acquisition of airborne laser scanning data is useful because of the limited area involved (UF, LS, VE) and the linear nature (SLD), corresponding to one flight line of the helicopter. Landslides and volcanic eruptions may be in areas hard to access, which can be problematic for setting up a GPS base station within a limited amount of time.

For forest fires (FF) and flooding (FL) the existence of a pre-event DTM is beneficial, which is to be acquired preferentially with airborne laser scanning due to its ability to penetrate the vegetation and measure the ground below. The high spatial resolution is especially indispensable for inhabited areas. After the Hurricane Katrina event in New Orleans 2005, such a DEM was used for computing the first inundation map and showing water heights in the city.

Change detection can be used in principle in EQ disasters, but the lack of existing building models typically hampers its applicability. With the current state of development the outcome of these algorithms has to be interpreted rather by trained personnel.

Draping is a technique that can be applied to any calibrated data source and makes sense in those cases where the relief is pronounced or such objects as houses are found. It is a method that is supporting the information that is being draped over the laser DSM or DTM, rather than the laser data. It can, therefore, be applied to all ER scenarios where calibrated images are of use.

4.3 *Thermal sensors*

Thermal remote sensing (TRS) results in image data similar to optical methods, yet the underlying approach differs fundamentally. Whereas *reflected* radiation is recorded in the former, the latter senses *emitted* radiation, the amount of which is a function of the surface temperature, among other variables. Since no illumination source is required, thermal RS is also possible at night, which, in particular for small thermal anomalies is also easier, since the contribution of day-time solar heating is reduced.

4.3.1 *Thermal systems*
TRS encompasses different elements, the radiation source and its characteristics, atmospheric attenuation, and the sensor type. Because of variable absorption by the atmosphere throughout the electromagnetic spectrum, thermal emissions are sensed in two distinct regions, 3–5 μm (Mid-Infrared, MIR) and 8–14 μm (Thermal Infrared, TIR). Higher temperature sources emit more energy, and the wavelength of the peak emission decreases with rising temperature. For example, a thermal feature of 50°C (323 K) has its peak at approximately 9 μm, while a low intensity grass

Table 2. Peak wavelength of different radiation sources. The object of interest can thus be used to select the appropriate sensor.

Radiation source	Temperature [°C/K]	Peak wavelength [μm]
Background	20/293	9.9
Fuel ignition	275/548	5.3
Grass fire	500/773	3.7
Forest fire	1200/1473	2
Lava flow	550–1400/823–1673	1.7–3.5

fire of 500°C (773 K) peaks at 3.7 μm (Table 2). While the MIR radiation is attenuated by smoke, longer wavelength TIR radiation still reaches the sensor, allowing imaging of the thermal anomaly as well as the surrounding ground. While the temperature can be calculated from the measured radiance, it is worth noting that temperature estimates will be lower then actual values. This is because fires do not emit IR energy perfectly, and also because some emitted energy is absorbed by heavy vegetation canopy if present, as well as by the atmosphere. Some temperature calculation algorithms are based on a bi-spectral technique, i.e. employing readings from two infrared channels, hence a multi-channel TIR system is advantageous. For an introduction to thermal remote sensing (TRS) see for example Quattrochi and Luvall (2004) or Lillesand et al. (2004).

Fuelled in particular by its suitability for forest fire applications, airborne TRS has undergone many developments since first experiments in the 1940s. Since 1966 it has been used operationally by the US Forest Service. While the focus in the beginning was on the detection of hotspots, fire perimeters and burned ground, there is great interest today to quantify fire intensity and fuel consumption, and to understand fire spread behaviour. This can only be achieved with detailed and frequent data. Airborne surveys serve this need for tactical information (as opposed to strategic, i.e. regular regional surveys that satellites sensors can best provide). A range of integrated systems has been constructed, comprising a sensor, GPS and IMU units, and data storage and processing equipment, all mounted on a suitable platform (see for example Greenfield et al., 2003; Riggan and Hoffman, 2003). Such systems allow the fast (digital) delivery of georeferenced products that are easily integrated with auxiliary spatial data, thus overcoming limitations of earlier technologies. In particular for the modeling of thermal phenomena the integration of additional data such as DEMs is critical, while knowledge on context and site access information is needed for operational use, for example by fire fighters.

4.3.2 Sensors and products

Existing TRS sensors are varied, but can be grouped into Forward Looking Infrared Cameras (FLIRs, also called thermographic cameras or thermal imagers), and multispectral radiometers/line scanners. FLIRs operate similar to video cameras, in that they can be handheld compact systems, recording in a certain spectral range to a single analogue or digital tape/file. They are available for both MIR and TIR, are versatile, inexpensive and easy to use. They can be handheld on a helicopter or aircraft platform, or gimbal-mounted for nadir- or forward-looking operations. Disadvantages are (i) the relative difficulty of interpreting and using oblique data, (ii) the provision of relative rather than absolute temperatures (depending on the camera type), and (iii) the lack of multi-channel data (e.g. for simultaneous availability of optical data or split-window processing). However, FLIR data can be quickly appraised, and merged in real-time with auxiliary information (e.g. from GPS). Work has also been carried out to georeference thermal video data (e.g. Wright et al., 2004).

Multispectral radiometers are nadir-looking scanners that record information in different channels. They are more sophisticated and expensive than FLIR, and have been primarily developed for operational fire monitoring from manned or unmanned airborne platforms. Examples are the Daedelus system (12 spectral channels), AIRDAS (4 channels, Ambrosia et al., 2003),

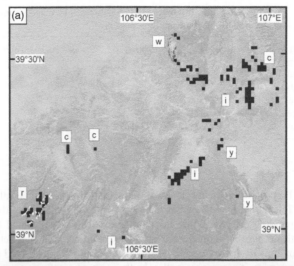

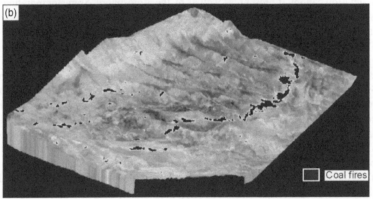

Figure 8. Thermal images of coal fires in Wuda and Ruqigou, China, respectively, (a) anomalous pixels superimposed on a Landsat image (Hecker et al., in press), and (b) draped over a DTM (courtesy of Xiangmin Zhang). Image (a) also highlights the potential problem of confusing coal fire related anomalies with those originating from industrial activity (marked *i*), or water bodies such as the Yellow River (*y*). *w* marks the Wuda coal field, while *r* demarcates Ruqigou.

FIREMAPPER (4 channels, Riggan and Hoffman, 2003), and PHOENIX (2 channels, Greenfield et al., 2003). These instruments are integrated into complete sensing systems, often with direct satellite or radio link for fast data processing. For more background information on TRS for forest fire mapping see Robinson (1991), Carlson and Burgan (2003) and Leblon (2005).

The simplest thermal sensor products are video or still images from a FLIR. Deployed on a police or media aircraft, they are rapidly collected and visually assessed. Data collected digitally can be further thresholded or ratioed to highlight thermal anomalies (Figure 8). If the sensor is properly calibrated, surface temperature calculations can be performed, otherwise relative change detection analysis of temperature patterns is possible for repeated data sets. Measuring the actual fire characteristics (dimensions, spreading rate, etc.) is difficult, since a fire site is composed of flames, hot ash, smouldering parts, unburned material etc. With integrated GPS/IMU information (Figure 9, Kerle and Stekelenburg, 2004), the thermal data can be also georeferenced, and integrated with auxiliary raster or vector data, as well as DEMs. For illustrations see http://www.firemapper.com/.

Figure 9.   Oblique thermal video frame captured with a Thermovision 1000 FLIR over the city of Enschede, Netherlands, following the explosion of a fireworks factory in May 2000, which incinerated an entire city block. Superimposed are GPS and relative camera orientation parameters. In the actual data (reproduced here in greyscale) blue hues indicate cold background, with a temperature increase towards red for intensely burning buildings.

Increasingly fire spread models are used in conjunction with TRS, meteorological and topographic information (e.g. Chuvieco, 2003). Since the accuracy of such models, as well as the disaster response efficiency, increases with frequently updated data, and for safety considerations, UAVs are increasingly used to acquire imagery over forest fires areas (Ambrosia et al., 2003).

4.3.3   *Applications*

Even though forest fires have been driving the fast developments in airborne TRS, there are other relevant applications. Underground coal fires are a common, costly and polluting occurrence. Even though TRS can only detect surface heat, shallow coal fires can raise the surface temperatures sufficiently for such fires to be detectable. Such mapping is often done at night to minimize the effect of solar heating (Figure 8a) (Torgersen et al., 2001; Zhang et al., 2003; Zhang et al., 2004). The same technology can be used to detect accidental warm, and potentially contaminated, water discharge from power stations and to monitor volcanic activities (Oppenheimer et al., 1993). Here it is possible to map thermal anomalies in surface or crater lake temperatures, or to detect magmatic sources (Leblon, 2005). Thermal data also aid the response to urban fires, and allow the most comprehensive assessment of a situation of a given crisis situation if properly integrated with additional sensors, as was the case after the World Trade Center collapse (Rodarmel et al., 2002).

4.4   *Radar sensors*

Imaging radar (*r*adio *d*etection *a*nd *r*anging) is a form of active microwave remote sensing. Despite its considerable utility for different disaster response situations, the number of available non-military airborne radar systems is very low (only some 45 systems, among them several experimental ones, have been built; see http://www.eoportal.org/documents/kramer/table.pdf). Compared to the 130 + laser scanners and thousands of optical and thermal systems it, therefore, remains a niche technology for emergency response purposes. Especially when considering the time restrictions, the limited availability of radar sensors reduces its practical use and the incentive to incorporate such technology in emergency response plans. In addition, the processing and interpretation of radar imagery is far more complex than the methods introduced above. Therefore, only a brief review of the potential of airborne radar is provided here, together with a summary of the main radar products, and examples of how they can be used in a crisis situation.

4.4.1 *Radar sensor and image characteristics*

Side-Looking Airborne Radar scanning (SLAR) was developed after the Second World War for military applications, and became declassified in the 1960s. Like laser scanning, radar is an active technique, but the wavelength of the signals employed ranges from 1 cm to 1 m, as opposed to the near-infrared wavelengths used by lasers. Since electromagnetic waves at those wavelengths are not scattered or attenuated by the atmosphere, radar signals penetrate clouds, haze and rain, giving it all-weather in addition to day-and-night imaging capabilities. The radar transmitter produces pulses of microwaves at a specific frequency that are focused by the antenna into a narrow beam, and emitted sideways perpendicular to the flight path. The strength and characteristics of the return signal allow discrimination between different surface features, while the time delay determines the distance (range) to the ground, similar to laser scanning. The return signal of each pulse is sampled and written out to one image line. Given the oblique viewing angle, the calculated direct distance to the ground (slant range) has to be converted to actual ground distances (ground range).

While the principle of Synthetic Aperture Radar (SAR) is straightforward, the actual image analysis process is rather complex, as the recorded signal contains two principal information parts. The first, *amplitude*, or its squared value, intensity, is used to create radar images (such as Figure 10a). Since radar radiation is coherent, the radar wave's *phase* is also recorded. Together with amplitude it is used in interferometric processing, i.e. to create interferograms or DTMs. However, the coherent nature of radar also leads to less desired effects. Since the return signals from individual elements within the ground resolution cell, the area imaged in one pixel, experience positive or negative interference, radar images show a characteristics grainy texture, called *speckle*, which gives images a salt-and-pepper appearance (Figure 10a). This effect can be removed by filtering procedures (Figure 10b), although at the expense of reduced spatial resolution.

The *Synthetic Aperture* in SAR refers to the way images are constructed, involving the processing of many separate return signals for a given point on the ground, as the scene is imaged many times while the sensor moves forward. This effectively synthesises a long antenna that increases the spatial resolution of the image. For an introduction to imaging radar see for example Henderson and Lewis (1998) or Stimson (1998).

Radar images are also difficult to interpret, as they result from a specific combination of (i) the radar system properties, (ii) the imaging geometry, and (iii) the interaction between signal and ground. Radar sensors can use a range of different wavelengths, or bands, ranging from the short K-band (ca. 1 cm) to the long P-band (1 m). Frequently used (also in satellite-based radars) are also X- (ca. 3 cm) and C-bands (ca. 5 cm). The wavelength determines how the signal interacts with surface features. For example, while P-band radar can penetrate vegetation and image the terrain (and in dry conditions even penetrate the ground for several meters), shorter bands lead to volume scattering within the foliage. In addition to wavelength, also the polarisation, i.e. the orientation of the electric field, can be changed. Both emission and reception of the signal can be either horizontal or vertical, giving a total of 4 possible combinations. Images corresponding to those different polarisation settings vary greatly. In addition to those system parameters the incident angle, i.e. the angle between radar beam and local nadir, is important, as it strongly influences the geometric distortions of the image. Radar images suffer distortions similar to oblique optical images, here termed foreshortening and layover. An additional problem is shadow, where the radar signal cannot reach the ground. Unlike in optical images, radar shadow areas contain no data. In addition to signal wavelength, the signal interaction with the ground is also influenced by the surface roughness and moisture. In summary, many variables in combination determine the nature

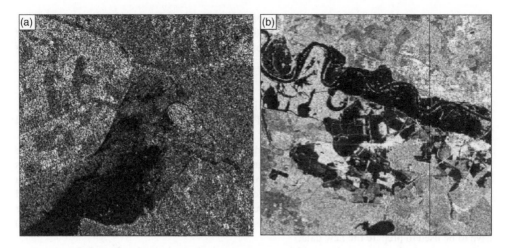

Figure 10.   (a) radar intensity image, and (b) flooded area highlighted by specular reflection.

of a given radar image, and only a good understanding of those parameters allows the appropriate setting to be determined for a given application.

4.4.2 *Applications*

Radar products can be 2- or 3-dimensional, and be analysed either individually or against existing data to detect change. The simplest visually interpretable radar product is an intensity image. Intensity (Figure 10a), the square of the amplitude, has been used to assess urban seismic damage, although, at least for spaceborne radar data, damage has been shown to cause both increase and decrease in intensity. More promising is the use of coherence, the additional property of a recorded radar signal, although that requires the use of images taken before and after a disaster (Arciniegas et al., 2007). Intensity images are most commonly used to map flooding, given that the obliquely arriving radar waves are reflected by water surfaces in a specular, i.e. mirror-like fashion, and water thus appears black (Figure 10b). Jiang and Cao (1994) reported on an operational airborne radar flood management system in China, where data were transmitted in real-time via satellite to an analysis command centre. Specular reflection also allows efficient detection of oil spills on water (Hühnerfuss et al., 1998). Terrestrial applications are more difficult, given the earlier mentioned parameter combinations. However, radar imagery has been used to map forest damage following storms, although this is also facilitated by available pre-disaster reference data. In particular the incident angle, radar band and polarisation are critical here.

Radar data can also detect 3-dimensional changes. For example interferometric radar (INSAR) data, which are based on the phase differences of corresponding pixels in 2 datasets, have been used to detect landslides (Singhroy et al., 1998), and can also be used to map other topographic changes such as subsidence or deformations due to earthquakes or on volcanoes (Rowland et al., 1999).

4.5 *Data integration*

In the previous sections different sensors and their products were introduced. Many of those, such as a laser- or stereophoto-derived DSMs or thermal images, are already of use without additional data. However, by integrating different information layers further knowledge can be gained. In this section we briefly summarise the main methods to do that, as well as prerequisites, and differentiate between simple overlays and multi-data information extraction. Note that we only address the use of airborne data here, and list methods that go beyond standard image enhancement such as filtering, or single-image classification approaches. For an introduction to such techniques see for example Kerle et al. (2004) or Mather (2004). In Chapter 3 a more in-depth discussion of multi-type data integration (e.g. raster and vector) in data base management systems (DBMS), as well as web-visualisation is provided.

(1) Simple overlay

Examples of this simple form of data integration were already given throughout this chapter. Recall how thermal data can be draped over DTM to provide topographic in addition to thermal information (Figure 8b). The same is possible with any of the data products discussed above. However, the principal requirement is that the datasets to be integrated are georeferenced to the same reference system (projection type, datum, units), although more sophisticated image analysis software packages can perform on-the-fly reprojections. Nevertheless, all of the airborne data discussed that are acquired without the direct integration of GPS and IMU data are initially devoid of reference information, and can thus neither be overlaid nor integrated with other data. In those cases ground control points (from GPS readings or other referenced map or image data sources) have to be used to prepare the new data.

Data overlay is similar to working with transparencies, in that information from different sources can be effectively related visually. It allows rough change detection operations, or quick appraisals of a given situation, for example when thermal data of forest fires are overlaid over older images of forest stands, or when radar images of a flood are integrated with optical imagery showing landcover or population centres. GIS technology is an ideal vehicle to perform such operations, but also facilitates more advanced processing, which we discuss in the following section.

(2) Multi-data information extraction

A wide variety of methods exists to combine different data sets, including both generic processing tools and highly specialised subject-specific operations. Some of them are similar to, or begin with, data overlays, but all are distinguished by the generation of new quantitative information. Here we summarise the principal groups of such methods.

4.5.1 *Increase in dimensionality*

With the exception of laser scanning data and some secondary DSM products, all data types discussed above are inherently 2-dimensional. Yet integrating such 2D with 3D information allows for more complex analysis. For example, flood extent outlines merged with topographic data allows the nature of further water spreading to be assessed. Integrating 3D data with earthquake epicentre location allows an assessment of intensity variations as local intensity increases on hills and mountains, as well as the likely location of landslide areas, especially if additional data on landcover or geology is integrated as well. Given frequent repeat data acquisition an extension to the 4th dimension – time – is also possible. This is of particular use in fire or flooding situations, but was also at the heart of the situation management following the World Trade Center collapse (Rodarmel et al., 2002).

4.5.2 *Spatial resolution increase*

Although more frequently used on satellite data, lower-resolution data (e.g. multispectral or thermal) can be enhanced through pan-sharpening, whereby pan-chromatic data, typically of higher resolution, are fused with lower resolution images. This results in data that retain much of the spectral information yet gain spatial resolution (e.g. Becker et al., 2005). In particular in urban emergency situations such added detail may be of use. See also Zhang and Kerle, this volume.

4.5.3 *Multi-criteria analysis and modelling*

Many emergencies result from a complex interaction of physical and social parameters that determine hazard and vulnerability. Multi-criteria analysis is thus required to assess any risk, but can also be used in the emergency response phase. It allows the nature of an event to be understood better, and, more importantly, further developments to be anticipated. Flooding was already named as an example. Others are the spread of diseases (combining information on pollution source, contamination pathways and susceptible population), fire evolution, where models integrate real-time fire parameter data with meteorological data and information on fuel availability, or coal fires, where thermal surface imagery is used to determine the underground extent of the fire, additional likely threatened areas, and optimal locations on which to concentrate fire suppression efforts. For examples see Mansor et al. (2004).

5 ASSESSMENT AND COMPARISON

In section 1.2 we explained that identifying the optimal geodata-based strategy to respond to an emergency is a function of possible sensor-platform combinations, the emergency type in question, and the utility of a given specific data product, in addition to given financial, environmental or logistic constraints. This precludes an absolute ranking of the value of a sensor or product for a given disaster situation. The purpose of this chapter was thus to introduce the available airborne platforms and sensors with respect to clearly defined assessment criteria, and to illustrate how airborne remote sensing technology has been used successfully in a range of emergency situations. The facts and observations given throughout the above sections are summarised in Table 3. One way to read the table is to begin with an emergency type. Useful data of a flooding situation, for example, can be acquired with optical data (both calibrated and non-calibrated), as well as with laser scanner and radar information. During a cloudy situation, however, optical sensors and laser scanners cannot be deployed. The likely availability of radar data, however, is very low, owing to the small number of operational systems. Hence a situation-specific compromise will be required. Note that no absolute costs are given, as those are highly variable among different providers. A non-calibrated optical photograph, acquired by a camera flown on a kite, might cost a few dollars, while an interferometric radar product likely exceeds 1000 US$.

6 OUTLOOK

In this chapter we have shown that airborne remote sensing provides a rich and varied arsenal of tools that can be employed in any type of emergency response situation. In particular given the continued relatively high cost and low temporal resolution of satellite data, aerial data should be strongly considered as a source of accurate and timely information. There are, however, important considerations and constraints as well – technical, economic, and environmental, which we discussed in this chapter.

Even though airborne remote sensing is a mature technology, it continues to develop rapidly. We can differentiate between developments in platform technology, sensors, and data processing tools. While there are clearly ongoing developments and refinements in airplane and helicopter technology, the most remarkable developments relate to alternative platforms. Over the next few years we can expect strong developments in drone, and especially HALE UAV development, with Pegasus expected to become operational in 2009. Smaller platforms, such as kites, balloons, or powered parachutes, will likely remain niche technologies, yet all are expected to become more sophisticated and reliable, and can thus well be considered in emergency response plans and operations.

There are currently no new airborne sensor types in development that are relevant in the ER context. There are sensors other than the ones discussed here, such as hyperspectral scanners, but those we omitted on purpose, given their very limited availability. Within the optical and thermal sensors we can expect a continued move to digital technology, and a faster data transfer and availability using radio or satellite links.

Data processing methods will also progress rapidly. We expect strong developments in the use of non-calibrated data (e.g. direct georeferencing, photogrammetric reconstruction), more automated

Table 3. Summary of the suitability of different data product characteristics, and their use for different emergency types.

Sensor	Calibr./non-calibr.	Envir. restriction	Product	Availability[10] 3 days	4–14 days	Accuracy	Analogue/digital	2D/3D	Cost[11]	State of development[12]	Need for expertise	Time requirement[13]	Specific disaster type[14]	Notes
Aerial camera	Non-calibrated (NC)	Daytime only, smoke, clouds, haze	Still/video images	H	H	L	A/D	2D	L	E	L	L	EQ, VE, LS, FL, SD, OS	– product availability is dependent on aircraft being quickly deployable
			Orthorectified photo	M	M	L	D	2D	M	PT	M	M	EQ, VE, LS, SLD, FL, SD	– products that go beyond simple imagery, i.e. surface models or orthorectified data, require extensive additional data, such as GPS information and ground control points (GCPs), and rapidly increase in required expertise
			DSM	L	L	L	D	3D	M	PT	M-H	M	EQ, VE, FL, LS, SD[15]	
			DTM (filtered DSM)	L	L	L	D	3D	M	PT	M-H	M		– to process non-calibrated data, only a few, relatively new technical solutions are available
	Calibrated (C)		Aerial photo	M	M-H	H	A/D	2D	H	E	M	L	EQ, VE, FL, LS, SLD, SD, OS	
			Orthorectified photo	L-M	M-H	H	D	2D	H	E	M	L-M	EQ, VE, LS, SLD, FL, SD	– while DSM generation is a largely automated process, DTM-filtering is more involved
			DSM	L-M	M-H	H	D	3D	H	E	M	M	EQ, VE, FL, LS, SD	
			DTM (filtered DSM)	L-M	M-H	M-H	D	3D	H	C	H	M-H	FF, EQ, UF, LS, VE, SD	
Airborne Laser Scanning	C	Clouds, smoke, haze, heavy rain	DSM	L	M	H	D	3D	M	E	M	L	EQ, VE, FL, SLD, LS, SD	– infrared lasers are reflected by water, while green laser light penetrates water and reflects bathymetry
			DTM (filtered)	L	M	H	D	3D	M	C	H	M	FL	
			Change detection	L	M	H	D	3D	M	C	H	M	EQ, VE, SLD, SD	
			Draping of other data	L	M	H	D	3D	M	E	M	L	EQ, VE, SLD, SD	

Thermal scanner/FLIR	C/NC	Clouds[16]	Thermal image	H	H	L-H	A/D	2D	L	E	L	L	FF, UF, VE
			Threshold/ratio	H	H	L-H	D	2D	L	E	L	L	FF, UF, VE
			Temperature calculation	M	M	M-H	D	2D	M	E	M	M	FF, UF, VE
			Integration with Met data	L	M	M-H	D	2D	M-H	C	H	M	FF, UF
Radar	C	None	Intensity image	L	L	H	D	2D	M	E	M	L	EQ, FL, OS, VE, LS, SD
			Coherence image	L	L	H	D	2D	H	C	H	M	EQ
			Change detection	L	L	H	D	2D	H	C	H	H	EQ, VE, SD
			Interferometric product	L	L	H	D	3D	H	C	H	H	EQ, VE, LS

Notes column (Thermal scanner/FLIR): – low cost FLIRs tend to lead to lower quality and non-calibrated data – availability of thermal real-time data integrated with meteorological information is limited by low number of system

Notes column (Radar): – low number of operational systems – processing of radar data is typically more complex and challenging

[10]L – low, M – moderate, H – high (likely availability within 3-day emergency response window).

[11]Relative cost for product generation, excluding sensor and platform cost, primarily a function of expertise, software and hardware required.

[12]PT – prototype, C – commercially available product, E – well established.

[13]L – low, M – moderate, H – high (for product generation after raw data acquisition).

[14]EQ – earthquake, FF – forest fire, UF – urban fire, FL – flooding, LS – landslides, VE – volcanic eruptions, SLD – surface lifeline damage (e.g. power cables, pipelines), OS – oil spill, SD – storm damage.

[15]The quality of DSMs, and therefore also DTMs, that can be generated today from uncalibrated photographs is not sufficient to make them very useful in disaster response. The DTM-category for calibrated data illustrates their potential use.

[16]For MIR.

processing of laser scanner data (e.g. automatic object identification and merging with vector data), as well as in damage assessment, using optical, laser scanner and radar data. It has to be kept in mind that Geoinformatics is a dynamic field advanced by many organisations and individuals, which makes accurate predictions of the future difficult. However, we attempted here to show that airborne remote sensing provides powerful tools, many of which are straightforwardly and usefully employed in an emergency response scenario.

REFERENCES

Ambrosia, V.G., Wegener, S.S., Sullivan, D.V., Buechel, S.W., Dunagan, S.E., Brass, J.A., Stoneburner, J., and Schoenung, S.M., 2003, Demonstrating UAV-acquired real-time thermal data over fires: Photogrammetric Engineering and Remote Sensing, v. 69, p. 391–402.

Arciniegas, G., Bijker, W., Kerle, N., and Tolpekin, V.A., 2007, Coherence- and amplitude-based analysis of seismogeic damage in Bam, Iran, using EnvisatASAR data: IEEETransactions on Geoscience and remote Sensing, v. 45, Special issue 'Reomote sensing for major disaster prevention, monitoring and assessment'. p. 1571–1581.

Becker, S., Haala, N., and Reulke, R., 2005, Determination and improvement of spatial resolution for digital aerial images, ISPRS Workshop High-Resolution Earth Imaging for Geospatial Information, Hannover, Germany, available at http://www.ifp.uni-stuttgart.de/publications/2005/becker05_hannover.pdf

Biesemans, J., Everaerts, J., and Lewyckyj, N., 2005, PEGASUS: Remote sensing from a HALE-UAV, ASPRS annual convention, Baltimore, USA.

Boike, J., and Yoshikawa, K., 2003, Mapping of periglacial geomorphology using kite/balloon aerial photography: Permafrost and Periglacial Processes, v. 14, p. 81–85.

Capel, D., and Zisserman, A., 2003, Computer vision applied to super resolution: Signal Processing Magazine, v. 20, p. 75–86.

Carlson, J.D., and Burgan, R.E., 2003, Review of users' needs in operational fire danger estimation: the Oklahoma example: International Journal of Remote Sensing, v. 24, p. 1601–1620.

Chuvieco, E., 2003, Wildland Fire Danger Estimation and Mapping: The Role of Remote Sensing Data, World Scientific Publishing, 280 p.

Dare, P.M., 2005, An innovative system for low cost airborne video imaging, 26th Asian Conference on Remote Sensing, ACRS 2005, Hanoi, Vietnam, 7p.

Dash, J., Steinle, E., Singh, R.P., and Baehr, H.P., 2004, Automatic building extraction from laser scanning data: an input tool for disaster management, Monitoring of Changes Related to Natural and Manmade Hazards Using Space Technology, Volume 33: Advances in Space Research: Kidlington, Pergamon-Elsevier Science LTD, p. 317–322.

Eisenbeiss, H., 2004, A mini Unmanned Aerial Vehicle (UAV): System overview and image acquisition, International Workshop on 'Processing and visualization using high-resolution imagery', Pitsanulok, Thailand, available at http://www.photogrammetry.ethz.ch/general/persons/henri/11.pdf

Greenfield, P., Smith, W., and Chamberlain, D., 2003, PHOENIX – The new Forest Service airborne infrared fire detection and mapping system, 2nd Fire Ecology Congress (American Meteorological Society), Orlando, Florida.

Haefner, S., 2004, Kite Aerial Photography, available at www.thehaefners.com/kap/?page5kites.

Hartley, R., and Zisserman, A., 2000, Multiple view geometry in computer vision: Cambridge, UK, Cambridge University Press, 624 p.

Hausamann, D., Zirnig, W., Schreier, G., and Strobl, P., 2005, Monitoring of gas pipelines – a civil UAV application: Aircraft Engineering and Aerospace Technology, v. 77, p. 352–360.

Hecker, C., Kuenzer, C., and Zhang, J., in press, Remote sensing based coal fire detection with low resolution MODIS data, in Stracher, G., ed., Geology of Coal Fires: Case Studies From Around the World; GSA Special Paper Series.

Henderson, F.M., and Lewis, A.J. (eds), 1998, Principles and applications of imaging radar: Manual of remote sensing, v. 2, p. 866.

Höfle, B., and Pfeifer, N., 2007, Correction of laser scanning intensity data: Data and model-driven approaches: ISPRS Journal of Photogrammetry and Remote Sensing, accepted for publication.

Hühnerfuss, H., Wismann, V., Gade, M., and Alpers, W., 1998, Radar signatures of marine mineral oil spills measured by an airborne multi-frequency radar: International Journal of Remote Sensing, v. 19, p. 3607–3623.

Jang, H.S., Lee, J.C., Kim, M.S., Kang, I.J., and Kim, C.K., 2004, Construction of national cultural heritage management system using RC helicopter photographic surveying system, XXth ISPRS Congress, Istanbul, Turkey.

Jelalian, A.V., 1992, Laser Radar Systems, Artech House Books, 308 p.

Jiang, J., and Cao, S., 1994, Real-time disaster monitoring system by using SAR, Microwave instrumentation and satellite photogrammetry for remote sensing of the Earth, Rome, Italy, 91–97.

Kerle, N., Janssen, L.L.F., and Huurneman, G.C. (eds), 2004, Principles of remote sensing: An introductory textbook: ITC Educational Textbook Series, p. 250 p.

Kerle, N., and Stekelenburg, R., 2004, Advanced structural disaster damage assessment based on aerial oblique video imagery and integrated auxiliary data sources, XXth ISPRS Congress, Istanbul, Turkey.

Kerle, N., Stekelenburg, R., van den Heuvel, F., and Gorte, B.G.H., 2005, Near-real time post-disaster damage assessment with airborne oblique video data, 1st international symposium on geo-information for disaster management Gi4DM, Delft, The Netherlands, 337–353.

Kramer, H.J., 2002, Observation of the earth and its environment: survey of missions and sensors: Berlin etc., Springer, 1510 p.

Kumar, R., Sawhney, H., Samarasekera, S., Hsu, S., Tao, H., Guo, Y., Hanna, K., Pope, A., Wildes, R., Hirvonen, D., Hansen, M., and Burt, P., 2001, Aerial video surveillance and exploitation: Proceedings of the IEEE, v. 89, p. 1518–1539.

Leberl, F., and Gruber, M., 2005, Ultracam-D: Understanding some noteworthy capabilities, Photogrammetric Week 2005, Stuttgart, Germany.

Leblon, B., 2005, Monitoring Forest Fire Danger with Remote Sensing: Natural Hazards, v. 35, p. 343–359.

LID, 2004, LiDAR Industry Directory 2004, available at www.airborne1.com.

Lillesand, T.M., Kiefer, R.W., and Chipman, J.W., 2004, Remote sensing and image interpretation: New York, John Wiley & Sons, 763 p.

Lindholm, S., 2004, Aerial digital photography from a balloon for fifty dollars. available at http://lindholm.jp/chpro_bal.html.

Mansor, S., Shariah, M.A., Billa, L., Setiawan, I., and Jabar, F., 2004, Spatial technology for natural risk management: Disaster Prevention and Management, v. 13, p. 364–373.

Mather, P.M., 2004, Computer processing of remotely – sensed images: an introduction: Chichester etc., Wiley & Sons, 324 p.

McGlone, J.C., Mikhail, E.M., and Bethel, J.S., 2004, Manual of photogrammetry: Bethesda, American Society for Photogrammetry and Remote Sensing (ASPRS), 1151 p.

Mikhail, E.M., Bethel, J.S., and McGlone, J.C., 2001, Introduction to modern photogrammetry: New York etc., Wiley & Sons, 479 p. + one CD-ROM.

Oh, P.Y., and Green, W.E., 2003, A kite and teleoperated vision system for acquiring aerial images, IEEE International Conference on Robotics and Automation (ICRA), Taipei, Taiwan, 1404–1409.

Oppenheimer, C., Rothery, D.A., Pieri, D.C., Abrams, M.J., and Carrere, V., 1993, Analysis of Airborne Visible Infrared Imaging Spectrometer (Aviris) data of volcanic hot-spots: International Journal of Remote Sensing, v. 14, p. 2919–2934.

Quattrochi, D.A., and Luvall, J.C., 2004, Thermal remote sensing in land surface processes: Boca Raton, Taylor & Francis, 440 p.

Rehor, M., 2007, Classification of building damages based on laser scanning data, ISPRS workshop on laser scanning 2007 and SilviLaser 2007, Espoo, Finland.

Ries, J.B., and Marzolff, I., 2003, Monitoring of gully erosion in the Central Ebro Basin by large-scale aerial photography taken from a remotely controlled blimp: CATENA, v. 50, p. 309–328.

Riggan, P.J., and Hoffman, J.W., 2003, FireMapper™: A thermal-imaging radiometer for wildfire research and operations, IEEE AerospaceConference, Big Sky, Montana, US, 12.

Robinson, J.M., 1991, Fire from space: Global fire evaluation using infrared remote sensing: International Journal of Remote Sensing, v. 12, p. 3–24.

Rodarmel, C., Scott, L., Simerlink, D., and Walker, J., 2002, Multisensor fusion over the World Trade Center disaster site: Optical Engineering, v. 41, p. 2120–2128.

Rowland, S.K., MacKay, M.E., Garbeil, H., and Mouginis-Mark, P.J., 1999, Topographic analyses of Kilauea Volcano, Hawai'i, from interferometric airborne radar: Bulletin of Volcanology, v. 61, p. 1–14.

Sandau, R., Braunecker, B., Driescher, H., Eckardt, A., Hilbert, S., Hutton, J., Kirchhofer, W., Lithopoulos, E., Reulke, R., and Wicki, S., 2000, Design principles of the LH Systems ADS40 airborne digital sensor: International Archives of Photogrammetry and Remote Sensing, v. 33, p. 256–263.

Schwarz, K.P., and El-Sheimy, N., 2004, Mobile mapping systems – state of the art and future trends, XXth ISPRS Congress, Istanbul, Turkey, 759 ff.

Singhroy, V., Mattar, K.E., and Gray, A.L., 1998, Landslide characterisation in Canada using interferometric SAR and combined SAR and TM images: Advances in Space Research, v. 21, p. 465–476.

Sithole, G., and Vosselman, M.G., 2004, Experimental comparison of filter algorithms for bare-earth extraction from airborne laser scanning point clouds: ISPRS Journal of Photogrammetry & Remote Sensing, v. 59, p. 85–101.

Stimson, G.W., 1998, Introduction to airborne radar, SciTech Publishing Inc, US, 576 p.

Thome, D.M., and Thome, T.M., 2000, Radio-controlled model airplanes: inexpensive tools for low-level aerial photography: Wildlife Society Bulletin, v. 28, p. 343–346.

Tielkes, E., 2003, L'oeil du cerf-volant : evalution et suivi des etats de surface par photographie aerienne sous cerf-volant: Weikersheim, Margraf Verlag, 113 p.

Torgersen, C.E., Faux, R.N., McIntosh, B.A., Poage, N.J., and Norton, D.J., 2001, Airborne thermal remote sensing for water temperature assessment in rivers and streams: Remote Sensing of Environment, v. 76, p. 386–398.

Toth, C.K., 2004, Future Trends in LIDAR, ASPRS Annual Conference, Denver, Colorado.

Tufte, L., Trieschmann, O., Hunsaenger, T., Kranz, S., and Barjenbruch, U., 2004, Using air- and spaceborne remote sensing data for the operational oil spill monitoring of the German north sea and Baltic Sea, XXth ISPRS Congress, Istanbul, Turkey, available at http://www.isprs.org/istanbul2004/comm7/papers/193.pdf

Tuong Thuy, V., Matsuoka, M., and Yamazaki, F., 2004, Employment of LiDAR for disaster management, Second International Workshop on Remote Sensing for Post-Disaster Response, Newport Beach, California, available at http://mceer.buffalo.edu/publications/workshop/05-SP03/206_T_Vu_paper.pdf.

Wehr, A., and Lohr, U. (eds), 1999, Theme issue on airborne laser scanning: ISPRS Journal of Photogrammetry and Remote Sensing, v. 54, p. 61–214.

Williamson, R.A., and Baker, J.C., 2002, Lending a helping hand: using remote sensing to support the response and recovery operation at the World Trade Center: Photogrammetric Engineering & Remote Sensing, v. 68, p. 870–875.

Wright, D.B., Yotsumata, T., and El-Sheimy, N., 2004, Real-time identification and location of forest fire hotspots from geo-referenced thermal images, XXth ISPRS Congress, Istanbul, Turkey, 6.

Zhang, X., Van Genderen, J., Guan, H., and Kroonenberg, S., 2003, Spatial analysis of thermal anomalies from airborne multi-spectral data: International Journal of Remote Sensing, v. 24, p. 3727–3742.

Zhang, X., Zhang, J., Kuenzer, C., Voigt, S., and Wagner, W., 2004, Capability evaluation of 3–5 micrometer and 8–12.5 micrometer airborne thermal data for underground coal fire detection: International Journal of Remote Sensing, v. 25, p. 2245–2258.

Geospatial Information Technology for Emergency Response – Zlatanova & Li (eds)
© *2008 Taylor & Francis Group, London, ISBN 978-0-415-42247-5*

Satellite remote sensing for near-real time data collection

Y. Zhang
University of New Brunswick, Fredericton, NB, Canada

N. Kerle
International Institute for Geoinformation Science and Earth Observation (ITC)
Enschede, The Netherlands

ABSTRACT: Satellite remote sensing has been applied to a variety of emergency responses for predicting, monitoring and/or managing natural or man-made disasters locally, regionally and globally. It has demonstrated that satellite imagery is a great data source for quickly responding to different emergency events. However, due to technical limitations, satellite remote sensing still faces certain challenges in real-time data collection. This paper discusses current satellite remote sensing technologies, on-going international initiatives, and future developments for rapid data collection. General discussions and examples with respect to spatial, spectral and 3D capacities of satellite imagery vs. effectiveness of information interpretation for certain natural and man-made disasters will also be given, to provide an overall guideline for data selection for a given emergency event.

1   INTRODUCTION

The previous chapter provided a detailed overview of the utility of airborne remote sensing for the provision of near-real time information in an emergency situation. Here we discuss the contributions satellite-based sensors can make. As in the previous chapter, by 'near-real time' we mean approximately the first 72 hours after an event.

Although satellite-based Earth observation only began in the 1960s, much later than airborne remote sensing, it has proved to be an excellent information source for emergency response locally, regionally and globally, because of its special advantages – (i) synoptic (i.e. large area) coverage, (ii) frequent and repetitive collection of data of the Earth's surface, (iii) diverse spectral, spatial and potentially three dimensional information, and (iv) relatively low cost for per unit coverage.

Examples of using satellite remote sensing for emergency responses can be found in many publications, such as Metternicht et al. (2005) on landslides, Tralli et al. (2005) on earthquakes, floods, and other disasters, Kerle et al. (2003) on mudflows, and Doescher et al. (2005) on satellite data for emergency support. For quick responses to forest fires nation-wide, the Forest Service of the U.S. Department of Agriculture (USDA) publishes an Active Fire Map on the Web to show weekly updated fire locations in the US (USDA Forest Service, 2006).

However, due to technical limitations, a range of challenges still affect real-time collection of imagery of areas of interest. This paper discusses state-of-the-art satellite remote sensing technologies, on-going international initiatives, and future developments for rapid data collection of the Earth's surface, to give readers an overview of current status and future development trends.

Spatial, spectral and temporal resolution (*repeat cycle*), as well as *spatial coverage* and *2D and 3D capacity* are critical technical aspects for selecting remote sensing imagery for a given emergency event. Therefore, this paper provides general information and examples on these aspects with respect to current and future Earth observation satellites and relevant natural and man-made disasters.

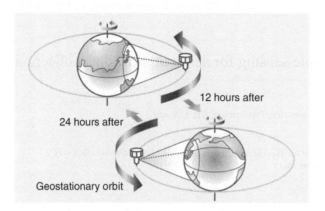

Figure 1.   Geostationary orbit (courtesy of the National Space Agency of Japan (NASDA)).

2   SATELLITE ORBITS, SENSORS AND IMAGES

2.1   *Satellite orbits*

For different remote sensing missions, different satellites carrying a range of sensors have been launched. The satellites follow different *orbits* around the Earth, acquiring images of the Earth's surface. A particular satellite is usually launched into a special orbit such that the satellite 'circles' the Earth repeatedly. The time used for completing one circle is called *period*. The time interval needed to collect repeatedly images of the same area is called *repeat cycle*. Depending on the characteristics of an orbit and the pointing capability of a satellite, the repeat cycle ranges from several days to approximately one month. The three principal satellite orbit types are introduced below. The fundamental distinction is whether a satellite revolves around Earth's major or minor axis, i.e. around its poles or the equator.

(1) Geostationary orbits
A geostationary orbit allows a satellite to appear stationary with respect to the Earth's surface (Figure 1). When positioned at approximately 36,000 km above the Earth, the period of the satellite will be 24 hours. When the satellite orbit is parallel to the equator in the same direction as the Earth's rotation, the satellite will rotate at the same speed as the Earth, appearing stationary above the same longitudinal position.

Geostationary orbits enable a satellite to view the same area of the Earth at all times. Given its large distance to the Earth, the ground coverage is very large (hemispheric). They are commonly used by meteorological satellites, such as the GOES series of NOAA, and Europe's Meteosat Second Generation (MSG). The permanent position over the same location allows a very high temporal resolution; in theory continuous, in case of MSG images are acquired every 15 minutes. However, given the large altitude, this comes at a cost of low spatial resolution, typically on the order of 3–4 km. This is sufficient for meteorological applications, but of less value in a disaster situation. Exceptions are thermal anomalies such as extensive forest fires or magmatic activity at volcanoes. Even small intense heat sources, such as 900°C lava covering only some 90 m², would be detectable with MSG's Seviri Sensor (see section 4.3 from the previous chapter). However, it is worth noting that the hemispheric coverage is not uniform, with spatial resolution decreasing steadily towards the periphery of the observed areas.

(2) Near polar orbits
A near polar orbit is one which passes close to the two poles of the Earth, in order for a satellite to collect images covering nearly the entire Earth surface in a repeat cycle (Figure 2). Polar orbits are

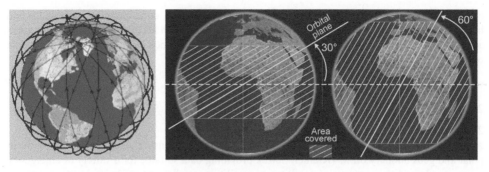

Figure 2.   Near polar orbit (left; courtesy of www.newmediastudio.org/DataDiscovery), and illustration of the effect of orbit inclination on ground coverage (right).

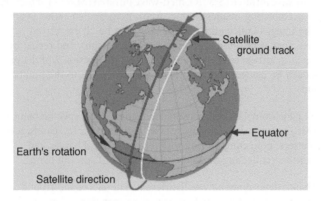

Figure 3.   Sun synchronous orbit (courtesy of Centre for Remote Imaging, Sensing and Processing).

much closer to the surface than those of geostationary satellites, typically between 500 and 800 km. A single satellite then travels in its orbit, while the Earth beneath rotates. Thus, upon return to the same point in the orbital cycle, typically some 80–100 minutes later, an area further west is imaged. Given that the Earth underneath rotates at a speed of 15 degrees per hour, the observation strips will not be contiguous.

While the spatial resolution of a polar orbit satellite is higher, the temporal resolution is generally lower compared to a stationary satellite. The current Landsat satellites, for example, have a repeat visit of only 16 days. For immediate disaster response this is clearly only sufficient if the satellite happens to be at a suitable position.

The alternative for increasing the temporal resolution is to use (i) several satellites, (ii) pointable sensors or platforms, or (iii) to incline the orbit. In the latter case the area near the poles is not observed, but the inclined orbit allows faster returns. This will, for example, be used by the FUEGO/Forest Fire Earth Watch constellation, whose purpose is forest fire monitoring in Mediterranean areas.

(3) Sun synchronous orbits
A sun synchronous orbit, a special case of a near-polar orbit, is one which allows a satellite flying over the ground track (vertical projection of the orbit on the ground) at the same local sun time, to achieve the same solar illumination condition (except for seasonal variation) for a given location (Figure 3).

Earth observation satellites usually have a sun synchronous and near polar orbit. This allows the satellites to collect images covering the entire world at a similar sun illumination condition. In a disaster response situation this is of value if pre-event data exist, thus facilitating the detection of change related to the disaster. To perform such change detection, corresponding datasets have to be tied to the same reference system, which is typically done with accurate ground control points acquired by GPS in the field, or from topographic maps. In absence of those, the ephemeris data of the satellite itself can be used, although errors will be larger.

2.2 *Remote sensing satellites and sensors*

2.2.1 *General classification of sensors*

Approximately 80 civilian imaging and non-imaging remote sensing satellites are currently in polar and geostationary orbits (Kramer, 2002), providing data suitable for a variety of emergency responses. In addition, many new satellites or satellite constellations are being built or planned. The satellites carry *optical* sensors that collect images within the wavelength of visible and near infrared (NIR), shortwave infrared (SWIR), med-wave infrared (MWIR) or thermal infrared (TIR) – and *radar* sensors – collecting images in microwave ranges, also called synthetic aperture radar (SAR) images. The images collected by the sensors can be characterized by spectral resolution (wavelength bands), spatial resolution, spatial coverage (swath), and repeat cycle, thus the same assessment criteria as in airborne remote sensing apply (see Section 2 of the previous chapter).

Optical sensors can be categorized into four types according to the number of spectral bands (spectral resolution):

- monospectral or panchromatic (Pan) sensors: collecting single spectral band, grey-scale images;
- multispectral (MS) sensors: collecting images with several spectral bands, forming MS imagery;
- superspectral sensors: collecting images with tens of spectral bands; and
- hyperspectral sensors: collecting images with hundreds of spectral bands.

The number of bands determines how well different materials on the ground can be distinguished. For broad discrimination of features such as water or vegetation, a low number of bands (3–5) is sufficient. However, to detect more subtle differences, e.g. disease damage or invading species within vegetation, or chemical contamination of soil or other surfaces, hyperspectral data would be needed.

Radar or SAR sensors can be categorized according to the spectral bands and the polarization modes (see Section 4.4 of the previous chapter).

- Wavelength:
  – L-band (15–30 cm), S-band (8–15 cm), C-band (4–8 cm), X-band (2.5–4 cm) or K-band (1.7–2.5, 0.75–1.2 cm).
- Polarization:
  – Single polarization modes: VV, or HH, or HV polarization (H: horizontal; V: vertical).
  – Multiple polarization modes: combination of two or more polarization modes.

The explanations given in Section 4.4 of the previous chapter for different radar properties, such as the effect of different polarizations or incident angles, are also applicable here. Principle differences are the large coverage achieved by satellite-based radar systems, and the regular repeat visits that facilitate interferometric applications and change detection.

In terms of **spatial resolution**, satellite sensors can further be classified into:

- low resolution sensors: >approx. 100 m ground resolution;
- medium resolution sensors: approx. 10 m to 100 m; and
- high resolution sensor: approx. sub-meter to 10 m.

The generally inverse relationship between spatial resolution and coverage discussed for airborne sensors also applies to satellites, as does the need to select an appropriate image resolution for a given emergency type.

2.2.2 *Overview of optical satellites and sensors*

An overview of the currently operational and future optical remote sensing satellites with the specifications of the corresponding sensors and images is given in Table 1, sorted by launch year. More detailed information on all operational and defunct satellites and sensors can be found at http://www.itc.nl/research/products/sensordb/searchsat.aspx.

From Table 1 it can be seen that the spatial, spectral and temporal resolution (repeat cycle), as well as spatial coverage (swath) follow certain patterns, which can provide some guidance for selecting suitable satellite imagery for a given emergency response:

Spatial resolution:

- The highest spatial resolution is 0.6 m for panchromatic images taken by the BGIS200 sensor onboard QuickBird 2, followed by Ikonos Pan with 1 m ground resolution.
- The highest spatial resolution for multispectral images is 2.5 m, also taken by QuickBird 2, followed by Ikonos MS and some other MS images with 4 m spatial resolution.
- Most high resolution satellite sensors collect Pan and MS images simultaneously, such as Quick-Bird, Ikonos and KOMPSAT-2. The resolution of Pan is usually four times higher than that of MS.

Spectral resolution:

- Most satellites collect both Pan and MS images in the spectral range of visible and near infrared wavelength (VNIR).
- Most MS images are taken within the visible and near infrared wavelength (VNIR) and are recorded into 3 or 4 multispectral bands.
- Nearly half of the sensors provide MS images with the spectral range from visible up to shortwave infrared (SWIR). However, these imagers fall mainly in the medium resolution category (20 to 30 m).
- Only a few medium resolution sensors collect MS images up to thermal infrared (TIR).

Temporal resolution:

- The repeat cycle of a single polar orbiting satellite is usually 2 to 3 weeks.
- Some satellites with body pointing capacity can reach a repeat cycle of 3 to 5 days, by turning the satellite off nadir and collecting oblique images (e.g. QuickBird, Ikonos).
- Only a few satellite constellations reach a daily repeat cycle by collecting images using many satellites (e.g. RapidEye A–E, and the Disaster Monitoring Constellation, both five-satellite constellations), others, such as SPOT, come close.

Spatial coverage:

- Usually spatial resolution and coverage are inversely related.
- High resolution satellite images usually have a swath of 10 to 20 km on the ground.
- Medium resolution images have a broader swath from 30 to 600 km.

Although in practice data are typically chosen according to what is available, the more appropriate strategy is to select imagery based on their characteristics as they pertain to a disaster situation in question. The principal question to ask is the purpose of the data/derived information. Is it to (i) obtain a general overview of the event, (ii) to aid in immediate crisis response, (iii) to assess the potential for secondary disasters (further building collapse, landslides after heavy rainstorms, etc.), or (iv) to guide cleanup and reconstruction? Beyond these questions the nature of the events, as well as its physical consequences (type, size, distribution, extent, etc.) has to be understood. Lastly, some image-derived products require special processing equipment or expertise, which also needs to be available in time.

Table 1. Main parameters of currently operational and future optical satellite systems with medium to high spatial resolution (adopted from Metternicht et al., 2005 and Earth Observation Resources, 2006).

Optical satellite	Sensor	Spatial resolution (meters) and (# bands)					Swath (km)	Repeat cycle (days)	Year launch
		PAN	VNIR	SWIR	MWIR	TIR			
Landsat 5	MSS		80 (4)				185	16	1984
	TM		30 (4)	30 (2)		120 (1)			
SPOT 2	HRV	10	20 (3)				60 (80)	26	1990
IRS-P2	LISS-II		36.4 (4)				74	22	1994
IRS-1C	LISS-III		23.5 (3)	70.5 (1)			142	24	1995
	PAN	5.8					70	24	
	WiFS		188 (2)	188 (1)			774	5	
IRS-P3	WiFS		188 (2)	188 (1)			774	5	1996
IRS-1D	LISS-III		23.5 (3)	70.5 (1)			142	24	1997
	PAN	5.8					70	24	
	WiFS		188 (2)	188 (1)			774	5	
SPOT 4	HRVIR	10	20 (3)	20 (1)			60 (80)	26	1998
	Vegetation		1000 (3)	1000 (1)					
Landsat 7	ETM+	15	30 (4)	30 (2)		60 (1)	185	16	1999
CBERS 1 and 2	HRCC	20	20 (4)				113	26	1999 & 2003
	IR-MSS	80			80 (2)	160 (1)	120	26	
	WFI		260 (2)				890	3 to 5	
Ikonos 2	OSA	1	4 (4)				11	3	1999
Terra	ASTER		15 (3)	30 (6)		90 (5)	60	16	1999
KOMPSAT-1	EOC	6.6					17	28	1999
	OSMI		1000 (6)						
EROS A1	PIC	1.9					14	2.5–4.5	2000
MTI	MTI		5 (4), 20 (3)	20 (3)	20 (2)	20 (3)	12		2000
Quickbird 2	BGIS 2000	0.6	2.5 (4)				16	3	2001
SPOT 5	HRG	2.5–5	10 (3)	20 (1)			60	26	2002
	HRS	10					120	26	
	Vegetation 2		1000 (3)	1000 (1)			2200	1	

Satellite	Sensor							
IRS-P6 (ResourceSat-1)	LISS-4	6	6 (3)			23.9 (70)	5	2003
	LISS-3		23.5 (3)	23.5 (1)		141	24	
	AWiFS		56 (3)	56 (1)		740		
DMC-AlSat1[a]	ESIS		32 (3)			600	4	2002
DMC-BILSAT-1[a]	PanCam	12				25 (300)	4	2003
	MSIS		28 (4)			55 (300)		
	COBAN		120 (4)					
DMC-NigeriaSat 1[a]	ESIS		32 (3)			600	4	2003
UK-DMC[a]	ESIS		32 (3)			600	4	2003
OrbView-3	OHRIS	1	4 (4)			8	3	2003
ROCSat-2/ FormoSat-2	RSI	2	8 (4)			24	14	2004
KOMPSAT-2	MSC	1	4 (4)			15	28	2004
China DMC + 4[a] (Beijing-1)	ESIS		32 (3)			600	4	2005
	CMT	4				24 (800)		
IRS-P5 (CartoSat-1)	PAN-F	2.5				30	5	2005
ALOS	PRISM	2.5				35 (70)	46	2005
	AVNIR-2					70		
RazakSat	MAC	2.5	5 (4)			20	13–15	2006
Resurs DK-1	ESI	1	3 (3)			28.3	N/A	2005
TopSat	RALCam1	2.5	5 (3)			25	4	2005
EROS B–C	PIC	0.7	2.8			11		2005–2008
RapidEye A–E[c]	REIS	6.5	6.5 (5)			78	1	2007
CBERS 3 and 4	MUX		20 (4)			120	26	2008 &
	PAN	5				60	1–26	2011
	ISR				80	120	26	
	WFI		40	40 (2)		866	5	
Plèiades[b]-1 and 2	HiRI	0.7	2.8 (4)			20	1–2	2008–2009

[a] DMC (Disaster Monitoring Constellation of 5 satellites) in sun-synchronic circular orbit, daily revisit cycle.
[b] Two-spacecraft constellation of CNES (Space Agency of France), with provision of stereo images.
[c] Five-satellite constellation.

2.2.3 *Overview of radar satellites and sensors*

Table 2 gives an overview of the current and future radar satellites and sensors with the specifications of individual images.

From Table 2 we can find that:

- The majority of radar satellites use C-band, while some use L- or X-band.
- The spatial resolution of most radar sensors is variable depending on the beam mode used (e.g. 1–100 m, 30–150 m, etc.), although it increases for modern systems.
- The spatial coverage (swath) of the sensors is also variable depending on the beam mode.
- Early radar satellites only have a single polarization mode, while new radar satellites tend to introduce multi-polarization modes.

This general information on radar remote sensing is also of use for users when considering radar data for emergency responses. For example, if high resolution is selected, the ground swath of image will be smaller; if radar data from an early satellite are used, multi-polarization will not be available.

2.2.4 *Spectral, spatial, stereo and orbit characteristics of selected satellites*
(1) Medium resolution optical satellites

Landsat 5 and 7

Landsat imagery has been a major source of optical images for regional and global observations in a variety of application areas, such as agriculture, forestry, geology, geography, land resources, water quality, oceanography, and global change. Table 3 presents the wavelength ranges of individual spectral bands and spatial characteristics of different Landsat sensors and orbit information of the satellites (Earth Observation Resources, 2006).

Landsat TM and ETM+ sensors have seven spectral bands and medium spatial resolution. Different spectral bands are sensitive to different materials of the Earth's surface. Therefore, different bands or band combinations will be suitable for different applications in emergency response. For example:

Band 1 (blue): coastal water mapping, soil/vegetation differentiation, deciduous/coniferous differentiation, chlorophyll absorption.
Band 2 (green): green reflectance, peak of healthy vegetation, plant vigor.
Band 3 (red): chlorophyll absorption, plant type discrimination.
Band 4 (NIR): biomass surveys, water body delineation.
Band 5 (SWIR): vegetation moisture measurement, snow/cloud differentiation.
Band 6 (TIR): plant heat stress, thermal mapping, soil mapping.
Band 7 (SWIR): hydrothermal mapping, geology.
The sensors on board of Landsat series do not have stereo capacity.

SPOT series

SPOT images are another frequently used information source from optical sensors for Earth observation and regional emergency response. Table 4 shows the wavelength ranges of individual spectral bands and spatial characteristics of different SPOT sensors and orbit information of the satellites (Earth Observation Resources, 2006).

SPOT sensors, compared to Landsat sensors, have a higher spatial resolution, but with fewer spectral bands. Due to the lack of a blue band in SPOT MS imagery, natural colour images are difficult to compose.

An advantage of SPOT is its capacity to collect panchromatic stereo images, which allows digital elevation models (DEMs) to be generated.

Table 2. Main characteristics of current and upcoming microwave satellites (Metternicht et al., 2005 and Earth Observation Resources, 2006).

Satellite	ERS-1	ERS-2	Radarsat-1	JERS-1	Envisat	Radarsat-2	Alos	TerraSAR-X	Cosmo/SkyMed[a]
Sensor	AMI	AMI	SAR	SAR	ASAR	SAR	PALSAR	TSX-1	SAR-2000
Space agency	ESA	ESA	RadarSat Int	NASDA	ESA	RadarSat Int	NASDA	DLR/Infoterra GmbH	ASI
Year of launch	1991	1995	1995	1992	2002	2006	2005	2006	2006
Out of service since	2000			1998					
Band	C	C	C	L	C	C	L	X	X
Wavelength (cm)	5.7	5.7	5.7	23.5	5.7	5.7	23.5	3	3
Polarization	VV	VV	HH	HH	HH/VV	All[b]	All	All	HH/VV
Incidence angle (°)	23	23	20–50	35	15–45	10–60	8–60	15–60	Variable
Resolution range (m)	26	26	10–100	18	30–150	3–100	7–100	1–16	1–100
Resolution azimuth (m)	28	28	9–100	18	30–150	3–100	7–100	1–16	1–100
Scene width (km)	100	100	45–500	75	56–400	50–500	40–350	5–100 (up to 350)	10–200 (up to 1300)
Repeat cycle (days)	35	35	24	44	35	24	2–46	2–11	5–16
Orbital elevation (km)	785	785	798	568	800	798	660	514	619

a Constellation of 4 satellites.
b All four polarization combinations HH, HV, VV and VH.

Table 3. Spectral bands and spatial parameters of Landsat sensors.

Satellite	Landsat 4 and -5	Landsat-4 and -5	Landsat-7
Orbit	Sun-synchronous polar orbit, altitude = 705 km, inclination = 98.2°, repeat cycle = 16 days, descending node at 9:30–10:00 AM (LS 4-5), 10:00–10:15 AM (LS 7)		
Sensor	MSS	TM	ETM+
Spectral bands (all bands in μm)	1) 0.5–0.6 2) 0.6–0.7 3) 0.7–0.8 4) 0.8–1.1	1) 0.45–0.52 VNIR 2) 0.52–0.60 VNIR 3) 0.63–0.69 VNIR 4) 0.76–0.90 VNIR 5) 1.55–1.75 SWIR 7) 2.08–2.35 SWIR 6) 10.4–12.5 TIR	P) 0.52–0.90 VNIR 1) 0.45–0.52 VNIR 2) 0.53–0.61 VNIR 3) 0.63–0.69 VNIR 4) 0.78–0.90 VNIR 5) 1.55–1.75 SWIR 7) 2.09–2.35 SWIR 6) 10.4–12.5 TIR
Swath width	185 km	185 km	185 km
Spatial resolution	80 m	30 m VNIR/SWIR 120 m TIR	15 m PAN 30 m VNIR/SWIR 60 m TIR
Radiometric resolution	6 bit	8 bit	9 bit (8 bit transmitted)

Table 4. Spectral bands and spatial parameters of SPOT sensors.

Satellite	SPOT-1,-2,-3	SPOT-4	SPOT-5
Orbit	Sun-synchronous polar orbit, altitude = 820 km, inclination = 98.8°, repeat cycle = 26 days, descending node at 10:30 AM		
Prime sensor	2 × HRV	2 × HRVIR	2 × HRG
Spectral bands PAN	PAN (0.51–0.73 μm)	PAN (0.61–0.68 μm) co-registered with B2	PA-1 (0.49–0.69 μm) PA-2 (0.49–0.69 μm)
Spectral bands MS	B_1 (0.50–0.59 μm) B_2 (0.61–0.68 μm) B_3 (0.79–0.89 μm)	B_1 (0.50–0.59 μm) B_2 (0.61–0.68 μm) B_3 (0.79–0.89 μm) SWIR (1.58–1.7 μm)	B_1 (0.49–0.61 μm) B_2 (0.61–0.68 μm) B_3 (0.78–0.89 μm) SWIR (1.58–1.7 μm),
Spatial resolution	10 m PAN 20 m MS	10 m PAN 20 m MS	5 m PA- and -2 10 m B_1, B_2 and B_3 20 m SWIR
Swath per sensor	60 km	60 km	60 km
Cross-track pointing	±27° about nadir	±27° about nadir	±27° about nadir (along-track/cross-track)
Radiometric resolution	8 bit	8 bit	8 bit

Terra Mission

ASTER is one of five sensors on board of the Terra satellite, launched in 1999. The sensor was build for collecting superspectral high-resolution imagery of the Earth's surface and clouds to better understand the physical processes affecting climate change (Yamaguchi et al., 1998; Yamaguchi et al., 1993). Table 5 gives detailed information on the spectral bands and spatial parameters of the ASTER sensor and orbit information of the satellite (Earth Observation Resources, 2006).

ASTER has 14 spectral bands with different ground resolutions from 15 m to 90 m. The main advantage of ASTER data is that the large number of bands allows substantially more detailed discrimination of different surface materials than Landsat TM/ETM+. A comparison of the spectral bands or ASTER and TM/ETM+ is provided in Table 6.

Table 5. Spectral bands and spatial parameters of ASTER sensor on board of Terra satellite.

Orbit	Sun-synchronous polar orbit, altitude = 705 km, inclination = 98.5°, repeat cycle = 15 days, descending node at 10:30 AM						

Parameter	Band No.	VNIR	Band No.	SWIR	Band No.	TIR
Spectral bands in μm	1	0.52–0.60	4	1.600–1.700	10	8.125–8.475
	2	0.63–0.69	5	2.145–2.185	11	8.475–8.825
	3N	0.76–0.86	6	2.185–2.225	12	8.925–9.275
	3B	0.76–0.86	7	2.235–2.285	13	10.25–10.95
	Stereoscopic viewing		8	2.295–2.365	14	10.95–11.65
	capability along-track		9	2.360–2.430		
Ground resolution	15 m		30 m		90 m	
Cross-track pointing	±24°		±8.55°		±8.55°	
Swath width	60 km		60 km		60 km	
Data quantization	8 bit		8 bit		12 bit	

Table 6. Spectral range comparison of ASTER and TM/ETM+ (Earth Observation Resources, 2006).

Sensor	ASTER		TM/ETM+ (Landsat 4/5)	
Wavelength Region	Band No.	Spectral Range (μm)	Band No.	Spectral Range (μm)
VNIR			1	0.45–0.52
	1	0.52–0.60	2	0.52–0.60
	2	0.63–0.69	3	0.63–0.69
	3	0.76–0.86	4	0.76–0.90
SWIR	4	1.60–1.70	5	1.55–1.75
	5	2.145–2.185	7	2.08–2.35
	6	2.185–2.225		
	7	2.235–2.285		
	8	2.295–2.365		
	9	2.360–2.430		
TIR	10	8.125–8.475	6	10.4–12.5
	11	8.475–8.825		
	12	8.925–9.275		
	13	10.25–10.95		
	14	10.95–11.65		

ASTER has two telescopes in the VNIR range, one nadir-looking with 3 spectral bands (1, 2 and 3N) and one backward-looking with a single band (3B). The band 3N (nadir) and 3B (backwards) are within the same spectral range, collecting along-track stereo images in NIR of nearly the same area on the ground as the satellite follows its orbit, thus allowing DEMs to be generated from a single ASTER data set. Such information is of use to study more voluminous landscape changes such as landslide scars (Figure 4).

(2) High resolution optical satellites

Ikonos and QuickBird
Ikonos-2, launched in 1999, was the first civilian high resolution optical satellite to reach a spatial resolution of 1 m. QuickBird-2, launched in 2001, surpassed this mark by acquiring data at a spatial resolution of 0.6 m. Both satellites collect Pan and MS images simultaneously at a resolution

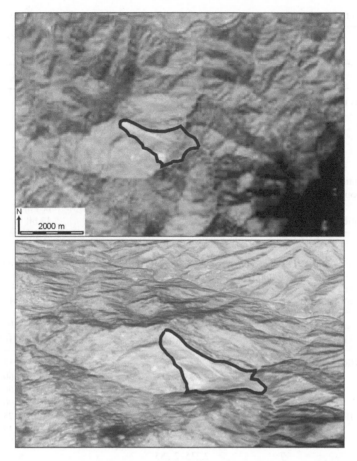

Figure 4.   ASTER image of the 2005 Pakistan earthquake area with an identified landslide outlined (top), and draped over a DEM created from the dataset. Length of scar in maximum dimension is appr. 2.5 km. Note that landslide identification is substantially facilitated when displayed in colour. (From U.S./Japan ASTER Science Team).

ratio of 1 to 4. To date, QuickBird-2 is still the satellite which provides commercially available images with the highest spatial resolution. Table 7 gives the information on spectral bands, spatial resolutions, 3D capacity and orbit information of the two satellites (http://www.spaceimaging.com/ and http://www.digitalglobe.com/).

The spectral bands of the two satellites are identical, but the spatial resolution and ground coverage are different. The capacities of body pointing and stereo imaging along or across the track are the special characteristics of the both satellites, which allow for quicker acquisition of ground images and effective 3D information acquisition and extraction.

The detailed data acquired by modern high resolution satellites, which, in addition to Ikonos and Quickbird, also include the Indian Cartosat (2.5 m Pan), Orbview-3 (1 m Pan, 4 m MS), EROS-A/B (1.9/0.7 m), Proba's HRC (5 m) and others have profound benefits for emergency response applications. These data facilitate detailed damage assessment at a house-level (Saito et al., 2004), and those with body pointing capability also allow DEM generation. However, data from these sensors tend to be very expensive, and the area covered small (Figure 5). For example, it would take more than 260 Ikonos scenes to cover the area observed in a single Landsat TM image.

Table 7. Spectral, spatial and orbital information of Ikonos and QuickBird.

Satellite	Ikonos	QuickBird
Orbit	Sun-synchronous polar orbit, altitude = 682 km, inclination = 98.1°, repeat cycle = 14 days, revisit cycle = 1–3 days (for observations at 40° latitude or higher), descending node at 10:30 AM.	Sun-synchronous polar orbit, altitude = 450 km, inclination = 98°, revisit cycle = 1–3.5 days depending on latitude (30° off-nadir), descending node at 10:30 AM
Sensor	OSA	BGIS 2000
Spectral range PAN (μm)	0.45–0.90	0.45–0.90
Spectral range MS (μm)	0.45–0.52, 0.52–0.60, 0.63–0.69, 0.76–0.90	0.45–0.52, 0.52–0.60, 0.63–0.69, 0.76–0.90
Spatial resolution	1 m PAN (0.82 m at nadir) 4 m MS	0.6 m PAN (at nadir) 2.4 m MS
Pixel quantization	11 bits	11 bits
Off-nadir pointing angle	±30° in any direction	±30° in any direction
Stereo capability	Along-track/cross-track	Along-track/cross-track
Swath width	11 km × 11 km (single image)	16.5 km × 16.5 km (single scene)
Nominal strips	11 km × 100 km (length)	16.5 km × 225 km Area (mosaic patterns): 32 km × 32 km (typically) Stereo: 16.5 km × 16.5 km typically; in along-track direction (single pass)

(3) Radar satellites

RADARSAT-1 and 2

First launched in 1995, RADARSAT acquires radar (or SAR) images in the microwave region. Because of the special advantages of active radar sensors, as described in Section 4.4 in the previous chapter, all radar satellites can take images of the Earth's surface under any weather condition and during day and night. This special characteristic makes radar sensors important for emergency response under bad weather conditions.

Compared to optical sensors the number of spaceborne radar instruments is low. After a first short-lived experiment in 1978 (SEASAT, operational for 106 days), the only civilian systems launched were Europe's ERS-1 (1991–2000) and ERS-2 (1995–), Japan's JERS (1992–1998), the Russian Cosmos-1870 and Almaz series (intermittently from 1987–1992), and Canada's Radarsat-1. After a long gap, ASAR was launched aboard Europe's ENVISAT, and in late 2005 Japan's Daichi (previously called ALOS) was deployed. The new Radarsat-2 and Germany's TerraSar-X are scheduled for launch in 2007.

Notable development trends are an increase in spatial resolution (up to 1 m for TerraSar-X) and more polarization settings. The principal disaster applications of spaceborne imaging radar are flooding (Tralli et al., 2005) and oil spill detection (Brekke and Solberg, 2005). Additionally, the utility of radar data to detect structural post-seismic damage has been explored (Bignami et al., 2004; Arciniegas et al., 2007). Remarkable success has been archived in mapping surface deformation on volcanoes, as well as following large earthquakes and landslides (e.g. Massonnet et al., 1995; Kimura and Yamaguchi, 2000). The capability to monitor ground deformation with interferometric techniques is one of the most important characteristics of SAR sensors, which can not be found in any optical remote sensing sensor.

Representative for other radar sensors, table 8 provides detailed information on RADARSAT-1 and -2 (Grenier et al., 2004; Earth Observation Resources, 2006).

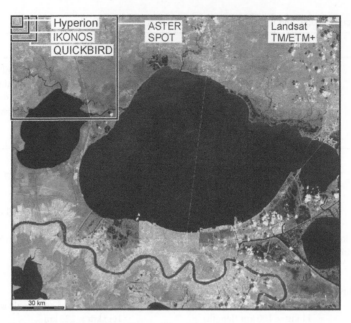

Figure 5. Ground coverage of a Landsat scene (185 × 172 km) of post-Katrina New Orleans, as well as of ASTER, SPOT, IKONOS, QUICKBIRD and Hyperion.

Table 8. SAR imaging modes of RADARSAT-1 and -2.

RADARSAT-1/2 modes (C-band, 5.5 cm center wavelength)

Orbit	Sun-synchronous polar orbit, altitude = 798 km, inclination = 98.6°, repeat cycle = 24 days, ascending node at 6:00 PM					
	Beam modes	Nominal swath width	Incidence angles to left or right side	Nr. of looks Range × Azimuth	Spatial resolution (m)	Swath coverage left or right (km)
Selective polarization:	Standard	100 km	20°–49°	1 × 4	25 × 28	250–750
Transmit H or V	Wide	150 km	20°–45°	1 × 4	25 × 28	250–650
Receive H or V or	Low incidence	170 km	10°–23°	1 × 4	40 × 28	125–300
(H and V)	High incidence	70 km	50°–60°	1 × 4	20 × 28	750–1000
	Fine	50 km	37°–49°	1 × 1	10 × 9	525–750
	ScanSAR wide	500 km	20°–49°	4 × 4	100 × 100	250–750
	ScanSAR narrow	300 km	20°–46°	2 × 2	50 × 50	300–720
New RADARSAT-2 modes (beyond those offered by RADARSAT-1)						
Polarimetric: transmit H, V on alternate pulses	Standard Quad polarization	25 km	20°–41°	1 × 4	25 × 28	250–600
Receive H, V on any pulse	Fine Quad polarization	25 km	20°–41°	1 × 1	11 × 9	400–600
Selective single polarization	Multi-look fine	50 km	30°–50°	2 × 2	11 × 9	400–750
Transmit H or V Receive H or V	Ultra-fine	20 km	30°–40°	1 × 1	3 × 3	400–550

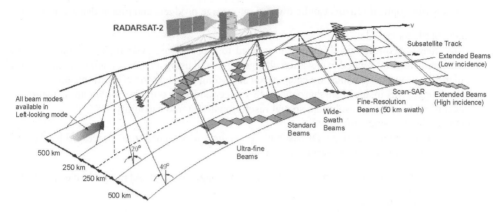

Figure 6.   Overview of different beam modes and ground coverage of RADARSAT-2 (Source: MacDonald
Dettwiler).

RADARSAT uses C-band with a center wavelength of 5.5 cm for image acquisition, which is
also being used by other radar satellites (Table 2). RADARSAT has the capability to generate 3D
information with two different techniques: (1) stereoscopic method for DEM generation; and (2)
interferometric method for DEM generation or ground deformation monitoring.

• For the stereoscopic method, stereo images taken from two different directions or orbits (Figure 6)
  are required.
• For the interferometric method, microwave phase information collected from two different orbit
  paths is required.
  – For DEM generation, it is desired that the two paths are apart from each other with a certain
    distance to allow the calculation of height information.
  – For deformation monitoring or measurement, it is desired that the two paths are close to each
    other, whiles the images are taken at different times.

(4) General orbit characteristics
In terms of orbit characteristics of all the above described polar orbiters, it can be seen that

• most satellites follow a sun-synchronous near polar orbit with inclination angles about 98° (with
  the exception of specialised satellites with inclined orbits for higher temporal resolution and
  limited coverage in higher latitudes);
• the satellite altitudes range from 450 km to 820 km;
• the temporal resolution for areas at the equator are between 2 and 3 weeks, while for regions in
  high latitudes (closer to the poles) it is significantly shorter since orbits begin to overlap;
• the temporal resolution of a satellite with body pointing capacity (e.g. off nadir ±30°) can be
  reduced to 1 to 3 days depending on the latitude of the region; and
• optical satellites usually take images around 9:30 to 10:30 AM, while radar satellites do not, for
  example RADARSAT at 6:00 AM or 6:00 PM (dawn-dusk orbit), since no sun illumination is
  needed.

3   IMAGE PRODUCTS AND IMPACT OF SPATIAL AND SPECTRAL CHARACTERISTICS

3.1   *Image products*

Image products from different providers may vary from each other and the product names may be
slightly different (e.g. Eurimage, 2006). But the essential products are generally the same both for
optical and radar images.

Products from optical sensors can be divided into image products and image-derived products:

- Image products include:
  - raw image, geo-referenced image (referenced to a coordinate system), and ortho-rectified image (distortions related to elevation changes removed using a DEM, creating an image with map-like properties), which are categorized according to geometric corrections; and
  - Pan image, MS image, Pan-sharpened MS image, and stereo image, which are categorized according to spectral resolution, spatial resolution and stereo capability.
- Image derived products include:
  - image maps, thematic maps, DEMs, and 3D images, which need to be processed by the end users or by remote sensing service companies.

Products from radar sensors can also be divided into image products and image derived products:

- Image products include:
  - raw data (unprocessed radar signals);
  - single look complex (stored in slant range, corrected for satellite reception errors, latitude and longitude information included, good for interferometric applications);
  - precision image (north oriented, corrected using ground control points); and
  - ortho-image (terrain distortions removed using a DEM and ground control points GCPs).
- Image derived products:
  - radar image maps, polarization colour image maps, DEMs, interferograms for surface deformation measurements, etc. (see for example Henderson and Lewis, 1998, or Stimson, 1998).

As was explained in the previous chapter (Table 3), the availability of image-derived products, which may require substantial processing and expertise, is generally low for the first 3 days after a disaster. While in principle an airborne platform can be launched quickly and flexibly to acquire imagery, fast data collection with satellite-based sensors is only possible if (i) a satellite happens to be in an appropriate orbit position, (ii) the satellite body can be pointed sideways, (iii) a constellation of several satellites is used, (iv) the area in question is at a high latitude, or (iv) geostationary satellites are used (although those provide data with low spatial resolution).

The availability of optical data has grown substantially in recent years. With some 80 systems in orbit, the possibility of acquiring data in an emergency situation is high. However, the time required for collected data to be downlinked, processed and distributed is typically still on the order of several days. This is a problem addressed by the International Charter 'Space and Major Disasters' (discussed below).

3.2 *Impact of spatial resolution*

Spatial resolution is one of the most important criteria for selecting remote sensing data for responding a specific emergency event. Figure 7 gives an example of spatial resolution vs. object detail. In an image with 30 m resolution, such as MS images of Landsat TM or ETM+, only large area objects can be recognized. Streets or highways are just single lines in the image, if they appear at all. However, in a QuickBird Pan image with a spatial resolution of 0.7 m, individual cars and parking lines can be clearly seen, even people in an open area can be identified.

Figure 8 shows a series of degraded images of a volcanic mudflow area. While coarse (low resolution) imagery is sufficient for general features related to the main flow, any detailed analysis involving smaller features such as houses demands very high resolution.

For emergency response, therefore, medium resolution images such as Landsat are more appropriate for macroscopic monitoring of areas affected by a natural disaster, such as flooded areas along a river or coast, or damaged areas of a forest fire or landslide. Conversely, high resolution images from Ikonos or QuickBird are more appropriate for detailed investigations within areas damaged by a natural or man-made disaster, in urban areas, or to examine individual objects damaged or in

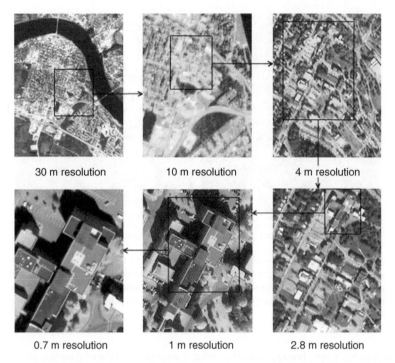

30 m resolution   10 m resolution   4 m resolution

0.7 m resolution   1 m resolution   2.8 m resolution

Figure 7.   Spatial resolution vs. object detail showing in Landsat TM (30 m), SPOT Pan (10 m), Ikonos MS (4 m), QuickBird MS (2.8 m), Ikonos Pan (1 m) and QuickBird Pan (0.7 m) (courtesy of the city of Fredericton).

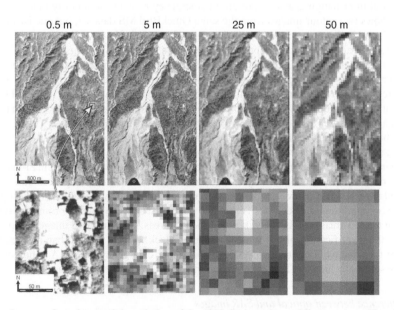

0.5 m   5 m   25 m   50 m

Figure 8.   Images of varying spatial resolution of the 1998 Casita volcano (Nicaragua) mudflow. Note that for characteristics of the actual flow, such as boundaries or main channels, even a comparatively low resolution is sufficient. For smaller features such as buildings, however, as illustrated in the bottom row, the imagery's utility declines rapidly.

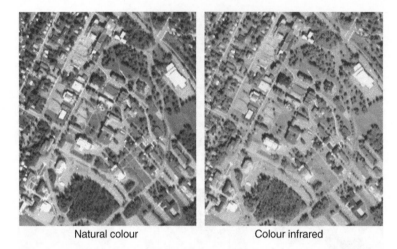

| Natural colour | Colour infrared |

Figure 9.   Natural colour image (left) and false colour composite (FCC) (right) of QuickBird 2.8 m MS data showing different landcovers on the campus of the University of New Brunswick. Note that in the original colour images, reproduced here in greyscale, buildings and paved areas are particularly apparent in the natural colour image (left), while vegetation is highlighted in red in the FCC image (right).

danger, such as buildings, houses, and other infrastructures. The rule of thumb is to use images as coarse as possible, but as detailed as necessary.

3.3   Impact of spectral resolution (or spectral bands)

Multispectral information is also important in emergency response to identify objects of interest. Figure 9 shows two colour images from the same QuickBird MS data set, but one is composed of bands 1, 2, and 3 forming a natural colour image (Figure 9 left), the other of bands 2, 3, and 4 forming a standard false colour composite (FCC) (Figure 9 right). Buildings and paved areas stand out in the natural colour image, while trees and other vegetations show up clearly in the FCC.

Another example of the importance of spectral information can be seen in Figure 10. The smoke of the forest fire can be clearly seen in both the natural colour image (left) and the FCC image (right), but in slightly different colour. However, it is difficult to identify burned areas in the natural colour image because forest (dark green) and burned areas (black) appear in very close colours. In contrast, burned areas (black) can be clearly differentiated from the forest areas (red) in the FCC due to the reduced near infrared signal of destroyed vegetation.

In general, natural colour images may be more effective for investigations of man-made objects in urban areas, because of the diversity of man-made objects and similar object colours in both the image and the real world. However, false colour infrared images may be better for investigations of forest fire or water flooding, because of clear colour contrast between vegetation and non-vegetation (including burned areas) and between flooded and dry areas. Near-infrared imagery is particularly suited for the identification of water, which appears black. Proper spectral band selection is also important for effective monitoring/investigation of other emergency events. A band combination should be selected which can best differentiate the objects of interest from each other.

3.4   Difference between optical and SAR images

Optical images are usually easier to interpret than SAR images, a difference shown in Figure 11. The grey values of an optical image represent the magnitude of the spectral reflectance of individual objects, making them similar in appearance to normal photographic images. This makes optical

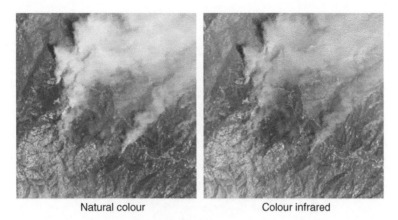

| Natural colour | Colour infrared |

Figure 10. The difference of forest fire shown in a natural colour image (left) and a colour infrared image (right). Note that in the original colour images, forest appears dark green in the natural colour image but red in FCC image, while burned areas appear black in both images. (Source: DigitalGlobe).

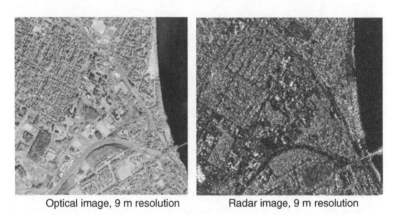

| Optical image, 9 m resolution | Radar image, 9 m resolution |

Figure 11. Difference between optical image (left, Ikonos pan resampled to 9 m) and radar image (right, RADARSAT-1 fine bean) (courtesy of the city of Fredericton).

images relatively easy to understand and interpret, although the unusual (near-) vertical viewing direction, as well as the low image scale, can also pose some problems. However, the amplitude or intensity values of a SAR image are a function of the roughness and moisture of the ground surface, as well as the radar type and incident angle used (see Section 4.4 of the previous chapter). Together with characteristic image distortions this can make for challenging image interpretation. An accurate interpretation of SAR images requires the operator to have the knowledge of the area of interest and knowledge of the principles of SAR imaging.

Normally, if optical images can be quickly acquired for an emergency area, they would be a more appropriate choice than radar due to their ease of interpretation. In case of bad weather conditions, however, as well as for specific damages such as oil spills, radar images are the only suitable option alternative for a quick response. In addition, radar images are good for certain special applications such as differentiation of flooded from dry areas and detecting ships in ice covered ocean (a challenge with optical images for typically white ships). For precise 3D measurements over larger areas, such as to detect ground deformation caused by an earthquake, interferometric SAR is uniquely suited.

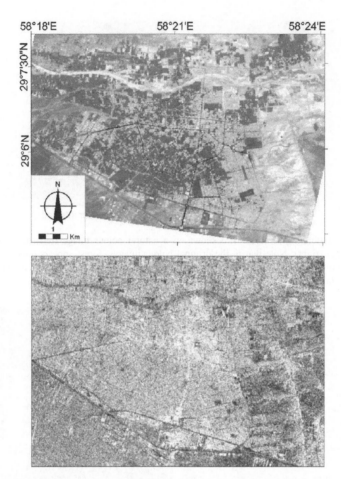

Figure 12.   Pre- and post-earthquake ASTER (top) and ASAR radar (bottom) images, respectively, of the city of Bam, Iran. Note that in particular vegetation would appear very prominently in optical data such as ASTER if displayed in colour. In this greyscale reproduction vegetation corresponds to the dark patches.

Radar data have also been used for urban damage assessment. However, Figure 12 illustrates how different radar data can be. The figure shows an ASTER image (left) and an ENVISAT ASAR scene (right) of the Iranian city of Bam. The latter was acquired following the December 2004 earthquake. While the build-up and vegetated areas are easily distinguished in the optical data, such interpretation is difficult and still subject to active research for radar images (e.g. Arciniegas et al., 2007).

3.5   Impact of image fusion

As discussed in section 2 (Table 1), most optical satellites simultaneously collect Pan and MS images at higher and lower spatial resolution, respectively. To obtain a high resolution MS image, image fusion is an important process, a technique to produce high resolution MS images by merging available high resolution Pan and low resolution MS Images. Many image fusion techniques have been developed to combine Pan and MS images for a high resolution pan-sharpened MS image. To date, the PCI Pansharp module produces the best Pan-MS fusion results among all commercially available software tools (Zhang, 2004; PCI, 2004; Gorin, 2005).

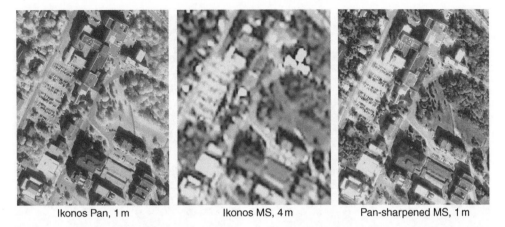

| Ikonos Pan, 1 m | Ikonos MS, 4 m | Pan-sharpened MS, 1 m |

Figure 13.   Spectral and spatial comparison between original Pan (left), original MS (middle), and pan-sharpened MS (right) images. Note that in the original colour images, the colour of the pan-sharpened MS (right) is almost identical to that of the Ikonos MS (middle) (engineering buildings and science buildings of the University of New Brunswick, image courtesy of the city of Fredericton).

Figure 13 illustrates the spatial and spectral differences between original Pan, original MS and pan-sharpened MS images from Ikonos-2. In the original Pan image (left) detailed information such as cars and trees can be clearly recognized, but due to the lack of spectral information it is not possible to identify their colours. On the other hand, the colours of cars and trees can be roughly seen in the original MS image (middle), but they cannot be clearly differentiated from each other due to the lack of spatial detail. However, in the pan-sharpened image (right), fused using the PCI Pansharp module, both colours and boundaries of individual cars and trees can be clearly seen. This example shows the importance of image fusion (or pan-sharpening) for detailed observation of the Earth's surface.

The impact of image fusion for detailed emergency investigation can also be seen in Figure 14, which shows a forest fire in southern California recorded by QuickBird on 27th October 2003. The left image is the original QuickBird MS image, 2.8 m, while the right one is the MS image after image fusion, 0.7 m. It is clear that the original MS image is not adequate for investigating individual houses in danger of the forest fire, but the fused image does provide much more adequate information for the investigation – individual family houses can be delineated and even cars in the area can be identified. Therefore, fused high resolution satellite MS images are a valuable information source for emergency response in residential areas or urban areas (also see Figure 15).

In addition to pan-sharpening, image fusion can also be used to integrate different data sources, such as GIS or map data with images, to support visual or quantitative analysis of change.

3.6   Impact of body pointing

Satellite body pointing is a relatively recent technique developed to increase the speed of image acquisition of areas of interest. It significantly enhances the agility of image acquisition and reduces the temporal resolution. Figure 15 shows two pan-sharpened Ikonos natural colour images of Manhattan before and after September 11, 2001. The right image was acquired on September 12th just one day after the man-made disaster. If body pointing technique had not been used to allow the satellite to take off-nadir images, it would not have been possible to observe the disaster site within such as short period, except with constellations of several satellites. However, it has to be remembered, the pointing of the satellite has to be specifically programmed. Such tasking is expensive, and may conflict with other observation priorities.

Grand Prix fire in southern California, October 27, 2003

QuickBird MS, 2.8 m   QuickBird pan-sharpened MS, 0.7 m

Figure 14. Difference of original natural colour image (left) and fused natural colour image (right) for investigating forest fire and residence in danger, both displayed at 1:1 scale. Original images were in
colour. (Credit to DigitalGlobe).

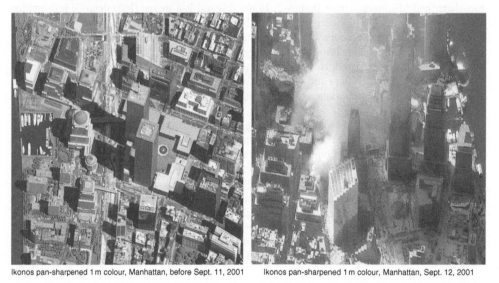
Ikonos pan-sharpened 1 m colour, Manhattan, before Sept. 11, 2001   Ikonos pan-sharpened 1 m colour, Manhattan, Sept. 12, 2001

Figure 15. High resolution oblique images collected by Ikonos before and after the man-made disaster in
New York on September 11th of 2001. Original images were in natural colour. (Credit to Space
Imaging).

From the two pan-sharpened natural colour images in Figure 15 it can be seen that buildings, streets and trees are clearly recorded, and cars can be clearly recognized. In the right image
the destruction of buildings and the impact to surrounding areas can be interpreted. Even the
chaotic traffic jam on the surrounding streets can be seen. This demonstrates the usefulness of high
resolution satellite images for urban emergency response.

However, due to the off-nadir viewing, which results in oblique imagery, many areas are occluded, i.e. blocked by taller buildings, making a comprehensive investigation of damages impossible. In particular precise mapping, as well as automated change detection based on pre-event data, suffer from such geometric distortion. Therefore, off-nadir pointing may also cause problems for emergency response, and should thus be avoided if possible.

4 CHALLENGES IN DATA AVAILABILITY

In the collection of remote sensing data for emergency response, one of the most challenging issues is to obtain image data of areas of interest within the 3 day emergency response window. From Tables 1 and 3–8 it can be seen that the normal repeat cycle of a single satellite is 2 to 3 weeks. If the satellite has the body pointing capabilities, the repeat cycle can be reduced to 3 to 5 days. However, in case of poor weather conditions etc., the actual time needed for collecting useful ground images often exceeds that time frame. On the other hand, oblique images taken by off-nadir pointing may not be suitable for many emergency events due to large blocked areas and geometric distortions. These limitations constrain the utility of using satellite remote sensing for emergency response. To overcome this limitation, some international initiatives have been proposed and developed, and joined by many countries and organizations to synergize the capacity of individual satellites and reduce the acquisition time.

There are technical solutions such as dedicated satellite constellations for disaster management, as well as more organisational approaches. The only currently operational satellite network solution is the Disaster Monitoring Constellation (DMC), which is described below. A constellation of 4 radar satellites, COSMO-Skymed, is being built by the Italian Space Agency, and is expected to be completed in 2008 (Metternich et al., 2005). The RapidEye constellation of 5 satellites, to be launched in 2007, also aims at supporting disaster response, although the principal focus is on agricultural remote sensing (http://www.rapideye.net/). For a list of other previously proposed constellations see Kerle and Oppenheimer, 2002.

4.1 Disaster Monitoring Constellation (DMC)

The Disaster Monitoring Constellation (DMC) is an international satellite program for rapid global response to natural or man-made disasters. DMC was initially proposed in 1996 and is led by Surrey Satellite Technology Ltd (SSTL), Surrey, UK. The objective of the constellation is daily global imaging capability by means of a network of five affordable micro-satellites, which collect medium resolution (28–32 m) images in 3 or 4 multispectral bands (corresponding to Landsat TM bands 2, 3, 4 and 1, 2, 3, 4, respectively) (Sweeting and Chen, 1996). As explained in section 2.2.2., such medium spatial resolution limits the utility of such data.

The DMC consortium consists of partners from Algeria, China, Nigeria, Turkey and the United Kingdom. Often involving engineers from these countries, SSTL built five low-cost Earth observation micro-satellites that were launched as the first Earth observation constellation (Table 9), and that provide daily images for global disaster monitoring (da Silva Curiel et al., 2005).

Already in March 2004, the first four DMC satellites, AlSAT-1, BILSAT-1, NigeriaSat-1 and UK-DMC, achieved the targeted orbit in the same orbit plane with a phase difference of 90° (Figure 16). This enables the DMC consortium to image anywhere on the Earth's surface with a revisit period of 24 hours (Earth Observation Resources, 2006). With the fifth satellite, Beijing-1, in orbit, the temporal resolution of the DMC constellation is even shorter. Another substantial strength of the DMC satellites is the large ground coverage in tiles up to 600 km wide (compare to figure 5).

4.2 International Charter 'Space and Major Disasters'

As detailed above, substantial space resources already exist, covering a wide range of technical specifications and hence specific utilities. This suggests that the best way forward may not be to

Table 9. The DMC satellite constellation for daily monitoring of global disasters.

Spacecraft	Spectral bands	Spatial resolution	Swath	Country	Launch date
AISAT-1	MS (3)	32 m	600 km	Algeria	Nov. 28, 2002
BILSAT-1	MS (4)	28 m	55 km (300)	Turkey	Sept. 27, 2003
	Pan	12 m	25 km (300)		
NigeriaSat-1	MS (3)	32 m	600 km	Nigeria	Sept. 27, 2003
UK-DMCSat-1	MS (3)	32 m	600 km	UK	Sept. 27, 2003
Beijing-1	MS (3)	32 m	600 km	China	Oct. 27, 2005
(China-DMC+4)	Pan	4 m	24 km within a FOR of 800 km		

Constellation orbit: Sun-synchronous polar orbit, altitude = 686 km, inclination = 98°, equator crossing at 10:30

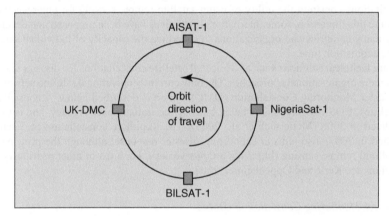

Figure 16. Orbit spacing of DMC satellite constellation for daily monitoring in 2004 (Earth Observation Resources, 2006).

equip every country with its own space technology, but rather to coordinate and optimise data acquisition by existing resources.

The International Charter 'Space and Major Disasters' is not a technical solution, but aims at improving the efficiency in the use of existing space technology during a disaster response phase. It was initiated by the European and French space agencies (ESA and CNES) after the UNISPACE III conference held in Vienna, Austria, in July 1999, and was declared formally operational on November 1, 2000. The objective of the Charter is to provide a unified system of space data acquisition and delivery for a rapid response to natural or man-made disasters. To date, the Canadian Space Agency (CSA), the National Oceanic and Atmospheric Administration (NOAA), the Indian Space Research Organization (ISRO), the Argentine Space Agency (CONAE), the Japan Aerospace Exploration Agency (JAXA), the United States Geological Survey (USGS), and the DMC-operator DMCII have joined the Charter (Table 10). Each member agency has committed resources to support the data collection of the Charter to improve the efficiency of data provision for emergency responses globally (disasterscharter.org).

To date the Charter has been activated more that 100 times, responding to a variety of natural and technological emergencies (see for example Bessis et al., 2004; www.disasterscharter.org). Once the Charter is activated, the most suitable and most quickly available space resources of the participating partners are used to obtain imagery of the affected areas. These data are then

Table 10. The current members of the Charter and their space resources.

Member	Space resources
European Space Agency (ESA)	ERS, ENVISAT
Centre national d'études spatiales (CNES)	SPOT
Canadian Space Agency (CSA)	RADARSAT
Indian Space Research Organisation (ISRO)	IRS
National Oceanic and Atmospheric Administration (NOAA)	POES, GOES
Argentina's Comisión Nacional de Actividades Espaciales (CONAE)	SAC-C
Japan Aerospace Exploration Agency (JAXA)	ALOS
United States Geological Survey (USGS)	Landsat
DMC International Imaging (DMC)	
Centre National des Techniques Spatiales (Algeria)	ALSAT-1
National Space Research and Development (Nigeria)	NigeriaSat
Tübitak-BILTEN (Turkey)	BILSAT-1
BNSC and Surrey Satellite Technology Limited (UK)	UK-DMC
Beijing Landview Mapping Information Technology Ltd (China)	BEIJING-1

further processed by Charter partners, such as the German Space Agency (DLR), UNOSAT or SERTIT, who provide map products that are made available via the Global Disaster Alert and Coordination System (GDACS, http://www.gdacs.org/), ReliefWeb (http://www.reliefweb.int) or Reuters's AlertNet (http://www.alertnet.org/). Such organisational improvements are arguably of greater value than the mere launching of more satellites. Typically the time-consuming aspect is not the data acquisition itself, but rather the transfer, processing and dissemination of useful products to emergency response personnel in the field, a requirement well served by Charter-like activities.

5 FUTURE DEVELOPMENTS

In parallel to improvements in data management and organisation, new satellites are also being built, either with new capabilities or to replace old ones. Satellite remote sensing has reached a point where it is of critical importance to many disciplines, from meteorology to the enforcement of non-proliferation treaties. It is thus an increasingly dependable and predictable tool that must be part of disaster management and crisis response plans. The future developments of remote sensing technologies are directed mainly towards the higher spatial, spectral, and temporal resolution, as shown below. It addition it must be realised that imaging satellites are only one part of relevant space infrastructure. Equally important are communication and navigation systems. The vital GPS system is satellite-based, as are emergency communication points set up in disaster areas.

5.1 *Higher spatial resolution*

WorldView. After the remarkable success of QuickBird-2 for acquisition of the highest resolution satellite images, DigitalGlobe is now in the development of the next generation satellite, WorldView, with an even higher spatial resolution of 0.25 m (Pan) and 1 m (MS), and 8 MS bands within the VNIR range. However, for non-US government customers the imagery must be re-sampled to 50-cm. The satellite is planned to be launched in 2007 (DigitalGlobe, 2004).

    OrbView-5. The company ORBIMAGE has also scheduled the launch of OrbView-5 for early 2007. The satellite will simultaneously acquire Pan images at 0.41 m resolution and MS images at 1.64 m. It will have the capacity of collecting more than 800,000 km^2 of imagery in a single day

Table 11. Future sensor development of the Pleiades program (after Earth Observation Resources, 2006).

Sensor	Resolution (m)	Swath width (km)	Nr. of bands	Revisit time (days)	Main applications
Wide field	2–5	40–100	3–4	3–7	Cartography, geology, agriculture, forest, hydrology
Optical HR	<=1	10–30	3–4	1–2	Cartography, risk, forest, geology
Superspectral	3–10	100–300	6–20	1–2	Agriculture, forest, geology
Hyperspectral	5–20	50–300	30–200	2–7	Geology
Thermal	1–40	100	TBD	<1	Forest fires, geology, ocean

and downlinking imagery in real-time to international ground station customers (ORBIMAGE, 2005).

TerraSar-X. This radar satellite will provide 1 m, 3 m and 16 m data in various polarization modes. In particular in spatial resolution this constitutes a substantial improvement over existing systems. Detailed studies such as urban damage assessment will benefit from the improvement (www.terrasar.de).

5.2 *Higher spectral resolution and response speed*

Pleiades-HR (High-Resolution Optical Imaging Constellation of CNES [French Space Agency]) is an example of the development towards higher spectral resolution and temporal resolution. A two-satellite constellation is planned for a higher repeat cycle. The planned first launch is in 2008 with the overall objectives for (Perret et al., 2002; Baudoin et al., 2001):

• high resolution panchromatic (0.7 m) and multispectral (2.8 m) imagery;
• a daily observation accessibility to any point on the Earth; and
• large coverage stereo imagery (up to 350 km × 20 km, or 150 km × 40 km).

Pleiades is a multi application, multi sensor and multi partnership program. The first generation will only carry an optical high-resolution imager (HiRI). Further sensor systems, such as wide-field, superspectral, hyperspectral and thermal (TIR) sensors, will be options for future Pleiades implementations to be launched after 2009 (Table 11).

6 OUTLOOK

Satellite remote sensing has demonstrated a tremendous potential in data collection for emergency response. Currently available satellite imagery has been applied to numerous applications in disaster prediction, investigation and/or management at global, regional and local scales. Due to technical limitations, it is still challenging to acquire quality imagery of emergency areas within a short time period (e.g. within 3 days). Therefore, the data have been mostly used for more detailed assessments of an event's aftermath and reconstruction, as well as in post-disaster scientific research. Unfortunately, bad weather conditions are often associated the natural disasters, causing further delay in the acquisition of suitable imagery and, hence, reduce the utility of satellite remote sensing for emergency response. Radar satellite images can overcome weather condition problems, although the nature of radar data limits their use to emergency aspects that entail characteristic changes in surface roughness or moisture (such as flooding), or to situations where subtle surface deformations occurred. In addition the number of operational radar sensors is very small.

The current limitations of satellite remote sensing for rapid data collection have been realized by the remote sensing community, including image providers and end users. To solve the existing problems and explore the potential of satellite remote sensing for emergency response, many international initiatives have been proposed, satellite constellations have been or are being constructed, new satellites with agile pointing capability have been developed and are being further improved, and new sensors with higher spatial or spectral resolution, or more polarization settings are under development. This will lead to faster data acquisition, and more diverse data types at higher resolutions (including multispectral, superspectral, hyherspectral, multipolorization and stereo images), which will significantly increase the utility of satellite remote sensing for emergency response. Currently the UN's Committee on the Peaceful Uses of Outer Space is also working on establishing an international space coordination body for disaster management that would link existing initiatives and mechanisms.

Importantly, there is increasing awareness of the need to integrate satellite infrastructure into a global observation strategies. To that end, groups such as the Committee on Earth Observation Satellites (CEOS) or the Integrated Global Observing Strategy (IGOS) seek to devise comprehensive and synergistic frameworks that maximise the benefit of existing and planned technology. Disaster management, including crisis response, features prominently in these efforts. An example is the Global Earth Observation System of Systems (GOESS), which is currently being developed, and for which the Global Monitoring for Environment and Security (GMES) is the European contribution. The processing of Charter data by DLR or SERTIT, as detailed above, is done under this umbrella.

Lastly, given the vast amounts of data being collected, some automation of image reception and processing is required. This is already being done successfully with lower resolution data, such as those collected by AVHRR (1.1 km resolution) and MODIS (250–1000 m resolution), to detect volcanic activity or forest fires automatically (see for example http://modis.higp.hawaii.edu/). Similar data mining approaches that comb large data sets (semi-)automatically can also be devised for other hazardous situations that allow disasters to be detected before or while they occur, and thus allowing a faster response.

REFERENCES

Arciniegas, G., Bijker, W., Kerle, N., and Tolpekin, V.A., 2007, Coherence- and amplitude-based analysis of seismogenic damage in Bam, Iran, using EnvisatASAR data: IEEETransactions on Geoscience and remote Sensing, v. 45, Special issue 'Remote sensing for major disaster prevention, monitoring and assessment', p. 1571–1581.

Baudoin, A., Boussarie, E., Damilano, P., Rum, G., and Caltagirone, F., 2001. Pléiades: a Multi Mission and Multi Cooperative Program, 52nd International Astronautical Congress (IAC), Oct. 1–5, 2001, Toulouse, France.

Bignami, C., Chini, M., Pierdicca, N., and Stramondo, S., 2004, Comparing and combining the capability of detecting earthquake damages in urban areas using SAR and optical data, IEEE International Geoscience and Remote Sensing Symposium, Alaska, USA.

Brekke, C., and Solberg, A.H.S., 2005, Oil spill detection by satellite remote sensing: Remote Sensing of Environment, v. 95, pp. 1–13.

Curiel, A.S., Boland, L., Meerman, M., Liddle, D., and Sweeting, M., 2003. Real Time, Regional Imaging Constellation, Third International Workshop on Satellite Constellations and Formation Flying, Pisa, Italy, Feb. 24–26, 2003, pp. 101–108.

Curiel, A.S., Gomes, L., Morgan, K., Butlin, T., Harding, J., Cooksley, J., Boland, L., and Sweeting, M., 2004. Small satellite constellations for Earth Observation, Proceedings of IAC 2004, Vancouver, Canada, Oct. 4–8, 2004.

da Silva Curiel, A., Boland, L., Cooksley, J., Bekhti, M., Stephens, P., Sun, W., and Sweeting, M.N., 2005, First results from the disaster monitoring constellation (DMC): Acta Astronautica, v. 56, pp. 261–271.

DigitalGlobe, 2004. DigitalGlobe's Next Generation System: WorldView. ASPRS Annual Conference, May 23–27, 2004, Denver, Colorado.

Disasterscharter.org, available at http://www.disasterscharter.org/main_e.html

Doescher, S.W., Ristyb, R., and Sunne, R.H., 2005. Use of commercial remote sensing satellite data in support of emergency response. ISPRS Workshop on Service and Application of Spatial Data Infrastructure, XXXVI(4/W6), Oct.14–16, Hangzhou, China.

Earth Observation Resources, 2006, available at http://directory.eoportal.org/res_p1_Earthobservation.html #note

Eurimage, 2006, available at http://www.eurimage.com/products/products.html

Gorin, B., 2005. Performance metrics for pan sharpening methods and comparison with ground truth, ASPRS Annual Conference, March 7–11, 2005, Baltimore, Maryland.

Grenier, C., Barnard, I., and Arsenault, P., 2004. The RADARSAT-2 synthetic aperture radar phased array antenna performance analysis methodology, Proceedings of EUSAR 2004, Ulm, Germany, May 25–27, 2004.

Henderson, F.M., and Lewis, A.J. (eds), 1998, Principles and applications of imaging radar: Manual of remote sensing, v. 2, 866 p.

Kerle, N., and Oppenheimer, C., 2002, Satellite remote sensing as a tool in lahar disaster management: Disasters, v. 26, pp. 140–160.

Kerle, N., Froger, J.L., Oppenheimer, C., and van Wyk de Vries, B., 2003, Remote sensing of the 1998 mudflow at Casita volcano, Nicaragua: International Journal of Remote Sensing, v. 24, pp. 4791–4816.

Kimura, H., and Yamaguchi, Y., 2000, Detection of landslide areas using satellite radar interferometry: Photogrammetric Engineering and Remote Sensing, v. 66, pp. 337–344.

Kramer, H.J., 2002, Observation of the earth and its environment: survey of missions and sensors: Berlin etc., Springer, 1510 p.

Massonnet, D., Briole, P., and Arnaud, A., 1995, Deflation of Mount Etna monitored by spaceborne radar interferometry: Nature, v. 375, pp. 567–570.

Metternicht, G., Hurni, L., and Gogu, R., 2005. Remote sensing of landslides: An analysis of the potential contribution to geo-spatial systems for hazard assessment in mountainous environments. Remote Sensing of Environment, vol. 98, pp. 284–303.

ORBIMAGE, 2005, available at http://www.orbimage.com/corp/orbimage_system/ov5/index.html

PCI 2004. Pan Sharpening in Geomatica 9, available at http://www.pcigeomatics.com/products/viewlets/ pansharpening/pansharp_viewlet_final.swf

Perret, L., Boussarie, E., Lachiver, J.M., and Damilano, P., 2002. The Pléiades System high resolution optical satellite and its performances, Proceedings of the 53rd IAC/World Space Congress, 2002, Oct. 10–19, 2002, Houston, TX.

Saito, K., Spence, R.J.S., Going, C., and Markus, M., 2004, Using high-resolution satellite images for post-earthquake building damage assessment: A study following the 26 January 2001 Gujarat earthquake: Earthquake Spectra, v. 20, pp. 145–169.

Stimson, G.W., 1998, Introduction to airborne radar, SciTech Publishing Inc, US, 576 p.

Sweeting, M., and Chen, F., 1996. Network of low cost small satellites for monitoring & mitigation of natural disasters, Proceedings of the 47th International Astronautical Congress, Beijing, China, October 7–11, 1996.

Tralli, D.M., Blom, R.G., Zlotnicki, V., Donnellan, A., and Evans, D.L., 2005. Satellite remote sensing of earthquake, volcano, flood, landslide and coastal inundation hazards. ISPRS Journal of Photogrammetry & Remote Sensing, vol. 59, pp. 185–198.

USDA Forest Service, 2006, available at http://www.fs.fed.us/eng/rsac/baer/status.htm

Yamaguchi, Y., Kahle, A.B., Tsu, H., Kawakami, T., and Pniel, M., 1998. Overview of Advanced Spaceborne Thermal Emission and Reflection Radiometer (ASTER), IEEE Transactions on Geoscience and Remote Sensing, Vol. 36, pp. 1062–1071, 1998.

Yamaguchi, Y., Tsu, H., and Fujisada, H., 1993. Scientific basis of ASTER instrument design, Proceedings of SPIE, Vol. 1939, pp. 150–160.

Zhang, Y. (2004). Highlight article: Understanding image fusion. Photogrammetric Engineering and Remote Sensing, Vol. 70, No. 6, pp. 657–661.

Geospatial Information Technology for Emergency Response – Zlatanova & Li (eds)
© 2008 Taylor & Francis Group, London, ISBN 978-0-415-42247-5

Terrestrial mobile mapping towards real-time geospatial data collection

J. Li
University of Waterloo, Waterloo, Ontario, Canada

M.A. Chapman
Ryerson University, Toronto, Ontario, Canada

ABSTRACT: Airborne and spaceborne remote sensing has proved their critical role in collecting geospatial data covering a relatively large geographic region in support of emergency response and disaster management. However, they fall short of capturing adequate urban details due to the limitations of spatial resolution and viewing angle. The advancement of navigation and imaging sensor technologies enabled land-based mobile mapping systems to rapidly and cost-effectively acquire detailed geospatial data on the ground. This paper provides a review of the operational principles and the state-of-the-art development of terrestrial mobile mapping technology. Following an overview of the existing systems, multi-sensor integrated mapping technology is described. The key concept of direct georeferencing is outlined. The sensors as well as their accuracy aspects are addressed. As an example, results of 2D and 3D object data collection obtained by ARAN – a Roadware terrestrial mobile mapping system are presented. The future research and development of mobile mapping towards ubiquitous mapping is summarized.

1 INTRODUCTION

In an emergency situation such as terrorist strikes and natural disasters, authorities need to quickly and effectively assess the situation and immediately decide on countermeasures. They need to answer some questions, such as, what has happened? Where has the greatest damage occurred? What kind of support systems are needed and to what degree are they needed? At the same time, there is a continuing need for actions that must be performed promptly, such as search and rescue missions, fire fighting, providing care and shelter for evacuees, and conducting assessment of building safety. Without real-time acquisition of accurate and reliable location-based information, these emergency response activities cannot be performed effectively and efficiently.

Hurricane Katrina of August 2005 that hit New Orleans, in the United States inflicted severe and extensive damage to highly built-up areas and the loss of terminals, pipelines, railroad lines, and bridges along the Gulf Coast, for example, had an immediate impact on the energy supply nationwide. Throughout the world, transportation is the most common target of terrorists, because people congregate in vehicles, terminals, and airports. The recent terrorist bombings of passenger trains in Madrid, Spain and of transit lines in London, England attest to the difficulty of protecting against such attacks. The U.S. federal government responded to the tragic events of September 11, 2001, by creating the Department of Homeland Security, which combined 22 federal agencies and entities. The amalgamation still faces significant challenges – as indicated when one of the incorporated agencies, the Federal Emergency Management Agency, struggled to respond to the devastation caused by Hurricane Katrina along the Gulf Coast. The slow and ineffective evacuations from Hurricanes Katrina and Rita in 2005 pointed to the importance of having plans that can be executed and of ensuring that intergovernmental collaborations are effective. In addition, the evacuations highlighted the need to plan and provide for transportation facilities that are adequate for response to, and recovery from, terrorist attacks and natural disasters. One of such plans is a well-established emergency response system. The emergency response system can be defined as a system for collecting and analyzing information in the event of such disaster situations, and for implementing appropriate and rapid response measures based on such information, in order to prevent or minimize damage (Tobita and Fukuwa, 2004). The emergency response systems are needed not only for implementing real-time response measures immediately following the occurrence of a disaster, but also for the recovery and reconstruction process. The emergency response system should make maximum use of information and communication technology, for example, mobile mapping systems for collecting location-based information and geospatial information systems (GIS) for handling, managing, analyzing, and representing spatial distribution of disaster and emergency situations.

In addition to standard 2D databases, the implementation of location-based services frequently requires 3D geospatial data sets. One example is personal navigation, which is facilitated by 3D visualizations and presentations in complex urban environments based on virtual 3D city models. Additionally, virtual city models are also required for architecture and town planning. Up to now, the growing number of applications resulted in great efforts in the development of tools for automatic and semiautomatic data collection (Baltsavis, 2002). Currently, the standard technique for creating large-scale city models is to use photogrammetric stereo vision approaches employing aerial and satellite images. Still, the reliable, fully automatic extraction of buildings in densely built-up areas has not yet been demonstrated (Grejner-Brzezinska et al., 2004). In recent years, advances in resolution and accuracy of airborne laser scanners (or LiDAR) have also rendered them suitable for the generation of 3D city models (Haala and Brenner, 1999; Maas and Vosselman, 1999; Rottensteiner and Briese, 2002; Schwalbe et al., 2005). However, airborne LiDAR can only capture building roofs but not the facades. This essential disadvantage prohibits their use in photo realistic walk- or drive-through applications such as training and simulation for urban disaster and emergency scenarios. Terrestrial mobile mapping systems have been used to acquire the complementary ground-level data and to reconstruct building facades.

The multi-platform and multi-sensor integrated technology has established a trend towards real-time spatial data acquisition. Multi-sensor systems can be mounted on various platforms, such as satellites, aircrafts or helicopters, land-based vehicles, water-based vessels, and even hand-carried by individual surveyors. As a result, every vehicle or individual surveyor becomes a potential data collector, responsible for globally integrated data acquisition. The recent development of terrestrial mobile mapping systems represents a typical application of this integrated technology. In fact, the development of mobile mapping systems was largely driven by the transportation applications and is being further inspired by the wide implementation of Intelligent Transportation Systems (ITS) and Geospatial Information Systems in Transportation (GIS-T). Furthermore, terrestrial mobile mapping systems have been widely used for collecting damage information, to support on-site investigation and emergency response and disaster management in urban areas.

2 STATE-OF-THE-ART TERRESTRIAL MOBILE MAPPING SYSTEMS

The evolution of mobile mapping systems from video-logging systems was mainly due to the efforts of two research groups in North America, The Center for Mapping at The Ohio State University, U.S.A. and the Department of Geomatics Engineering at The University of Calgary, Canada (Bossler et al., 1991; and Schwarz et al., 1993). Compared to video-logging systems, mobile mapping systems are able to offer full 3-D mapping capabilities that are realized by using an advanced multi-sensor integrated data acquisition and processing technology (Novak, 1995; Schwarz and El-Sheimy, 1996; Li, 1997). An overview of the mobile mapping technologies can be found in Grejner-Brzeeinska et al. (2004) and Schwarz and El-Sheimy (2006). A recent collection of the mobile mapping literature can be found in Li and Chapman (2005) and Tao and Li (2007).

A common feature of mobile mapping systems is that more than one camera is mounted on a mobile platform, allowing for stereo imaging and 3-D measurements. Direct georeferencing of

Figure 1. LD2000-RH system from Leador Spatial Co. Ltd. Wuhan, China.

digital image sequences is accomplished by the multi-sensor navigation and positioning techniques. Multiple positioning sensors, GPS, Inertial Navigation System (INS) and dead-reckoning (DR), can be combined for data processing to improve the accuracy and robustness of georeferencing. The ground control required for traditional mapping is eliminated. The systems can achieve centimeter accuracy of vehicle positioning and meter or sub-meter 3-D coordinate accuracy of objects measured from the georeferenced image sequences.

Another advantage of mobile mapping systems is that the data link to a geospatial database is easy and straightforward. The collected geometric and attribute information can be directly used to build and update a database. With the development of fast communication and image compression technologies, real-time image data link from a field mobile mapping system to an office GIS can be realized. Furthermore, such data can be disseminated and accessed through widely distributed Internet and even wireless networks. With the development of direct-georeferencing and multi-sensor integration, the technology has matured to the operational level with a number of commercial systems developed and in operation. Currently, there are many private companies who offer the commercial mobile mapping system development, integration and services. One of those is Leador Spatial Co. Ltd. Wuhan, China, which exhibited their latest mobile mapping system LD2000-RH (see Figure 1) at the conference venue of the 4th International Symposium on Mobile Mapping Technology (MMT2004) held in Kuming, China, March 29-31, 2004. Table 1 lists some available terrestrial mobile mapping systems based on the sources from the literature and the websites (up to January 2006).

3 TERRESTRIAL MOBILE MAPPING: OPERATIONAL PRINCIPLES

3.1 Definition of mobile mapping

The first formal definition of the term "mobile mapping" was offered in 1995 by Dr. John Bossler, director of the Center for Mapping, The Ohio State University, USA, "Mobile mapping is a technique used to gather geographical information, such as natural landmarks and the location of roads, from a moving vehicle" (Bossler and Toth, 1995). The research on mobile mapping dates back to the late 1980's in North America. It was mainly driven by the need for highway infrastructure mapping and transportation corridor inventory. Cameras along with navigation and positioning sensors, e.g., the Global Positioning System (GPS) and inertial navigation devices such as inertial measurement unit (IMU), were integrated and mounted on a mobile vehicle for mapping purposes. Objects can be measured and mapped from images that are directly georeferenced by the navigation and positioning sensors. In early years, the research community had used various terms to characterize this evolving research area. Terms like kinematic surveying, dynamic mapping, vehicle-based mapping, etc., appeared in literature. In 1997, the First International Symposium on Mobile Mapping Technology was held at the Ohio State University, Columbus, Ohio and the term of 'Mobile Mapping' then became a popularly accepted term used by both the research and development communities. Today, a commonly accepted definition on mobile mapping systems is given below:

A mobile mapping system can be defined as a kinematic platform, upon which multiple sensors have been integrated and synchronized to a common time base, to provide three-dimensional

Table 1. Some examples of terrestrial mobile mapping systems (updated based on Tao, 2000).

System	Developer	Positioning sensors	Mapping sensors	Website or references
ARAN	Roadware Corp., ON, Canada	Accelerometers/IMU/GPS	1 VHS, 2 or more CCD cameras, 1 laser scanner	www.roadware.com
CDSS	RWTH Aachen, Germany	GPS, 2 odometers, 1 barometer	2 monochrome CCD cameras	Bennin and Aussems, 1998
GeoVAN	GeoSpan Corp., USA	GPS, dead reckoning	10 VHS cameras, voice recorder	www.geospan.com
GEOVAN	Cartographic Institute of Catalonia, Spain	GPS/IMU	2 CCD camera, 1 laser scanner	Talaya et al., 2006
GI-Eye	NAVSYS Corp., USA	GPS/IMU	1 CCD	www.navsys.com
GPSVan™	The Ohio State University, USA	GPS/Gyro/wheel counter	2 CCD camera, voice recorders	Novak and Bossler, 1995
GPSVision	Lambda Tech. WI, USA	GPS/INS	2 color CCD cameras	www.lambdatech.com
KiSS™	University FAF Munich ikv, Germany	GPS/IMU, odometer, barometer, inclinometer, compass	2 monochrome CCD cameras, 1 colour VHS camera	www.ikv-kiss.de
LD2000	*Leador, China	GPS/INS, dead-reckoning, odometer	4 CCD colour cameras,	www.leador.com.cn Li et al., 2004
MoSES	**ikv, Germany	GPS/IMU	2 CCD cameras 2 laser scanners	www.ikv-kiss.de
MTLS	Terrapoint Ottawa, Canada	GPS/INS	Laser scanner	www.terrapoint.com
ON-SIGHT™	***TransMap Corp., USA	GPS, IMU	4 colour CCD camera	www.transmap.com
OrthoRoad	University of Calabria, Italy	GPS, IMU	1 CCD camera with a timer	Artese, 2006
Photobus	ETH Lausanne	GPS/IMU	1 CCD camera	Gontran et al., 2006
ROMDAS	Data Collection Ltd., New Zealand	GPS/IMU	Video camera, laser profilometer	www.romdas.com
ScanVan	DelftTech bv	N/A	Laser scanner	www.delfttech.com
UMMD	Sao Paulo State University, Brazil	GPS	2 CCD cameras	Silva et al., 2003
VISAT™	Univ. of Calgary and Geofit Inc., Canada	GPS/INS/Anti-lock Brake System	8 monochrome CCD cameras, 1 color VHS camera	Schwarz et al., 1993
VLMS	University of Tokyo, Japan	GPS/INS	6 linear CCD cameras, 3 laser scanners	Zhao and Shibasaki, 2005

* Leador in Wuhan, setup in 1999, conducted the commercialization of LD2000 series, originally developed at Wuhan University and is providing mapping and spatial data collection services.

** ikv in Munich is carrying out the commercialization of KiSS and MoSES, originally developed at the University FAF Munich in Germany and providing the services in street data collection.

*** Transmap Inc., established in 1994 carried forward the commercialization of GPSVan™ technology, now has been redesigned, modernized and implemented on several vehicles and serves mapping needs of the government and private sector.

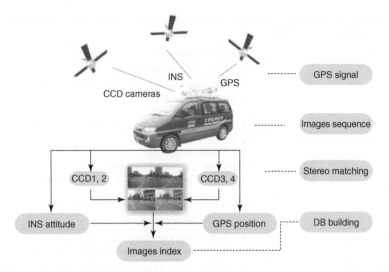

Figure 2.   The operational principle and major components of a terrestrial mobile mapping system (Source: www.leador.com.cn).

near-continuous and automatic positioning of both the platform and simultaneously collected geospatial data (Grejner-Brzezinska, 2001).

In parallel, there has been an impressive development in airborne digital frame cameras (e.g. Leica ADS40, Z/I Imaging DMC, Vexcel Ultracam), three-line scanners (e.g., STARLABO STARIMAGER), hyperspectral sensors (or imaging spectrometers, e.g., AVIRIS), laser scanners (or Lidar, e.g., Optech ALTM, TopoSys Falcon, IGI LiteMapper), and interferometric synthetic aperture radar (IfSAR or InSAR, e.g., Intermap STAR-3i) mapping systems. In the last five years, spaceborne sensors have played a significant role in mapping. In particular, high-resolution commercial imaging satellites (e.g., IKONOS, QuickBird, Orbview-3) present a new way of carrying out mapping from space. Remotely sensed imagery becomes a critical source for mapping and information extraction. Mobile mapping is all about the integration of multi-sensor configuration, multi-platform constellation, multi-data fusion, and multi-level information processing. As such, the conventional definition of mobile mapping may not be applicable to the growing evolution of the technology (Tao and Li, 2007).

3.2   *Major components of a mobile mapping system*

Mobile mapping systems are most commonly designed as modular systems that can be installed on various terrestrial and airborne platforms, and their components can be easily replaced by more advanced counterparts as technology progresses. A typical mobile mapping system consists of four key components: (1) positioning/navigation module, (2) imaging/ranging module, (3) system control module, and (4) data processing/feature extraction module (see Figure 2). The modules or subsystems can be integrated together to create a multi-task system for handling various concurrent operations in real-time and/or post-processing mode as well as for providing automatic acquisition of directly georeferenced digital data for mapping and geospatial data collection.

In a typical terrestrial mobile mapping system, multiple CCD digital cameras are mounted on the top of the land vehicle (a van or SUV) permitting stereo imaging and 3D measurement (some systems also have single or multiple down-looking digital cameras to take images of the pavement conditions of road surfaces). The stereo vision system provides 3D measuring capacity to obtain object coordinates without object space control. Two forward and two backward looking cameras that capture image pairs, are mounted on the well-defined bases which establish the model scale.

107

Table 2. Main sensors of a typical mobile mapping system and their functionality (modified according to Grejner-Brzezinska, 2004).

Sensor type	Sensor functionality
GPS receiver	<ul><li>Image geopositioning in 3D</li><li>Time synchronization between GPS and IMU</li><li>Image time-tagging</li><li>IMU error control</li><li>Furnishes access to the 3D mapping frame through WGS84</li></ul>
Inertial Measurement Unit (IMU)	<ul><li>Image orientation in 3D</li><li>Support image georeferencing</li><li>Provide bridging of GPS gaps<ul><li>Provide continuous, up to 256 Hz, trajectory between GPS measurement epochs</li><li>Support ambiguity resolution after losses of lock, and cycle slip detection and fixing</li></ul></li></ul>
CCD digital camera	<ul><li>Collect images to derive object positions<ul><li>Two or more cameras provide 3D coordinates in space</li></ul></li></ul>
Laser range finder	<ul><li>Support feature extraction from the imagery by providing precise distance measurements</li></ul>
Laser scanner or Lidar	<ul><li>Collect 3D point clouds to derive 3D object models</li><li>Collect both object's geometry and surface texture information</li></ul>
Voice recording, touch screen, Barometers, Gravity gauges	<ul><li>Document attribute information</li></ul>

The relative position and orientation of the cameras are constant, thus 3D spatial coordinates can be computed in a local coordinate system, while the final object coordinates can be determined by integrating this system with the GPS/IMU navigation subsystem. Digital imaging connects the moving platform to the object or environment being mapped. Features are identified in the images captured and their coordinates are derived from the positioning and orientation info rmation of the vehicle. Currently, multiple imaging sensors, such as CCD cameras and laser scanners are mounted on the same vehicle to support multi-purpose geospatial data collection. Table 2 presents the main sensors of a typical mobile mapping system and their functionality.

Multi-sensor integrated systems that combine direct positioning and imaging sensors are rapidly becoming a standard data acquisition tool for various mapping applications. The availability of mobile Internet in a distributed environment facilitates the extension of the operational environments to real time, enabling change detection, emergency response, and real-time decision making. An optimal fusion of multi-sensor data, supported by the geometric fusion facilitated by GPS/INS provides complementary information and higher fault resistance. This, in turn, translates to a more consistent scene description, enabling an improved scene interpretation and enhanced knowledge content. However, proper individual sensor calibration and inter-calibration become crucial in providing the required mapping accuracy, because no provision can be made for incorrect or varying sensor models (inner orientation) when direct platform orientation is used. For example, a laser scanner can deliver range information with an accuracy of a few mm for terrestrial applications. Thus, in order to properly utilize this high quality information, the sensor has to be positioned and oriented with a comparable accuracy. The quality and stability of calibration and time synchronization are especially important for mobile mapping systems (Grejner-Brzezinska et al., 2004).

GPS provides accurate position data to the mobile mapping system. Because of low data rates (1 Hz) and the requirement of viewing at least four satellites, the use of GPS alone is limited. In contrast, INS provides high rate position (X, Y, Z coordinates) and attitude (direction) information (up to 256 Hz), but its sensor errors tend to grow with time. By integrating GPS and INS, the accurate GPS position is used to update the INS, while the INS then produces the high data rate

Table 3. Summary of integrated GPS/INS navigation systems (modified according to Grejner-Brzezinska, 2001).

Characteristics	GPS	INS	GPS/INS
Advantages	• High-accurate position and velocity estimation • Practically time-independent error spectrum	• Self-contained, independent system • Almost provides continuous positioning. • Provide both position and orientation data. • High data sampling rate up to 256 Hz	• Combine all advantages of both system • INS serves as a backup in the case of GPS signal loss of lock due to platform maneuvers and/or nearby structures.
Disadvantages	• GPS readout is only available at discrete intervals (1–10 Hz). • Losses of lock causing gaps in positioning	• Sensor error growth with time leading to positioning error divergence	• No significant limitations • Precise time synchronization

thus producing accurate position and attitude data of the mobile mapping system, even when the GPS signals are lost. The integration of GPS and INS information has a number of advantages over the use of either system alone (see Table 3).

From a photogrammetric perspective, a terrestrial mobile mapping system is a fixed-based stereo vision system with known position and attitude provided by the GPS/INS component. Just as a person uses two eyes to determine the distance of an object, every infrastructure feature that is "seen" by two cameras can be triangulated into a three dimensional local coordinate frame and then transferred into a global coordinate system (e.g., latitude, longitude, height). From an application perspective, a terrestrial mobile mapping system can be used to collect digital images along highways, public roads, residential streets and/or railroads while traveling at posted speed limits. The image sequences acquired are tagged with their position and attitude information determined by the GPS/INS component. These georeferenced digital images are used for videolog applications. Most importantly, the stereo images are also designed to be accessed by the feature extraction software to position visible physical features, such as curb lines, traffic signs, manhole covers, fire hydrants and building locations. In addition, the GPS/INS positioning component creates base maps of the street network for GIS base map applications.

The georeferenced digital image data and the position and attribute data of the features are stored in a simple format that can be used in standard GIS systems. Once the processed data are loaded into the target GIS, the data are easily displayed in map format, analyzed and/or manipulated utilizing GIS database query functions. A typical client can use this data to accurately position items of infrastructure, develop base maps or view image data directly from an operator's personal computer in the vehicle or in the office.

3.3 Direct georeferencing

The most important concept of mobile mapping is *direct-georeferencing*. The conceptual layout of direct georeferencing is shown in Figure 3. Direct-georeferencing refers to the determination of the exterior orientation of the mapping sensor without using ground control points and the photogrammetric block triangulation. For example, if a camera sensor is used, any captured image can be "stamped" with the georeferencing parameters, namely three positional parameters and three attitude parameters. As a result, 3-D object measurements can be achieved directly via photogrammetric intersection.

The concept of direct georeferencing is based on the premise that auxiliary information other than traditional ground control is available to transform data sets into a well-defined reference frame.

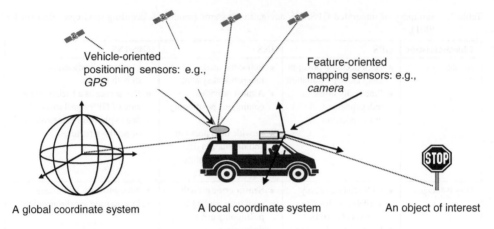

Figure 3.   The concept of direct georeferencing (Tao, 2000).

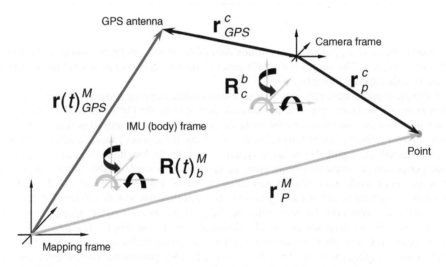

Figure 4.   Relationships between sensors in a direct georeferencing system.

These reference frames are generally defined by datum such as WGS84, GRS80, and NAD83. With the advent of GPS it became possible to precisely determine the location of an exposure station at the instant of exposure. This was primarily due to the availability of the precise time offered by GPS (Mostafa et al., 1998). In addition, the inclusion of data from an IMU introduced the possibility of observing the angular orientation of the camera at the instant of exposure. When such systems are used to the exclusion of control points on the ground, the process of direct georeferencing is invoked.

The physical relationship between a camera, an IMU, and a GPS antenna in a mobile mapping system is shown in Figure 4. In this figure, $\mathbf{r}_P^M$ indicates the coordinates of a point in the mapping coordinate system, and $\mathbf{r}_p^c$ indicates the coordinates of the same point in the camera coordinate system. Mathematically, these coordinates are related by

$$\mathbf{r}_P^M = \mathbf{r}(t)_{GPS}^M - \mathbf{R}(t)_b^M \mathbf{R}_c^b \left(\mathbf{r}_{GPS}^c - \mu_p^P \mathbf{r}_p^c \right) \tag{1}$$

where $\mathbf{r}_{GPS}^M$ are the co-ordinates of the GPS antenna, $\mathbf{R}_b^M$ is the rotation matrix between the IMU (body) and mapping coordinate frames, μ_p^P is the scale between the image and mapping frames for

110

the point, $\mathbf{r}_{GPS}^c$ is the vector connecting the GPS antenna and the camera, and $\mathbf{R}_c^b$ is the rotation matrix between the camera and IMU frames. The t indicates the time changing quantities – specifically, the GPS position and IMU orientation.

The significant terms in Equation (1) are $\mathbf{r}_p^c$, $\mathbf{r}_{GPS}^c$, and $\mathbf{R}_c^b$. The former term – the co-ordinates of the point in the camera coordinate system – has the effects of camera calibration embedded in it. The latter two terms describe the relative position and orientation between the navigation sensors and the camera. Determination of the $\mathbf{r}_{GPS}^c$ vector between the camera and the GPS antenna is known as lever-arm calibration and determination of the $\mathbf{R}_c^b$ rotation matrix between the camera and IMU is known as boresight calibration. It should be noted that some authors refer to both calibrations as the boresight calibration (or the calibration of the boresight parameters). However, to clearly demarcate between the two calibrations the terminology presented above will be followed herein.

The effect of calibration errors on the total system error can be found by performing a first-order error analysis on the Equation (1). The result, shown in Equation (2), illustrates that errors in calibration have the same effect as measurement errors. As a corollary, a more accurate calibration can mean that less accurate – and less expensive – navigation sensors are required.

$$
\begin{aligned}
\delta\mathbf{r}_P^m \quad & \delta\mathbf{r}\,(t)_{GPS}^M && \text{GPS position error} \\
- \quad & \delta\mathbf{R}\,(t)_b^M\,\mathbf{R}_c^b\left(\mathbf{r}_{GPS}^c - \mu_p^P\mathbf{r}_p^c\right) && \text{IMU Attitude error} \\
- \quad & \mathbf{R}\,(t)_b^M\,\delta\mathbf{R}_c^b\left(\mathbf{r}_{GPS}^c - \mu_p^P\mathbf{r}_p^c\right) && \text{Calibration error in IMU/Camera} \\
& && \text{misalignment angles} \quad\quad (2) \\
- \quad & \mathbf{R}\,(t)_b^M\,\mathbf{R}_c^b\,\delta\mathbf{r}_{GPS}^c && \text{Calibration error in Camera/GPS offset} \\
+ \quad & \mathbf{R}\,(t)_b^M\,\mathbf{R}_c^b\,\delta\mu_p^P\mathbf{r}_p^c && \text{Scale error} \\
+ \quad & \mathbf{R}\,(t)_b^M\,\mathbf{R}_c^b\,\mu_p^P\,\delta\mathbf{r}_p^c && \text{Image point measurement and camera} \\
& && \text{calibration error} \\
+ \quad & \delta t\left[v(t) - \omega(t)\mathbf{R}_c^b\left(\mathbf{r}_{GPS}^c - \mu_p^P\mathbf{r}_p^c\right)\right]. && \text{Synchronisation error}
\end{aligned}
$$

Employing these equations with photogrammetric data captured using a mobile system can yield point coordinates of features with accuracies of 10–15 cm. Such accuracies are generally more than sufficient for most GIS applications.

Mobile mapping systems such as the Automatic Road Analyser produced by Roadware GRP of Paris, Ontario, Canada used an array of sensors for mapping features such as pavement conditions, road marking luminosity, and road surface roughness (see Figure 5).

3.4 *Accuracy assessment of sensors*

The choice of sensors and the effective integration of sensors is a key to the development of a mobile mapping system. Accuracy, cost, reliability, data rate, portability, power consumption as well as integratibility are the most important factors to be considered. Among of them, accuracy and cost are the primary factors for a system design and implementation. However, it is beyond the scope of this paper to provide a full analysis of both positioning and imaging sensors available in the market and used in various mapping applications. A summary of some of the sensor aspects is given in the following sections.

3.4.1 *Accuracy of positioning sensors*

GPS is the most viable choice for the determination of position information due to its wide acceptance and proved performance. GPS can be operated in a variety of modes. The associated accuracies of each mode are summarized in Table 4 (Ellum and El-Sheimy, 2002).

With respect to the attitude determination, there are a number of options. Table 5 gives a list of possible sensors for attitude determination. The accuracies stated are for tilt angles (roll and pitch) below 20 degree. For more information regarding these specifications, one may refer to (Ellum and El-Sheimy, 2002; Schwarz and El-Sheimy, 1996).

It is in fact that only the IMU can directly provide all three attitude angles. For the other systems to provide all three attitude angles, they must be combined with additional sensors. Alternative to

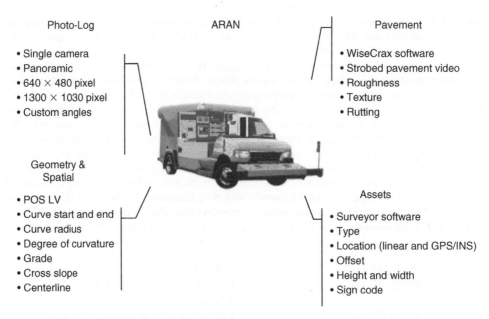

Photo-Log

- Single camera
- Panoramic
- 640 × 480 pixel
- 1300 × 1030 pixel
- Custom angles

ARAN

Pavement

- WiseCrax software
- Strobed pavement video
- Roughness
- Texture
- Rutting

Geometry &
Spatial

- POS LV
- Curve start and end
- Curve radius
- Degree of curvature
- Grade
- Cross slope
- Centerline

Assets

- Surveyor software
- Type
- Location (linear and GPS/INS)
- Offset
- Height and width
- Sign code

Figure 5. The Automated Road Analyser (ARAN) van.

Figure 6. Digital colour image acquired by ARAN.

Figure 7. Panoramic image collected by ARAN.

the IMU approach is to use GPS multi-antenna systems. It has been used in the airborne and ship-borne environments where its accuracy for attitude determination has been acceptable. However, for the land vehicle based mobile mapping applications, such a multi-antenna based GPS system can not reach the acceptable accuracy level due to the fact that the baseline between the antennas is

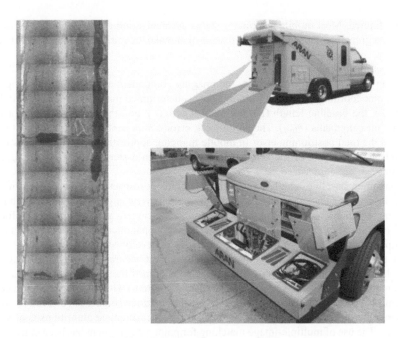

Figure 8.   Image sequences acquired by ARAN for pavement condition assessment.

Table 4.   Position accuracy of GPS.

GPS mode	Horizontal accuracy (2D RMS)	Vertical accuracy (RMS)
Code differential (Narrow Correlator, Carrier-phase smoothing)	0.75 m	1.0 m
L1 Carrier-phase RTK (Float ambiguities)	0.18 m	0.25 m
L1/L2 Carrier-phase RTK	0.03 m	0.05 m
L1 and L1/L2 Post-mission Kinematic	0.02 m	0.03 m
L1 Precise ephemeris (with Ionospheric Modelling)	1.0 m	3.0 m

Table 5.   Accuracy of sensors for determining attitude.

Sensor type	Accuracy in roll and pitch	Accuracy in Azimuth	Cost (USD)
Navigation grade IMU	<0.01°	<0.03°	>$100,000
Six-axis tactical grade IMU	0.25°	2°	$12,000–20,000
Twin antenna GPS	0.5°–1.0°	0.75°	$2,000–6,000
High-accuracy tilt sensor	0.05°	–	$3,500
Low-accuracy tilt sensor	0.25°	–	$700
Magnetic azimuth sensors	–	1.0°	$250
3-axis magnetometer integrated with 2-axis tilt sensor	0.25°	1.0°	$700–1,200

physically limited. Most mobile mapping systems use navigation grade IMU in order to meet the accuracy requirements. This is one of the reasons that make the initial development cost very high.

3.4.2 *Accuracy of mapping sensors*

Considering the error contribution from the imaging component alone, the positioning accuracy of object point coordinates derived from imagery is determined mainly by four factors: the object distance (Y), the baseline length (B), the focal length (f), and the mean square error of image coordinate measurements (m_{pxl}). The along-track error component is the predominant error contributing to the total positioning error in mobile mapping applications. Tao (1999) has conducted a comprehensive analysis of the achievable accuracy from a stereo imaging system. The main conclusions can be summarized as:

The maximum baseline length between two cameras is restricted by the desired overlap percentage; the overlap percentage is affected by the field of view angle of the camera; and the field of view angle is determined by the focal length and the camera sensing area. Therefore, a best trade-off is required for the configuration of these imaging parameters. In Tao (1999), a method to find an optimal combination of imaging parameters was developed.

Under the assumption of that the manual measurement accuracy for a single image point is 0.29 pixel, the maximum camera-to-object distance is 35m if the object positioning accuracy of 30 cm is required (with a standard CCD stereo camera system). However, if the image coordinate measurement accuracy of 0.2 pixel can be achieved, the positioned object can be 50 m removed from the camera. Therefore, development of sub-pixel image measurement algorithms is of significant importance. The use of multiple-image matching for point measurement has been considered as an effective approach (Tao, 1999).

The choice of camera types is critical to overall system performance. The camera parameters such as pixel spacing, sensing area, electronic noise, data capture rate and storage requirement must be considered carefully. Cameras with small pixel spacing, pixel synchronization unit and built-in A/D converter allow for obtaining more accurate image coordinate measurements. The use of large sensor cameras will improve the overall system accuracy and permit flexibility on the configuration of imaging parameters.

An accurate calibration of interior orientation parameters, rotation angles φ (rotation around the axis Z) and the baseline is required. Based on an analytical analysis (Tao, 1999), the calibration accuracies of 0.3 pixel of interior orientation parameters, 0.02° for the rotation angles φ, 0.03° for the relative orientation angle and 3.5 mm for the camera baseline have to be achieved, so that the effects of the calibration errors onto the total system positioning accuracy are at the same level as that of the errors arising from image coordinate measurements. However, calibration accuracy for the rotation parameter ω and κ is not as stringent.

Accuracy for time referencing is also important but is technically achievable. The tests showed that if the error of 10^{-3} second in synchronization can be controlled, the resulting position error is less than 4 cm at a vehicle speed of 60 km per hour (Schwarz and El-Sheimy, 1996).

3.4.3 *Accuracy improvement*

The accuracy from an individual sensor component is discussed above. However, effective combination of these sensors and integrated processing of the sensor data will be able to further improve the total system accuracy. There are basically two levels of the integrated processing: (1) integrated processing of multiple positioning sensor data; and (2) integrated processing of positioning and mapping sensory data. For example, if GPS and INS are combined for positioning, the INS drifts with time can be largely controlled by using GPS updates while GPS outages and cycle slips can be corrected by using INS data (Wei and Schwarz, 1990; Schwarz and Wei, 1994; Toth and Grejner-Brzezinska, 1998; and El-Sheimy et al., 1999). As for the second level of integrated processing, image-based sequential triangulation using the tie points from the overlapped images can be used to determine the orientation parameters of each image. This technique can be used to augment the georeferencing accuracy derived from the positioning sensors or to bridge gaps where the GPS signals are lost. However, this technique is not as effective as that being used in the airborne photogrammetry. This is mainly due to the poor triangulation geometry of tie points obtained from the terrestrial images. The accuracy evaluation and the automatic determination of these tie points can be found in Chaplin and Chapman (1998). One good example on the use of this technique for georeferencieng of terrestrial digital images can be found in Silva et al. (2000).

3.4.4 *Automated data processing in mobile mapping*

The implementation of real-time processing requires considerable computing power, including fast data transfer, sizeable memory, and rapid CPU time. Therefore, terrestrial mobile mapping systems are more appropriate for initial implementation of real-time processing, because the volume of data is substantially less than that of the airborne case (Grejner-Brzezinska et al., 2004. There are several advantages of using mobile mapping technology in highway applications. Because it employs dynamic data acquisition, mobile mapping technology can be directly used in highway-related applications such as traffic sign inventory, monitoring of speed limits and parking violations, and generation of road network data-bases. When laser technology is jointly applied, road surface condition inspection can also be performed. As long as traffic velocity is no less than approximately 70 km per hour, data can be acquired without disturbing the traffic flow. In addition, a single data collection mission provides diverse information for multiple purposes. Moreover, because data can be both collected and processed in a short time period, frequent repetition of road surveys and updating of databases are feasible.

4   TERRESTRIAL MOBILE LIDAR SYSTEM

However, these approaches which mainly use a few digital camera images are typically limited to one or few buildings. A terrestrial lidar-based automated data acquiring system has been developed at the University of California at Berkeley, USA (Früh and Zakhor, 2005). The system is capable of acquiring 3D geometry and texture information for an entire city at the ground level by using a combination of a horizontal and a vertical Sick LMS laser scanners and a digital color camera with a wide-angle lens. The system is mounted on a rack approximately 3.6 m high on the top of a truck, in order to obtain measurements that are not obstructed by pedestrians and cars. The scanners have a 180° field of view with a resolution of 1°, a range of 80 m and an accuracy of ±3.5 cm. Both scanners face the same side of the street and are mounted at a 90° angle. The real time software has also been developed in order to handle the incoming data stream of more than 80 Mbit/sec.

Similarly, a mobile mapping system called the Vehicle-borne Laser Measurement System (VLMS) has been developed in a joint research project between Asia Air Survey Co. Ltd and the University of Tokyo, Japan. VLMS is designed to collect detailed spatial data in central urban areas. Three single-row laser scanners and six line CCD cameras are mounted on a van to map object's geometry and texture along the streets, a GPS/INS-based navigation unit is also available. A framework was developed to reconstruct textured 3D models of buildings, roads, and trees automatically using vehicle-borne laser range and line images (Zhao and Shibasaki, 2005). An interface for extracting a broad range of urban objects, such as commercial signboards, road boundaries, traffic signs and signals, and telephone poles with cables in a semi-automatic manner (Zhao and Shibasaki, 2003) was designed. The efficiency of the system for generating a database of urban details was demonstrated by many experiments in central Tokyo, Japan.

Table 6 lists the commercially available terrestrial laser scanners which could be used for acquire high-accuracy 3D objects.

5   FROM MOBILE MAPPING TO UBIQUITOUS MAPPING

There has been an impressive development in airborne digital camera sensors, laser scanners and InSAR mapping systems. In the last five years, spaceborne sensors have played a significant role in mapping. In particular, high-resolution commercial imaging satellites present a new way of carrying

Table 6. Available terrestrial laser scanners.

System	Manufacturer	Field of view	Sampling rate	Range	Resolution/Spot size
GS200	Trimble www.Trimble.com	60°(V) × 360°(H)	5,000 pts/s	200 m	3 mm at 100 m accuracy 1.5 mm at 50 m
GX	Trimble www.Trimble.com	60°(V) × 360°(H)	5,000 pts/s	350 m	3 mm at 50 m accuracy 2.5 mm at 100 m
ILRIS-3D	Optech www.optech.ca	−20°~90°(V) × 360°(H)	2,000 pts/s	3–1000 m	3 mm
LMS	Riegl www.riegl.com/	0°~90°(V) × 360°(H)	12,000 pts/s	1–200 m	5 mm
I-SiTE 4400	I-SiTE www.isite3d.com/	80°(V) × 360°(H)	4,400 pts/s	3–400 m	50 mm
IMAGER 5003	Z+F www.zf-laser.com	310°(V) × 360°(H)	125,000 pts/s	1–53.5 m	1 mm
HDS4500	Leica www.leica-geosystems.com	310°(V) × 360°(H)	100,000–500,000 pts/s	0.75–53.5 m	8.5 mm at 25 m
HDS3000	Leica www.leica-geosystems.com	270°(V) × 360°(H)	1,800 pts/s	1–100 m	6 mm at 50 m accuracy 1.5 mm
FARO LS 880 HE80	FARO www.iqvolution.com	320°(V) × 360°(H)	120,000 pts/s	<80 m	3 mm at 25 m
FARO LS 880 HE40	FARO www.iqvolution.com	320°(V) × 360°(H)	120,000 pts/s	<40 m	3 mm at 25 m

out mapping from space. Remotely sensed imagery becomes a critical source for mapping and information extraction. Mobile mapping is all about the integration of multi-sensor configuration, multi-platform constellation, multi-data fusion, and multi-level information processing. As such, the conventional definition of mobile mapping may not be applicable to the growing evolution of the technology.

With the increasing availability of abundant, cheaper and miniature sensors; and wireless, mobile, and nomadic network access; ubiquitous mapping becomes a reality. Ubiquitous mapping refers to a new way of collecting geospatial data using various sensors and platforms and their ubiquitous combinations. A large sensor network can be formed with massive sensors mounted on different platforms or carried by individual data collectors. Moreover, the collected data can be transmitted to a central server or distributed servers in real-time for registration, integration and distribution.

There three primary characteristics of ubiquitous mapping: (1) ubiquitous sensing with mobile and portable sensors, (2) combined sensing with flexible deployment platforms, and (3) collaborative sensing with networked and controllable sensor nodes. Given the improved capacities in telecommunication bandwidth and distributed computing power, collaborative data collection is no longer a technical hypothesis. Figure 1 illustrates an exciting and comprehensive framework for the ubiquitous and collaborative data acquisition schema. This schema is also called "Sensor Web", a connected web or a network that offers a multi-level data acquisition (Tao, 2003).

In fact, real-time information processing and integrated management is also central to the ubiquitous mapping. For example, real-time image data link and communication from a terrestrial mobile mapping system to an airborne mobile mapping system will allow an efficient emergency mapping at different resolutions. Ground-based mobile sensors can be integrated with the spaceborne sensors for real-time in-situ ground truthing and sampling. Furthermore, collaboratively collected data can be disseminated and accessed through widely distributed Internet and even wireless networks. As a result, data acquisition, processing, and management are all integrated into a seamlessly integrated workflow.

6 CONCLUDING REMARKS

Recent advances in direct georeferencing, imaging sensor technology, increasingly enhanced computing power, and easy access to relatively inexpensive telecommunication services have dramatically affected the trends in modern mobile mapping system developments. Building a cost-effective terrestrial mobile mapping system by integrating off-the-shelf hardware and software components is becoming a reality. Cost-effective, high-quality CCD cameras, commercially available GPS/INS systems, and new map matching algorithms contributed substantially to the increasingly low development costs for systems that do not target at the top level of accuracy. More new sensor types have been added to the mobile mapping sensor family, including lidar for 3D object modeling acquisition and short-range laser for road surface inspection. One example of a high-end mobile mapping system is the mapping and vision systems on board rovers in the 1997 Mars Pathfinder Mission and the 2003 Mars Exploration Rover mission (Li et al., 2002). The new development of real-time mapping and easy access to wireless communication provide mobile mapping technology with an opportunity to expand its uses to many areas which were not traditionally considered part of the "mapping" business, for example, inventory and maintenance of highways, buildings, and communication facilities; mobile environmental monitoring; emergency management; and location-based services.

In this paper, an overview of the terrestrial mobile mapping technology and its evolution over the past decade was presented. Sensor technology and relevant algorithmic solutions that contributed to the success and expansion of this technology were discussed.

Several examples of mobile applications were also presented. Some examples of state-of-the-art mobile technology and applications presented here clearly illustrate the potential of multisensor systems as important mapping/GIS tools, enabling the much-desired automation of photogrammetric data collection and interpretation. The aspect of real-time mobile mapping/mobile computing, based on GPS/INS, automatic image processing, and telecommunication networks, was highlighted as the newest trends in mobile mapping technology. Although some tasks related to calibration and image data processing can be done automatically, in many cases a human operator is still needed. Thus, more research in the area of automatic interpretation and image data fusion are expected in the near future.

REFERENCES

Artese, G., 2007. OrthoRoad: a low cost mobile mapping system for road mapping, In Tao, C.V. and J. Li (eds.): Advances in Mobile Mapping Technology, ISPRS Book Series, Vol. 4, Taylor & Francis, London, pp. 31–41.

Baltsavias, E., 2002. Object extraction and revision by image analysis using existing geospatial data and knowledge: state-of-the-art and steps towards operational systems, International Archives of Photogrammetry and Remote Sensing, 34(2), pp. 13–22.

Benning, W. and T. Aussems, 1998. Mobile mapping by a car-driven survey system (CDSS), In Kahmen (ed.): Proceedings of the Symposium on Geodesy for Geotechnical and Structural Engineering, Eisenstadt (A), available at http://www.gia.rwth-aachen.de/Forschung/Strassen/CDSS/english/

Bossler, J. D., C. C. Goad, P. C. Johnson, and K. Novak, 1991. GPS and GIS – Map the nation's highways. GeoInfo Systems, March, pp. 27–37.

Bossler, J. D., and C. Toth, 1995. Bossler, J. D., and C. Toth, 1996. Feature positioning accuracy in mobile mapping: Results obtained by the GPSVan™, International Archives of Photogrammetry and Remote Sensing, 31(part B4): 139–142.

Accuracies obtained by the GPSVan™, Proceedings of GIS/LIS'95 Annual Conference, 14–16 November, Nashville, Tennessee, Vol. 1, pp. 70–77.

Chaplin, B. and M. A. Chapman, 1998 A procedure for 3D motion estimation from stereo image sequences for a mobile mapping system, Proceedings of the ISPRS Commission III Symposium on Object Recognition and Scene Classification from Multispectral and Multisensor Pixels, July 6–10, Columbus, Ohio, pp. 17–22.

Ellum, C. and N. El-Sheimy, 2002. Land-based mobile mapping systems, Photogrammetric Engineering & Remote Sensing, 68(1): 13–17.

El-Sheimy, N., M. Chapman, and C. V. Tao, 1999. An intelligent mobile mapping system, Proceedings of the 2nd International Symposium on Mobile Mapping Technology, Bangkok, Thailand, pp. 21–23.

El- Sheimy, N. and M. Lavigne, 2004. VISAT – Achieving decimeter accuracy with mobile mapping, Proceedings of the 4th International Symposium on Mobile Mapping Technology (MMT2004), Kunming, China, March 29–31.

Früh, C., S. Jain, and A. Zakhor, 2005. Data processing algorithms for generating textured 3D building façade meshes from laser scans and camera images, International Journal of Computer Vision, 61(2): 159–184.

Grejner-Brzezinska, D. A., 2001. Direct Sensor Orientation in Airborne and Land-based Mapping Applications, Report No. 461, Geodetic and Geoinformation Science, Department of Civil and Environmental Engineering and Geodetic Science, The Ohio State University, Columbus, OH 43210-1275, June 2001, 52p.

Grejner-Brzezinska, D. A., R. Li, N. Haala, and C. Toth, 2004. From mobile mapping to telegeoinformatics: paradigm shift in geospatial data acquisition, processing, and Management, Photogrammetric Engineering & Remote Sensing, 70(2): 197–210.

Gontran, H., J. Skaloud and P.-Y. Gilliéron, 2007. A mobile mapping system for road data capture via a single camera, In Tao, C.V. and J. Li (eds.): Advances in Mobile Mapping Technology, ISPRS Book Series, Vol. 4, Taylor & Francis, London, pp. 43–49.

Haala, N., and C. Brenner, 1999. Extraction of buildings and trees in urban environments, ISPRS Journal of Photogrammetry and Remote Sensing, 54(2–3): 130–137.

Li, R., 1997. Mobile mapping: An emerging technology for spatial data acquisition, Photogrammetric Engineering & Remote Sensing, 63(9): 1085–1092.

Li, D., S. Zhong, X. He, and H. Zheng, 1999. A mobile mapping system based on GPS, GIS and multi-sensor, Proceedings the 3rd International Symposium on Mobile Mapping Technology, April 21–23, Bangkok, Thailand.

Li, D., S. Guo, C. V. Tao, H. Sun, Q. Hu, and Z. Chen, 2004. The development and applications of the LD-2000RH system, Proceedings of the 4th International Symposium on Mobile Mapping Technology, March 29–31, Kunming, China.

Li, J. and M. Chapman (eds.), 2005. Mobile Mapping Systems, Special Issue of Photogrammetric Engineering & Remote Sensing, Vol. 71, No. 4.

Li, R., F. Ma, F. Xu, L. H. Matthies, C. F. Olson, and R. E. Arvidson, 2002. Localization of Mars rovers using descent and surface-based image data, Journal of Geophysical Research-Planet, 107(E11): FIDO 4.1–4.8.

Maas, H.-G., and G. Vosselman, 1999. Two algorithms for extracting building models from raw laser altimetry data, ISPRS Journal of Photogrammetry & Remote Sensing, 54(2–3): 153–163.

Mostafa, M. M. R., K. P. Schwarz, and M. A. Chapman, 1998. Development and testing of an airborne remote sensing multi-sensor system, Proceedings of the ISPRS WG II/I Conference on Data Integration: Systems and Techniques, Cambridge, UK, 13–17 July, pp. 217–222.

Novak, K., 1995. Mobile mapping technology for GIS data collection. Photogrammetric Engineering & Remote Sensing, 61(5): 493–501.

Novak, K. and J. D. Bossler, 1995. Development and application of the highway mapping system of Ohio State University, Photogrammetric Record, 15(85): 123–134.

Rottensteiner, F. and C. Briese, 2002. A new method for building extraction in urban areas from high-resolution LiDAR data, International Archives of Photogrammetry and Remote Sensing, 34(3A): 295–301.

Schwalbe, E., H.-G. Maas, and F. Seidel, 2005. 3D building model generation from airborne laser scanner data using 2D GIS data and orthogonal point cloud projections, Proceedings of ISPRS WG III/3, III/4, V/3 Workshop on Laser Scanning 2005, Enschede, the Netherlands, September 12–14, pp. 209–214.

Schwarz, K. P., M. Chapman, M. E. Cannon, and P. Gong, 1993. An integrated INS/GPS approach to the georeferencing of remotely sensed data, Photogrammetric Engineering & Remote Sensing, 59(11): 1667–1674.

Schwarz, K. P. and N. El-Sheimy, 1996. Kinematic multi-sensor systems for close range digital mapping, International Archives of Archives of Photogrammetry and Remote Sensing, 31(B3), pp. 774–784.

Schwarz, K. P. and N. El-Sheimy, 2006. Digital mobile mapping systems – state of the art and future trends, In Tao, C. V. and J. Li (eds.): Advances in Mobile Mapping Technology, ISPRS Book Series, Taylor & Francis.

Schwarz, K. P. and M. Wei, 1994. Aided versus embedded: A compar-ison of two approaches to GPS/INS integration, Proceedings of IEEE Position Location and Navigation Symposium, 11–15 April, Las Vegas, Nevada, pp. 314–321.

Silva, J. F. C., P. O. Camargo, and R. A. Oliveira, 2000. A street map built by a mobile mapping system, International Archives of Archives of Photogrammetry and Remote Sensing, 33(B2), pp. 510–517.

Silva, J. F. C., P. O. Camargo, and R. B. A. Gallis, 2003. Development of a low-cost mobile mapping system: a South American experience, Photogrammetric Record, 18(101): 5–26.

Talaya, J., E. Bosch, R. Alamús, A. Serra, and A. Baron, 2006. GEOVAN: The mobile mapping system from the ICC, In Tao, C. V. and J. Li (eds.): Advances in Mobile Mapping Technology, ISPRS Book Series, Taylor & Francis.

Tao, C. V. 1999. Error analysis and modeling of terrestrial stereo imaging systems, Surveying and Land Information Systems, 59(3): 187–196.

Tao, C. V., 2000. A review of mobile mapping technology for road network data acquisition, Journal of Geospatial Engineering, 2 (2): 1–13.

Tao, C. V., 2001. Database-guided automatic inspection of vertically structured transportation objects from mobile mapping image sequences, Photogrammetric Engineering & Remote Sensing, 67(12):1401–1409.

Tao, C. V., 2003. The smart sensor web, GeoWorld, September, pp. 28–32.

Tao, C. V. and J. Li (eds.), 2007. Advances in Mobile Mapping Technology, ISPRS Book Series, Vol. 4, Taylor & Francis, London, 176p.

Tobita, J. and N. Fukuwa, 2004. Emergency Response Systems, In Karimi, H. A. and A. Hammad (eds.) 2004: Telegeoinformatics, CRC Press, pp. 263–286.

Toth, C. and D. A. Grejner-Brzezinska, 1998. Performance analysis of the Airborne Integrated Mapping System (AIMS™), International Archives of Photogrammetry and Remote Sensing, 32(2): 320–326.

Wei, M. and Schwarz, K.P., 1990. Testing a decentralized filter for GPS/INS integration, Proceedings of IEEE PLANS, Las Vegas, USA, pp. 429–435.

Zhao, H. and R. Shibasaki, 2005.Updating a digital geographic database using vehicle-borne laser scanners and line cameras, Photogrammetric Engineering & Remote Sensing, 71(4): 415–424.

Zhao, H., and R. Shibasaki, 2003. Special Issue on Computer Vision Systems: Reconstructing textured CAD model of urban environment using vehicle-borne laser range scanners and line cameras, Machine Vision and Applications, 14 (1), 35–41.

Part 3
Data management and routing in 3D

Geospatial Information Technology for Emergency Response – Zlatanova & Li (eds)
© 2008 Taylor & Francis Group, London, ISBN 978-0-415-42247-5

Real time and spatiotemporal data indexing for sensor based databases

S. Servigne & G. Noel

LIRIS: Research Center for Images and Information Systems, INSA-Lyon, Villeurbanne Cedex, France

ABSTRACT: The most important aspect of a system for emergency response is time. So it is necessary to organise fast storage of newly arriving data into databases, fast search of data, maintenance of time sequences, robustness of the systems. All of these processes have to be real time. Another important aspect for environmental risk monitoring for example, is to allow real-time queries based on spatiotemporal attributes. Fastest accesses to the most recent data are also necessary. In addition, real-time data collected into real time database from sensors require fast and efficient management to avoid main memory saturation. The chapter discusses real-time spatiotemporal indexing schemes and real time management of main memory. Some possible solutions for real time spatiotemporal data indexing are presented to allow real time and fast data structuring. The proposed index methods support queries privileging recent data. One of the presented indexing solutions is dedicated to real time spatiotemporal data issued from fixed sensors. Another one is dedicated to real time spatiotemporal data collected from agile sensors. Finally, an outline solution concerning management of real time memory saturation according to the significance of data is presented.

1 INTRODUCTION

Sensor networks are now common and usually used for many applications, especially when it comes to risk monitoring. The monitoring of natural phenomena is indeed a key element of any situation likely to provoke some disasters, such as flooding, volcanic eruptions, landslides and so on. Sensor networks often share the same global architecture, as defined in Figure 1. Different sensors measure various data. From here on, they have to transmit these data to a monitoring system for processing. The transmission policies must take into account the nature of the sensors, their autonomy and their processing and storage capacities. A balance between energy saving and data accuracy must be reached.

When data reach the monitoring system, data first go through a main-memory sub-system (database,...) for short-term analysis. Main-memory is faster than disk-based memory and therefore allows faster processing of recent data. This is particularly interesting when the monitoring system is linked to a risk management system. After the main-memory, data can be stored into a secondary memory (disk-based memory, a data warehouse...) for longer term-analysis.

This whole process means that the system architecture may change according to various elements. As a matter of fact, the type of sensors used may change the network structure. Data location (in the network, in a database...) has an impact on data querying. The use of a database means that data have to be correctly indexed.

First, an example of a risk monitoring system for environmental phenomena is presented in order to better explain the need of real time querying and structuring with spatiotemporal criteria. Real time spatiotemporal query requires specific data indexing method. After a state of the art on data

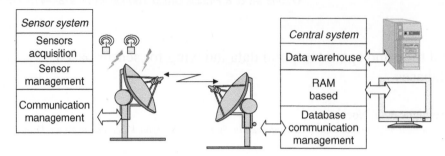

Figure 1.   Sensor network architecture.

structuring according to real time and spatiotemporal specifications, some possible solutions for real time spatiotemporal data indexing are presented.

2   NATURAL PHENOMENON MONITORING SYSTEM

After a presentation of the general system architecture, an attention is given on an existing system: a volcanic activity monitoring system. Data location and data updating in sensor networks, as well as data storage are also tackled. Systems have to offer various and new real time queries for natural phenomenon monitoring. These issues are finally addressed.

2.1   General architecture

Different monitoring systems are used to collect and process data issued by sensor networks. Among these, the Earthworm system, developed by the USGS (USGS, 2006) appears as a good example. It was designed to replace older systems for seismic activity notification. While the first version of the specifications did not take into account the storage and use of older data, the later ones tried to meet these new requirements. The main developments are split up into two parts: the real-time notification (Automatic Earthworm) and the longer term processing of data (Interactive Earthworm). Different databases are used for the second part.

The Automatic Earthworm is a message passing system. The different components broadcast announces labelled according to the kind of message sent and the source of the message. In that way, the different listening systems can select the relevant messages. It must be noted that even though the system has been defined as passing message, it can be implemented on a single computer. In that case, a part of the memory can be shared among different processing modules.

The Interactive Earthworm uses a database for storing old data. That is, data that have exited the Automatic Earthworm system and data that may be eligible for medium to long term analysis. Different systems can access those data through querying or using specific messages. However, even though the specifications were set for the Earthworm architectures, different variations have been developed for specific activities.

As a whole, the Earthworm system was primarily created for notification purpose but has been improved to add long term analysis capacities, by utilising a database. This approach has been used in many natural phenomena monitoring systems, including systems for monitoring volcanic activity.

2.2   Instance of system: a volcanic activity monitoring system

Data collected by the monitoring network of Popocatépetl volcano (seen in Figure 2. Puebla, Mexico) is heterogeneous, coming from various types of sensors. While the sensors are mainly spatially fixed, some data-processing could generate aggregates, which can be regarded as *agile*

Figure 2.   Measuring stations around the Popocatépetl volcano.

or even mobile data (epicentre locations). A sensor is considered *agile* if it has the capacity of changing the spatial location between two data acquisition phases from a specific sensor. A sensor that continually changes its location is labelled as mobile. A sensor that sometimes changes its location is considered agile.

Developments in sensor technologies have lead to the use of agile or even mobile sensors to obtain data on specific geographical sectors. However, even though sensors are able to move, measurements are usually considered more important as location changes.

As for now, collected data are indexed through their timestamps and sensor identifiers. Following the Earthworm specifications, data are first sent to an in-memory turnpike system for current data notification, pre-processing and aggregation and to a database for short to medium-term analysis. For longer-term data analysis, the database sends the measurements to a data warehouse. We shall focus on the database, with the idea that data can later on be sent to a data warehouse, possibly using a compatible indexing system.

2.3 *Data location and data updating in sensor networks*

Storage and transmission limits, data location and updating schemes can be affected by the potentially high number of sensors. In networks consisting of micro sensors, the continuous transmission of data is usually unrealistic, due to the transmission cost. Different studies tend to agree on the fact that before transmission, data aggregation at sensor level is advisable, despite the scanty calculation capabilities of the sensors (Intanagonwiwat, 2001) (Ratnasamy, 2002). However, aggregating data leads to temporal delays between data measurement and data availability for the users.

For the same reasons of data cost, several studies focused on data location. Two main approaches can be defined (Eiman, 2003).

The first one, the *local approach*, aims at storing data at the local sensor. This approach minimizes updating costs while increasing querying cost. Furthermore, to keep tracks of previous records, it is necessary to send data to a storage point.

The second approach, the *external approach*, often found in some Geographical Information Systems, uses for storage databases or data warehouses, which are outside of the sensor network. The updating cost is obviously higher while the querying cost is noticeably lower compared to the first approach. Currently, most decision support and monitoring systems rely on such solutions, combining the benefits of an in-memory storage system for the most recent data and a data warehouse for archive data. This approach also offers a relative level of safety against sensors failures (Satnam, 2001). If a sensor suddenly fails during a critical period, the data previously collected by this sensor, transmitted to the centralized database, are still available for the users.

Various approaches coexist between these two extremes, allowing solutions for data replication between specific nodes of the network. These schemes also have their constraints. Data-centric storage patterns are used to spatially determine specific nodes in the network. Therefore, data can be accessed from these replicas. By distributing data between different network nodes, it is possible to achieve a compromise between updating / querying costs. However, sensor mobility can lead to data relocating problems, limiting the real-time efficiency of such techniques.

Natural phenomenon monitoring, in which the number of sensors is limited and the environment is particularly hostile, should focus on external approaches, at least until better communication protocols are developed. External approaches limit the risk of loss data due to sensor failure, which is a common problem of some natural phenomenon monitoring systems. However, the problem of data storage is not completely solved yet.

2.4 *Data storage*

As stated before, the key idea in sensor monitoring systems is storing the maximum amount of recent data into the main memory, and to flush ancient data into an archive (disk, data warehouse). This however, does not explain either how, or where data should be kept. Three approaches for storage are discussed in the literature (Laurini 2005).

The first possibility is to keep data, at least for some time, by the sensor. In general it would be too expensive to directly send data. Therefore most sensors aggregate data before a transmission. This means that most recent data are kept at sensor level, and therefore they are unavailable to system for analysis. Some recent systems tend to favour this approach even for older data.

Due to the large amounts of sensor data, current analysis systems rely on in-memory data. Therefore data must be sent towards a central point, where it is kept in main-memory (directly or in main-memory databases). Main-memory is faster than disk and is therefore interesting for short term analysis. For longer term analysis, data are sent to disks or secondary-memory (data warehouses...). While organising data several issues have to be taken into consideration. If data are kept in a database, they must be indexed correctly, taking into account the amount of data issued from a sensor network and their real-time specificities. Furthermore, when the memory is filled in, memory saturation may bring the whole system down, which requires solutions to free database memory without loses. A specific real-time spatiotemporal database management must be coupled with a specific archiving management (Data Warehouse), as seen in Figure 3.

2.5 *Queries*

Even though spatiotemporal data are definitely useful in monitoring systems, most actual implementations are sensor-identifier-based systems.

While current indexing patterns use only timestamps and sensor identifiers, specialists' needs for queries based on spatial attributes are increasing (Figure 4). Querying patterns usually focus on finding data from a specific sensor within a given temporal interval (ending at the present time). When the number of sensors increases, users often fetch data from specific reference sensors to estimate the global state of the monitored system before querying other more specific sensors to refine their estimation. Other kinds of queries are possible but less frequent.

Sample common queries could be: "Fetch the data issued from sensor number 'IIB' in the last 5 minutes." "Fetch the data from the sensors 'IPP' between time T1 and time T2."

Due to the impact of GPS technologies and alike, specialists now demand systems that can answer queries based on spatial components. **Therefore, future systems should meet real-time and spatiotemporal requirements**. For example, scientists would like to add spatiotemporal

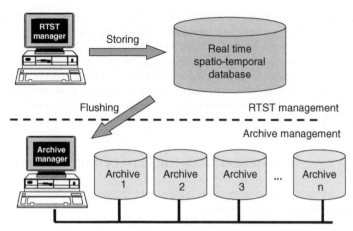

Figure 3. Structure of a real time system for geographic information with regular flushing of data into archive (RTST meaning real time spatiotemporal).

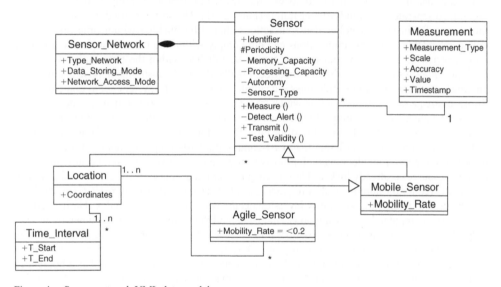

Figure 4. Sensor network UML data model.

constraints to answer queries such as "Fetch the data issued from the sensor at point $<X, Y>$ issued in the last 5 minutes."

The Popocatépetl monitoring system shows some specific characteristics that can be extended to other monitoring systems. In this system, a significant number of updates are sent from a sensor network to a centralized in-memory database. Because of the number of sensors and their measurement frequencies, soft real-time constraints emerge. Measurements from a sensor at timestamp Ti must be committed, copied into the database, before a measurement with a timestamp Tj > Ti reaches the database to be processed. For example, under normal volcanic activity, update queries must be considered as more important than other lookup queries. While the processes using the collected data run as background tasks, the updates must be carried out with higher priority. Usually, the most recent data are much more critical than the older data. In other words, volcanologists are more interested by the current state of the volcano than by its former states (provided these ones are not typical of a significant activity phase).

One last specific issue must also be considered, i.e. the sensors are usually referred to through spatial positions as well as through their identifiers. This requires a finer management of sensor agility and an addition of new querying patterns while preserving older ones.

An access method for monitoring systems should meet those requirements. Therefore, we focus on real-time indexing methods for spatiotemporal data into real-time main-memory databases to allow fast accesses to real-time data. After an overview concerning spatiotemporal indexes, two new methods for real time databases are presented. A solution for real time memory saturation is also addressed.

3   OVERVIEW OF REAL TIME SPATIOTEMPORAL DATA INDEXING METHODS

A state of the art on various existing methods is essential to better understand the pros and the cons as well as the main concepts of the situation.

3.1   *Soft real-time*

The main idea in this approach is to answer queries within time constraints (Noel, 2004). It is possible to separate three kinds of constraints (Lam, 2001). The soft one, used in case of volcanic monitoring for instance, implies that transactions should be fulfilled within time limits, yet it is understandable that some transactions can not comply within the limits. The firm constraints, more restrictive, allow some transactions not to be fulfilled within the time limits, yet in this case the whole system can be slightly impaired. Finally, the hard constraints impose that under no circumstances a transaction should miss a deadline. Otherwise, the system could come to a halt. Priorities are generally used to define which transaction is more important than another; although this is not equivalent to defining which one should occur before another. Different techniques can be used to assign priorities: Earliest Deadline First, Rate Monotonic and other variants (Lam, 2001).

Real-time computing is different than Fast-computing. Fast-computing does not prevent a low priority transaction from blocking high priority transactions (priority inversion) because the access to some resources is already locked. The current paradigm in databases is to keep the index in main memory to reduce the number of slow disk accesses. Index Consistency Control (ICC) methods can then be used to make sure no priority inversion occurs while accessing the index (Haritsa, 2001).

Currently computers' processing power increases and the memory cost drops down quite regularly. However, the memory access cost does not decrease with processors improvements. Real-time meant memory access for more than 20 years. Only since the late 90's, researchers have reached the conclusion that this was not enough. Main memory access is now considered as a bottleneck. Some researchers have provided models for a better management of memory buffers. While classical indexing structures such as the B+tree were ill-considered in the early 90's, assessments of performance have shown that the processing capacity improvements could give them a new life in the real-time applications. At least if some changes were made to their initial structure. As such, the CSB+tree (Rao, 2000), a cache-conscious B-tree limits the number of pointers in the structure, optimizes the node length in order to fit the memory line length. . . It has proven to be a real improvement over older Real-time structures, which were meant to outlast the B-trees. Some researchers are now working on extensions of the ideas set up for the CSB+tree so as to provide other indexing systems.

3.2   *Spatial indexing*

The classical spatial approaches in indexing often tend to linearise data in order to use known "fast" structures. This is the case for quadtrees, kd-trees (Samet 1984; Ooi, 1997) as well as for other indexing methods used for spatial objects, or more accurately spatial points. Kd-trees are related to binary trees. A reference point is taken, along with a reference dimension. Every other point that falls below the reference point for this dimension shall branch to the left, all points with higher values branch to the right. At the next level, a new reference point is taken in each branch and the next dimension is used as a reference. It is relatively fast, can be updated on-line but the final shape of the tree depends on the insertion order of data, which can lead to unbalanced trees.

Another widely accepted approach is the use of rectangles, bounding structures to match the location of objects. These bounding rectangles can then be regrouped within bigger rectangles to create a balanced tree. The R-tree, and its sibling R*-tree (Ooi, 1997) are examples of this approach. While R-trees allow to work with complex objects (approximated as rectangles and not points), their higher building and querying time make the use of lighter structures appealing to index points.

A special version of the R-tree, named CR-tree (Kim, 2001) uses compression to limit the node size in memory, in order to make it fit in a buffer line. Some works now try to mix this approach with the patterns developed for the CSB+tree so as to further reduce the memory access cost to spatial data.

3.3 *Temporal indexing*

In general, different notions of time can be used for databases (Ooi, 1997). The Transaction Time allows users to perform "rollbacks" in order to find past-values. It does not allow to modify previously entered values, or to enter future values. One can only append new data issued at the present moment. The Valid Time represents the time when a fact is considered true. It allows users to modify past data, and to enter future data. However, it does not allow rollbacks. The Bi-Temporal Time is a mix between the other two, allowing rollbacks, post-modifications and future updates.

There are mainly two ways to look at the temporal aspect. The first one is to consider that time is monotonous (time goes in one direction) and to use B-trees as index structures. An interesting variation of this is to consider that data flows constantly; therefore it becomes possible to link the root of the tree more closely the last leaf, this leaf containing the most recent data. Such an idea has been developed in AP-trees (Gunahdi, 1993).The other way is to consider time just as a spatial dimension and to use R-trees, with, on one dimension, the timestamps, and on the other one, the validity duration.

3.4 *Spatiotemporal indexing*

Spatiotemporal indexing has to be applied for different types of data: points, ranges, intervals (Wang, 2000). According to the type of data, families of indexing trees can be distinguished: those which work with objects in continuous movement, those for discrete changes and finally those for continuous changes of movements. Another way of differentiating the index families has been brought by (Mockbel, 2003). The author has focused on approaches aiming at indexing past, present and future locations.

Many trees have been developed to answer specific needs. Some of these trees tend to consider the temporal aspect as yet another spatial dimension, which has led to 3DR-trees (Theodoridis, 1996). However, they do not take into account the monotonicity of time and usually need to have a previous knowledge of data. Another family of trees makes a difference between spatial and temporal dimensions. In HR-tree (Nascimento, 1998), snapshots of spatial R-trees are linked in a time indexed balanced tree. The main problem of this kind of trees is their size. While nodes that do not change between two snapshots are shared among the R-trees, only minor changes force to duplicate some of the data. Furthermore, they tend not to be optimal for interval queries.

While most of these approaches focus on the data as a whole, some specific indices focus on a particular aspect of system monitoring. As an example, structures such as the SETI (Chakkar, 2003) or SEB-tree (Song, 2003) divide the global space in sub-zones for a zone-based spatiotemporal indexing. Other structures such as the FNR-tree (Frentsos, 2003) or MON-tree (Almeida, 2005) are dedicated to indexing data from an existing road or access network. With such structures, the focus is no longer the data themselves but more conceptual entities that can be used for more efficient querying or monitoring.

Recently, researches have lead to different structures based on mobility. Objects in the physical world can often move. Monitoring the movements of these objects has proved an interesting issue, particularly for systems trying to predict the future location of objects. TPR-trees (Saltenis, 2000) use velocity vectors to estimate the future location of an object or its future expansion. While these structures can be efficient in forecasting locations, they are usually not perfect for keeping tracks of past data. Therefore analysis based on specific past data can be impaired by a structure otherwise interesting for forecasting.

It should be noticed that there is difference between mobility and agility. Mobility is linked to the constant movement of objects. A car, bus or plane is an accurate example of what a mobility-based indexing structure usually has to deal with. Agility is linked to a more restrictive notion. Data sources can move, but usually stay in the same location for long times. A portable measurement station, set up at a location for one week then set up elsewhere qualifies for the definition of agility. Most current monitoring systems rely on some kind of sensor agility. Sensor agility management is usually more important to these systems than mobility management. Even though newer sensors tend to be lighter and therefore more 'mobility-oriented', most systems as still widely based on fixed or agile sensors.

3.5 Real time indexing and memory saturation management

The faster development of processor technologies over memory technologies during the 90's has lead to great changes in main-memory and real-time databases. Actually, it has appeared that memory access should be considered as a major bottleneck to system performances. This has lead to the development of structures based on B+trees, such as the CSB+tree (Rao, 2000) or structures akin to the CSB+tree (Bohannon, 2001) (Raatikka, 2004). These structures focus on limiting the cost of memory access by a better use of memory buffers. In the spatial and spatiotemporal domains, new indices have also been developed. The CR-tree (Kim, 2001) and other indexes (Sitzmann, 2002) (Min, 2004), based on modified R-trees limit heap of memory to store and to access data through compression algorithms or other processor intensive computations.

Even though these structures prove to be interesting assets for real-time indexing, they do not comply with all the requirements of sensor databases. The CR-trees are based on R-trees. Therefore, they tend to store data regardless of their origin. Once again, resolving a sensor specific query involves taking into account the whole dataset, thus slowing down the whole process. The specificities of real-time, sensor related data indexing are not within the parameters of the actual indexing solutions. Still, there is no appropriate method for management of spatiotemporal data.

In the next section, two new methods of indexing for real time spatiotemporal databases are detailed, giving a solution to the first three specifications detailed above. Section 5 will detail a solution integrating the management of main memory saturation.

4 REAL TIME SPATIOTEMPORAL DATA INDEXING

One method, the PoTree is designed for structuring masses of spatiotemporal data collected from fixed sensors. The second method, the PasTree is dedicated to indexing spatiotemporal data issued from agile sensors.

4.1 An indexing structure for a "fixed real-time sensor" database

The following sections describe the PoTree, an indexing method (Noel, 2004a). This centralized main-memory database spatiotemporal index can store real-time data issued from a fixed sensor network. It also allows real-time querying while focusing on the most recent data.

4.1.1 Overview of a solution
As mentioned above, the difference between temporal and spatial data can be used to segment the index tree. Then, a variation of B+-tree, the AP tree, suggests that a direct link between the root

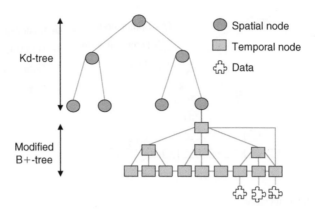

Figure 5.   A possible tree structure for a real time spatiotemporal index for fixed sensor database.

of the temporal data and the latest node can be established. These ideas have been associated to provide a solution of a real time indexing structure for fixed sensor database, i.e. the PoTree.

The PoTree is based on the difference between temporal and spatial data, with a focus given to the latter. This way the notion of information sources is linked to a specific spatial location. It has been devised so as to deal with systems such as volcanic activity monitoring, which involves a number of different sensors with measurement frequencies going as high as 100 Hz. The spatial aspect is indexed through a Kd-tree, while the temporal aspect uses modified B+-trees (see Figure 5). This combination of structures and, more than anything else, of the modifications of the B+-tree are the major innovations of the structure. As for now, mobility is not managed by the structure. However the specificities of both of these trees allow on-line and batch updates: it is possible to update the structure either on real-time or by using batch files.

The monitoring stations being immobile, this structure does not allow mobile sources of information. In this way, every spatial location, akin to spatial object (sensor) is directly linked to a specific temporal tree. Requests shall first determine the spatial nodes concerned and later on determine the temporal nodes.

4.1.2   Details and discussion

Kd-trees are simple spatial structures, but they are not perfect structures. One of their main problems is the fact that they rely on the order of inserted data. If data are entered in different orders, the final trees may have different shapes. Another issue is the fact that they are not perfectly suited for mobility, which is not part of our needs in this section. However, ICC methods, originally designed for B-trees can easily be adapted to cover Kd-trees. Different tests have also proven that these trees behaved reasonably well compared to R-trees for small number of data (Paspalis, 2003). As each B+-tree is linked to one object it is possible to develop a secondary structure so as to access directly the temporal data of specific objects, without the need to first determine their location. This can be useful for the notion of hierarchy of information sources.

Furthermore, it has been noticed that the most recent data are considered of higher interest than the older ones. It has also been noticed that inserts are generally held at rightmost of the structure, where are found the newest nodes. Therefore, the temporal tree has been modified to add a direct link between the root and the latest node. While maintaining this link requires minimum work for the system, a simple test prevents being forced to traverse the whole tree so as to append or to find the requested data. This direct link is useful to save processing time.

As most, if not all, of the updates take place to the rightmost part of the temporal tree, the fill factor of leaf nodes can be placed higher than usual. Deleting should be somehow rare under normal conditions, and updates that do not concern the newest data should be even rarer, unless the system experiences lag time due to network problems between the sensors and the database.

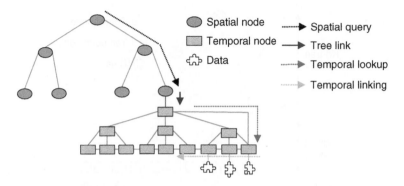

Figure 6.   A Point/Interval Query.

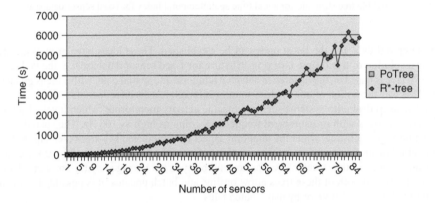

Figure 7.   Comparing R*-tree and PoTree construction time.

Therefore the split and merge procedures can be changed so that the nodes can be filled almost at their maximum capacity.

Figure 6 shows an example of point / interval query. In order to process queries, it is first needed to determine which information source it is relating to. Therefore, queries first browse the spatial sub-tree to determine the relevant node. Once this node has been determined, we can access to its temporal sub-tree. For interval queries, the next step is to look for the end of the interval. Optimally, it should be on the last node of the sub-structure. From this point, it is possible to use the links between the leaf nodes to fetch all the data of the requested interval.

This configuration implies that this tree is more specifically designed for queries on the most recent data. Spatial range/temporal interval queries that do not end at the present time, do not take any advantage of the specificities of the tree.

4.1.3   *Experimental results*

Various tests have been conducted to compare the PoTree and R*-tree structures (thanks to Hadjieleftheriou's implementation, Hadjieleftheriou 2004). As a matter of fact, every indexing solution is designed to focus on some features, some applications. While some structures have been designed to store data related to mobile sensors (evolution of positions and alike, cf. TPR-tree (Saltenis, 2000)), or to store locations and timestamps of special events, so far the focus has not been on data issued from sensors. As a consequence the PoTree has been compared to the most widespread indexing structure in the spatiotemporal database world: the R*-tree.

Randomly generated data have been sequentially issued to a fixed number of random points acting as information sources. Tests have been conducted while changing the total number of data

132

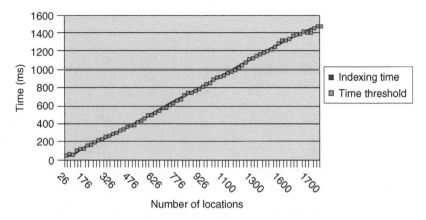

Figure 8.   Influence of the number of station locations on Tree construction time.

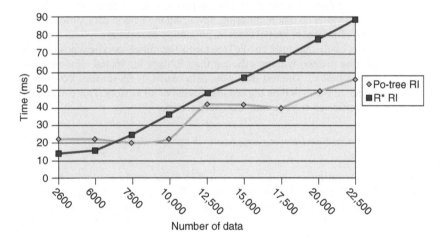

Figure 9.   Range-Interval queries.

to index (1000–200,000), the number of information sources (10–100) used and the portion of the base to scan for interval queries. The tests have been conducted on a 1.6 G Hz, 128 MB RAM computer, running Linux. The programming language used was Java.

Due to the differentiation of spatial and temporal components, and due to the fact that data were coming from a finite set of spatial points, the PoTree built time has been greatly reduced compared to the R*-tree (Figure 7). Please note that for a better readability concerning time, scale on figure 7 uses seconds instead of milliseconds (Figure 8). While 25,000 points stemming from 100 different locations were indexed in less than one second with the PoTree, it took nearly 45 seconds with a R*-tree. Other tests have shown that the construction time of the PoTree evolved linearly with the number of stations, the number of different spatial locations (Figure 8).

Various queries (Noel, 2004b) have shown interesting properties as well. Interval queries took an advantage of the linking of the temporal nodes of the PoTree. For point-interval queries, the PoTree can be up to 8 times faster the R*-tree. While for interval queries the difference has proved to be much lighter, it still remains in favour of PoTree solution, as shown in Figure 9. On this figure, the last 10% of the collected data were fetched. The spatial range covered the whole of possible locations. It is visible than for a low number of data the R*-tree fares better, yet when the amount of data rises past 6000, it is the PoTree is more efficient.

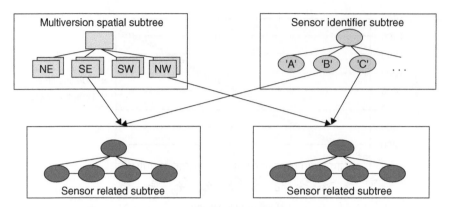

Figure 10.   A possible multi-tree structure for a real time spatiotemporal index for agile sensor database.

Results obtained have shown that the PoTree is compatible with the constraints set by application cases concerning environmental risk monitoring: favouring the newest data, processing of masses of data in a given time, fixed set of spatial sources, possibility of real-time system use. Even though the mobility is not yet easily managed, the PoTree meets the initial specifications defined in 5.1.

The PoTree indexing method is more efficient than R*trees for spatial windows/time interval queries (see Figure 13 below). The test data were data issued by 68 sensors, covering the whole studied area. The queried was based on the last ten percent of all sensors. This can be explained by the use of sensor specific sub-trees to limit data access cost.

4.2   Spatiotemporal real time indexing structure with sensors agility (location changes) managing

The PoTree offers interesting results for data issued by fixed sensor networks, but it lacks finer agility management. The next sections describe the PasTree, another main-memory database indexing structure. It also manages real-time data and real-time querying. However it also offers sensor agility management and multi-dimensional access to the data. Queries can be based on spatiotemporal aspects or on sensor identifiers.

4.2.1   Overview of a solution

While the index previously presented uses two structures to index spatiotemporal data, it does not allow sensor location changes. If a location was to change, a new temporal sub-tree would have to be created. This could be troublesome if a specific location became irrelevant, or in order to index some specific data. For example, seismic epicentres are usually determined by specialized processes but can be considered as data to be indexed. Those epicentres can be considered as data collected from evolving sensors. So, a new tree is needed. The PasTree has to deal with location changes, yet remain focused on prioritizing the newest data and update transactions.

Sensors can move from time to time. The spatial sub-tree should be able to track these modifications. Different approaches exist, yet we shall aim at introducing a multi version approach. A given node records the presence of past sensors (with an end-time) and of actual sensors. So as to offer other querying options, quadtrees could be used instead of Kd-trees, as seen in Figures 10 and 11.

The temporal sub-trees should also keep tracks of the location changes. Each of these sub-trees is related to a specific sensor. Keeping track of their location allows following the movement of sensors through a time interval without querying the spatial sub-tree. The temporal sub-trees of the PoTree can be suitably adjusted to differentiate two kinds of entries: measurement data and location changes.

A tertiary structure, based on B+ tree keeps record of the sensor IDs. As a matter of fact, some queries do not need spatiotemporal properties. Scientists are used to use sensor identifiers, and

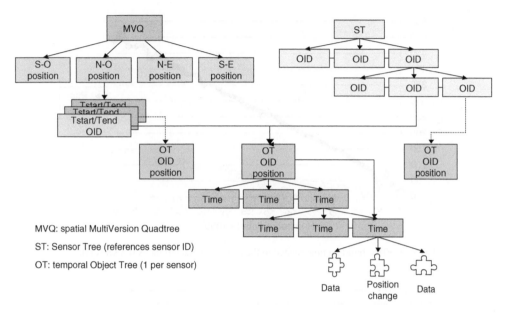

Figure 11.   Detailed structure of the PasTree.

do not always rely on spatiotemporal properties. Therefore this structure is directly linked to the temporal, sensor related sub-trees without using the spatial sub-tree.

4.2.2   Details and discussion

The PasTree suffers from data duplication, yet it also allows more request types than the PoTree. As a matter of fact, queries can be based on the sensor ID as well as on spatiotemporal properties. It allows users to follow a specific sensor through its location changes or to have a look at different sensors passing through a region during a lapse of time. It stills focuses on update transactions and on the newest data.

4.2.3   Experimental results

The PasTree was implemented in Java (Sun VM) on a 1.3 GHz Athlon XP based system with 512 MB of memory. Different tests have been carried out on the PasTree. Randomly generated data, as well as real-world data (from the K-Net) were used.

The PasTree has been tested against the R*Tree. Once again, this choice is linked to the lack of purely sensor-based indexing structures focusing on data and not the spatiotemporal properties of the sensor. It was also tested against the PoTree with static sensors. As the PasTree does not specifically aims at managing mobility, it has not been tested against mobility-centered access methods, such as the TPR-tree.

Figure 12 shows the impact of the number of data on the PasTree building time. This test has been carried out with real data issued by 68 sensors. A semi-linear raise in building time can be noted. As a matter of fact, during the updates, the spatial and identifier sub-trees did not change.

Only the sensor sub-trees had to be really updated. Even then, the structure made use of the direct link to the last node. This sub-tree needs only further processing when the last node has to split. The spatial sub-tree is only updated if agility becomes a factor. Otherwise, it is only used to reach the relevant sensor sub-tree (which takes $\theta(n \log Nl)$, with Nl the number of locations used).

Other tests have been carried out to understand the impact of agility on the building cost. Randomly generated data simulating data issued by 100 sensors with different agility levels have

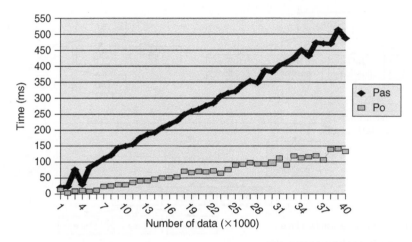

Figure 12.   PoTree/PasTree building time related to the number of data to index (68 sensors).

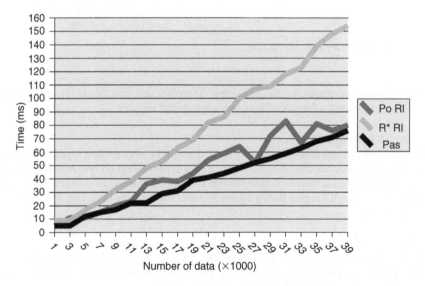

Figure 13.   Spatial window/Time interval query solving time related to the number of data within the base.

shown that the PasTree can process data issued by 600 fixed sensors and 300 agile sensors (agility of 0.5). The PoTree could have dealt with 1200 fixed sensors in similar conditions. However, it could not have dealt with agility.

Comparisons with the PoTree have shown that this structure was 3 to 4 times faster in building the index. However, the limitations on the answerable queries limit the potential of the PoTree. The various data access patterns provided by the PasTree are proposed at the cost of construction time.

Even then, the PasTree is still far more efficient than the R*tree. As a matter of fact, it uses the notion of information source, gathering data into specific sub-tree to limit updating costs.

The PasTree is, as the PoTree, more efficient than R*trees for spatial windows/time interval queries (Figure 13). The test data were data issued by 68 real sensors, covering the whole studied area. The queried was based on the last ten percent of all sensors. This can be explained by the use of sensor specific sub-trees to limit data access cost.

From these different results it appears that the PasTree is a suitable database indexing method for data issued from a set of real-time sensors. It offers spatiotemporal as well as sensor-identifier-based data access patterns. Furthermore, it can manage sensor agility. It uses more resources than the PoTree in processing queries. However, it also adds new features, linked to sensor agility and users needs to alleviate the cost of using the PasTree.

5   OUTLINE OF A METHOD OF INDEXING FOR MANAGEMENT OF MEMORY SATURATION: StH

The StH method that we outlined is an indexing method for spatiotemporal real time database in main memory. This index is able to manage real time measurements collected by a network of agile sensors.

5.1   *Specifications*

The StH is like the PasTree. It is dedicated to answer problems related to the real time spatiotemporal indexing of data resulting from a network of agile sensors. The StH method is focused on the initial role of sensors namely collecting data resulting from one sensor in only one indexing substructure. It aims at allowing the resolution of queries based on sensor identifiers as well as on spatiotemporal attributes.

The StH is more particularly dedicated to the databases into main memories, connected to a data warehouse to store the least important data. In order to prevent memory saturation of the database, the StH index takes into account the need for "emptying" the database by selecting data to be transferred downwards the data warehouse according to the importance of data. The importance of data is determined according to several criteria. The most recent data are regarded as priority, but it is advisable to consider other criteria. Variation of value between two data can impact the importance of the measurements. Small variation involves that the new measurement is less important.

Moreover, concerning environmental risk monitoring, it appeared that some processes could include data corresponding to observations on particular areas in addition to the data resulting from sensors. Thus, in the case of the volcanic monitoring for example, the passage of one lava flow can be regarded as an observation relative to a zone, and not at a located point.

The StH index uses the global structure of the PasTree adding development of particular methods of management of memory saturation.

5.2   *Description of the structure of the StH index*

The total structure organisation of the PasTree index based on a set of substructures is used again (Figure 14) to define the StH index. The difference between the StH and the PasTree concerns the central substructure. The central substructure of the StH is related to sources of information (sensors), contains all the data relative to one sensor and is able to ensure a management of memory saturation. One of the two access structures of the PasTree is re-used just as it is: the tree based on sensor identifiers. The other access structure, the spatial tree, is modified.

As stated before, the tree for sensor identifiers results from the PasTree index. This tree allows database queries based on sensors identifiers.

In the StH index, the spatial tree, based on a multiversion quadtree in the PasTree, is now based on a multi-version R-tree. This structure allows to record data resulting from specific located points (sensors) as well as data resulting from observations on areas (difficult to support by a quadtree). It also allows intersection or nearest neighbours spatial queries.

The major change comes from the sensor based substructure. It is no more a tree but it is now a dynamic staircase. This sensor staircase behaves like an escalator (Figure 15). The structure is filled by the left and initializes procedure of dumping by the right. The staircase is composed of steps (cf. Figure 15). Each step is a pile of scales. In each scale, pointers towards the data are

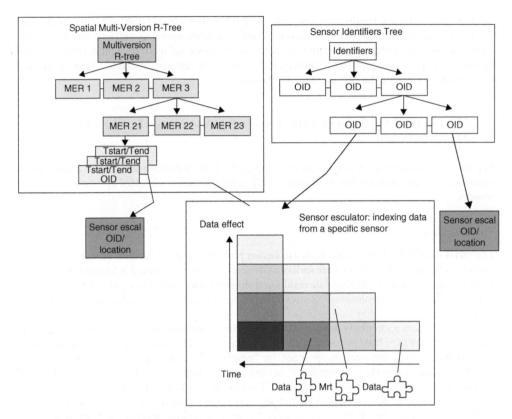

Figure 14.   Structure of the StH index.

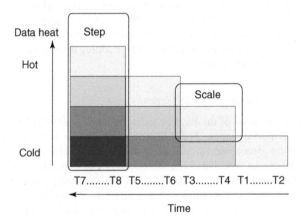

Figure 15.   Structure of the StH escalator of sensor.

stored. The choice of the storage scale is determined by the significance of data (according to the application). A function calculates the heat of data which characterizes the significance of the data. When a number of data by step is reached, the procedure of dumping is carried out. During the dumping procedure (Figure 16), the lowest scale of each step (corresponding to the coldest data) "is flushed" towards a data warehouse and removed from the StH structure. A new step is then

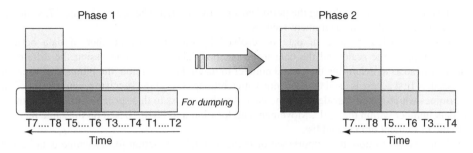

Figure 16.   StH index dumping.

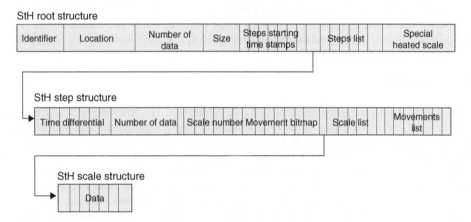

Figure 17.   Data structure of the escalator of the StH index.

created (on the left of the structure) to collect the new data. The structure is dumped gradually, preserving the most significant data for a longer time.

Concerning the global structure of the StH index, several attributes allow the resolution of the most current queries (Figure 17, root). In addition to the identifier of the sensor associated with one staircase, the number of data contained in the staircase, the memory size and the last known position of the sensor, a pointer on a table containing the temporal markers of each step is stored. This table is used to determine the step to access so as to answer a query. The root of the StH preserves also a pointer on a table containing the headings of different steps. Bound to the staircase is also a special step, which contains only one scale. In this scale are preserved the data explicitly qualified "hot data" by the user. These data are copied from the staircase and are preserved explicitly in this structure. Only an explicit declaration of the user can empty this special step, created in order to force the conservation of certain data.

A pointer on a table of differential time is associated to each step (Figure 17). The difference in time since the origin of creation of the step is preserved there. It is consequently possible to use shorter temporal records in order to limit the size of the structure. A notation of the differentials on two bytes instead of four is used. Concerning the step, a pointer on a table of bitmaps is stored. A bitmap is associated to each step. They are used in order to determine in which scale is a particular data as well as a number of records of a particular measure in the scale. Another bitmap is also preserved indicating the changes of locations of the sensor. If a change of location takes place, the bit corresponding to the data is placed at the value TRUE. The step also contains a pointer towards a table of pointers on scales. Finally, it contains a bond towards a list of locations corresponding to

139

the movements carried out during the period covered by the step. The scales (Figure 17, scale) are finally made up only of measurement records.

Finding a particular piece of data is resolved by counting the number of data between the beginning of the scale and the data to be reached. This involves accessing the differential time record so as to determine the record number to reach. From here on, the bitmaps table can be accessed. A first access is used to determine which bitmap is set at the specific record number. This determines which specific scale holds the data. Another access to this specific bitmap determines the data number within the scale. Movement and location queries are based on the same methods with the movement bitmap and list.

It is necessary to note the importance of using a suitable function to determine data heat. If the function is too complex, the cost will be too high. Badly defined, the function will not offer a homogeneous distribution of data during phases of normal system activities. This will lead to bad performances of functions of "emptying". It is quite obvious that at the time of particular activities, functions must classify a greater number of hotter data.

Actually, the structure of the StH is currently under experimentation and validation. The experimental phase already emphasized the role of the function of heat attribution, which must be adapted to each case of specific application.

6 CONCLUSION

Time is crucial for emergency response and risk monitoring. The latter is practically based on collections of sensor data. In this chapter, a discussion is presented concerning real time spatiotemporal indexing and real time main memory management for sensor databases. As an example, a global architecture of environmental risk monitoring system is outlined. The difficulties concerned real time data management in database and also spatiotemporal data management are addressed in detail. After a state of the art on data structuring according to real time and spatiotemporal specifications, some possible solutions for real time spatiotemporal data indexing are presented to allow real time, fast data structuring and queries. One of the presented indexing solutions is dedicated to real time spatiotemporal data issued from fixed sensors. Another one is dedicated to real time spatiotemporal data collected from agile sensors. This second solution offers multidimensional accesses: according to purely spatiotemporal criteria or sensor identifiers. In addition, it allows managing saturation of main memory according to the significance of data which depend on applications.

The indexing solutions presented here are new and efficient as they permit to:

– store spatiotemporal data collected in real-time from sensor network
– index spatiotemporal data in real time allowing real time spatiotemporal queries and facilitating rapid access to recent data (most of existing methods do not privilege last entered data)
– allow real time spatiotemporal queries taking into account sensor location and type of movement (fixed sensor, agile sensor)
– manage in real time main memory saturation using data significance (application based) to elect data to store on long period.

REFERENCES

Almeida V.T., Guting R.H. 2005. Indexing the Trajectories of Moving Objects in Networks, in: GeoInformatica, vol. 9(1), pp33–60
Barbará D. 1999. Mobile Computing and Databases – A Survey. In: IEEE Transactions on Knowledge and Data Engineering, vol. 11 (1), pp. 108–117
Behr F.J. 1995 Mobile GIS: Contributing to Corporate Benefits. DA/DSM Seminar, 20 November 1995, Rome, Italy, available at http://www.graphservice.de/papers/mobile_g.htm
Bohannon P., McIlroy P., Rastogi R.. Main-Memory Index Structures with Fixed-Size Partial Keys, In SIGMOD Conference , 2001

Buelher K., Mckee L. (eds). 1996. The OpenGIS? Guide, Introduction to Interoperable Geoprocessing. Open GIS Consortium, available at http://www.opengis.org

Chakka V.P., Everspaugh A., PATEL J.M. 2003. Indexing large Data sets with SETI, Proceedings of the Conference on Innovative Data Systems Research (CIDR 2003), Asilomar, january 2003

Eiman E. 2003. Research Directions in Sensor Data Streams: Solutions and Challenges, technical report DCIS-TR-527, Rutgers University

Frentsos. 2003. Indexing Objects Moving on Fixed Networks, Proceedings of 8th Int. Symp. On Spatial and Temporal Database (SSTD)

Gunadhi H., Segev A. 1993. Efficient indexing methods for temporal relation, In IEEE Transactions on knowledge and Data Engineering, 5(3), 496–509

Guting R.H., Bohlen M.H, Erwig M., Jensen C.S., Lorentzos N.A, Schneider M., Vazirgiannis M. 2000. A Foundation for Representing and Querying Moving Objects. ACM Transactions on Database Systems, Vol. 25 (1) pp. 1–42

Hadjieleftheriou M. 2006. Spatial Index Library, available at http://www.cs.ucr.edu/~marioh/ spatialindex/

Haritsa J.R., Seshadri S. 2001. Real-time index concurrency control. In Real Time Database System – Architecture and Techniques, Kluwer Academic Publishers, Boston, ISBN: 0-7923-7218-2, 60–74

Intanagonwiwat C., Estrin D., Govindam R. et al. 2001. Impact of Network Density on Data Aggregation in Wireless Sensor Networks, in Proceedings of the International Conference on Distributed Computing Systems, Vienna, 2001

Kim K., Cha S.K., Kwon K. 2001. Optimizing Multidimensional Index Trees for Main Memory Access, Proceedings of 2001 ACM SIGMOD Int. Conf. On Mnagement of Data, USA May 2001, pp. 139–150

Lam K.Y., Kuo T.W. 2001. Real time database systems: an overview of systems characteristics and issues. In Real Time Database System – Architecture and Techniques, Kluwer Academic Publishers, Boston, ISBN: 0-7923-7218-2, 4–16

Laurini R., Servigne S., Noel G. Soft Real-time GIS for Disaster Management, GI4DM, In: Proceedings of International Symposium on Geo-information for Disaster Management. Delft, The Netherlands March 21–23, 2005. 10 p

Long M., Murphy R., Parker L. 2003. Distributed Multi-Agent Diagnosis and Recovery from Sensor Failures, IEEE/RSJ International Conference on Intelligent Robots and Systems(IROS), Vol. 3, pp. 2506–2513

Min Y.S., Yang C.Y., Yoo J.S., SHIM J.M., SONG S.I., PCR-Tree: An Enhanced Cache Conscious Multi-dimensional Index Structures, proceedings of DEXA 2004, pp. 212–221

Mokbel M., Ghanem T.M., Aref W.G. 2003. Spatiotemporal Access Methods, IEEE Data Engineering Bulletin, Vol 26, n° 2, pp. 40–49

Nascimento M.A., Silva J.R.O. 1998. Towards Historical R-trees, In: Proceedings of ACM Symposium on Applied Computing (ACM-SAC) Atlanta, USA, p. 235–240

Noel G., Servigne S., Laurini R. 2004a. Real-time spatiotemporal data indexing structure, Proceedings of 2004 AGILE 7th conference on Geographic Information Science, Heraklion, 2004, pp. 261–268.

Noel G., Servigne S., Laurini R. 2004b. The Po-tree: a soft real-time spatiotemporal data indexing structure, Proceedings of 11th SDH International Symposium on Spatial Data Handling, Leicester, 2004, 10p.

Noel G., Servigne S., Laurini R. 2005. Spatial and Temporal Information Structuring for Natural risk monitoring, GISPlanet'2005, Lisbonne 30 mai-2 Juin 2005. 10 p

Ooi B.C., Tan K.L. 1997. Spatial Databases. In Indexing Techniques for Advanced Database Systems, Kluwer Academic Publishers, Boston, ISBN 0-7923-9985-4, 39–75

Paspalis N. 2003. Implementation of Range searching Data-Structures and Algorithms, available at http://www.cs.ucsb.edu/~nearchos/cs235/cs235.html

Prichard J., Fortier P. 1997. Real-Time SQL. In: Second International Workshop on Real-Time Databases September 18–19, 1997 Burlington, Vermont, USA, pp. 289–310.

Rao J., Ross K. 2000. Makinb B+ trees Cache Conscious in Main Memory, Proceedings of ACM SIGMOD 2000, pp. 475–486

Raatikka V. Cache-Conscious Index Structures for Main-Memory Databases, Master's Thesis of university of Helsinki, Departement of computer Sciences, 2004

Ratnasamy S., Estrin D., Govindan R., et al. 2002. Data-Centric Storage in Sensornets, in Proceedings of the First ACM International Workshop on Wireless Sensor Networks and Applications, Atlanta, 2002

Samet H. The Quadtree and Related Hierarchical Data Structures,1984 , ACM Computer Survey , Vol 16 (2) , pp. 187–260

Saltenis S., Jensen C., Leutenegger S., Lopez M. 2000. Indexing the Positions of Continuously Moving Objects, In Proceedings of ACM SIGMOD 2000

Satnam A., Agogino A.M., Morjaria M., A. 2001. Methodology for Intelligent Sensor Measurement, Validation, Fusion, and Fault Detection for Equipment Monitoring and Diagnostics, Journal of Artificial Intelligence for Engineering Design, Analysis and Manufacturing, , vol 15, n. 4, 2001, 307–320

Servigne S., Tanzi T., Noel G. Telegeomatic System and Real Time Spatiotemporal Database. Proceedings of Urban Data Management Systems, UDMS'06. Aalborg, Denmark, May 15–17, 2006. 12 p

Sitzmann I., Stuckey P.J., Compacting discriminator information for spatial trees, proceedings of the Thirteenth Australasian Database Conference, 2002, pp. 167–176

Song Z., Roussopoulos N., (2003), SEB-tree: an Approach to Index Continuously Moving Objects, in Mobile Data Management (MDM), January 2003, pp. 340–344

Theodoridis Y., Vazirgiannis M., Sellis T. (1996) Spatiotemporal indexing for large multimedia application, In Proceedings of the 3rd IEEE conference on multimedia computing and systems (ICMCS)

Wang X., Zhou X., Lu S. (2000) Spatiotemporal Data Modeling and Management: A Survey, In Proceedings of the 36th International Conference on Technology of Object-Oriented

USGS, Earthworm Overview, available at http://folkworm.ceri.memphis.edu/ew-doc/OVERVIEW/1_History.htm

Wolfson O. (1998) Moving Objects Databases: Issues and Solutions. In: the Proceedings of the 10th International Conference on Scientific and Statistical Database Management (SSDBM98), Capri, (Italy), July 1-3, 1998, pp. 111–122

Geospatial Information Technology for Emergency Response – Zlatanova & Li (eds)
© 2008 Taylor & Francis Group, London, ISBN 978-0-415-42247-5

A 3D data model and topological analyses for emergency response in urban areas

J. Lee
The University of Seoul, South Korea

S. Zlatanova
Delft University of Technology, The Netherlands

ABSTRACT: 3D geospatial information has always been a challenge due to a variety of data models, resolutions and details, ways of representation (b-reps, voxel, SCG), etc. After 9/11 the interest in 3D models (buildings or undergrounds) for emergency responses is progressively increasing. Such models are mostly available from the design phase (as CAD models). Design CAD models are in most of the cases too detailed for computing, for example evacuation routes. Therefore, this chapter is motivated by the need of a new data model to represent and to analyze 3D geospatial data in emergency management systems for field workers and decision makers. This chapter reviews 3D data models developed for geometric or topological representations of 3D objects and proposes a 3D Data Model for emergency response to represent urban built environments in multi-levels. The proposed data model is a composite model integrating: (1) 3D geometric model to measure and represent 3D spatial objects geometrically only, (2) 3D topological model to represent only the topological relationships among the 3D objects using a network-based model, and (3) 3D city model to visualize the 3D objects in multi-views.

1 INTRODUCTION

Human-induced disasters such as fires and the terrorist attacks on September 11th, 2001 (WTC, New York), March 11th, 2004 (Madrid) or July 7th, 2005 (London) usually occur on the micro-space of multi-level structures (such as buildings) in urban areas. Such disasters not only affect multi-level structures in urban areas, but also impact upon their immediate environment at the street level in ways that considerably reduce the speed of emergency response. The complex internal structures of built environments and the traffic bottlenecks at the street level also make speedy escape or rescue particularly difficult in any emergency situations (Kwan & Lee 2005). Reducing rescue time can have a significant impact on evacuation in disaster environments. Geospatial researchers have learned that the availability, management and presentation of geospatial information play a critical role in disaster management, especially in 3D urban space such as large public buildings, shopping centers, underground metro (subways) and garages. However, most current GIS-based emergency management systems for earthquake, floods and other disasters have been developed using 2D GIS with 3D visualization systems. The systems have limitations in representing the micro-scale urban areas in 3D space, such as the complex internal structure of buildings, as well as in analyzing human movements during emergency situations in micro-scale environments.

With the goal being to achieve a real-time emergency response system for evacuations in 3D GIS, this chapter focuses on developing a 3D data model to represent urban built-environments including the interior structures of buildings and on 3D spatial analysis functions used for emergency responses such as 3D navigation and 3D buffering.

Section 2 of this chapter reviews 3D data models developed for geometric or topological representations of 3D objects, and Section 3 proposes an 3D Data Model for emergency response, which is a composite model integrating three data models: (1) 3D geometric model to measure and represent 3D spatial objects geometrically only, (2) 3D topological model to represent only the topological relationships among the 3D objects using the network-based topological model, and (3) 3D city model to visualize the 3D objects in micro-view. The following section describes the algorithms of 3D spatial analysis functionalities. In the fifth section, the output from implementing the emergency response system is discussed. And, the final section discusses several significant substantive insights derived from this study.

2 REQUIREMENTS OF EMERGENCY RESPONSE SYSTEM

In general, emergency management is described in terms of how societies respond to disasters. These responses are a 4-stage cycle of emergency response phases: mitigation, preparedness, response and recovery (Cutter 2003). The mitigation phase is related to activities leading to a reduction of occurring emergency situations, and the second phase is the active preparation for any following unexpected events. Response is an acute phase after an emergency, while recovery is a phase after the acute emergency including all arrangements to remove arisen detriments and long-term supply of irreversible detriments (Zlatanova & Holweg 2004). Geospatial technologies have been used throughout all phases of the emergency response cycle, although more in some phases than others. Especially, systems to support decision makers in the phases of mitigation, preparedness and recovery are in use, but the number of systems for technical support in the response phase is quite limited (Zlatanova & Holweg 2004), which requires time-critical response. The emergency response system, one of the time-critical applications (TCA), is related to decisions that have to be made by a human decision maker in emergency situations. The geospatial technology supports the decision maker in getting several rescue strategies derived from the highest quality and quantity of spatial data. The GIS based decision support system in areas of TCA requires appropriate data management and efficient data discovery and integration to facilitate the decision makers whenever they need to make a decision in real-time.

In order to respond to emergencies in real-time, Kwan & Lee (2005) proposed GIS-based Intelligent Emergency Response System (GIERS) and evaluated the potential benefit of a 3D GIS for improving the speed of emergency response. The experiment demonstrates that response delay within multi-level structures due to the indoor route uncertainty can be much longer than delays incurred in ground transportation in terms of the street network uncertainty. The results express that extending conventional 2D GIS to 3D GIS representing the internal structures of high-rise buildings can significantly improve the overall speed of rescue operations. Such an output motivates geospatial scientists to develop an intelligent emergency evacuation system of complex buildings using 3D GIS (Meijers et al. 2005) integrated with Intelligent Transportation System (ITS) technologies, called an Intelligent Building Evacuation (IBE) System.

In terms of TCA, the 3D GIS-based emergency response system has to fulfill requirements similar to 2D GIS but and some specific for the 3D domain (Cutter 2003, Zlatanova and Holweg 2004):

- Dynamic and multi-dimensional representation of physical and human processes – the system incorporates important geospatial data about the emergency situations, represented by combined indoor (3D models) and outdoor (more traditional geospatial data) data models, which also deal with dynamically changing and uncertain disaster environments such as current availability of exits, stairs and the characteristics of the evacuees (age, gender, disability).
- Spatial data acquisition and integration – data updated with newly collected data from the field can be very critical for both a) monitoring the disaster events and b) giving instructions to the involved people. From a database point of view, this process requires 3D position utility to determine the event locations, and strict consistency rules for integration with existing models and immediatly propagating the information to all the users;
- Interoperability for data integration and semantic/data discovery – the integration of multiple systems and databases become a critical issue to develop the emergency response system, in

order to access data from multiple data sources. In this respect, integration of CAD (3D indoor models) and 3D GIS (3D outdoor models) is becoming of critical importance

- 3D spatial analysis – incorporate dynamic geospatial data about the emergency situation, as well as have spatial-temporal analytical and modeling capabilities to facilitate better planning and decision making on emergency responses, for example, 3D topological analytical methods (3D buffering, overlap, intersect, etc.) and 3D shortest route analysis functions.
- Mobile and wireless communication – the ability to provide updated information (3D graphics, images, etc) to rescue units, decision makers and citizens fast, almost in real-time, through communication technologies to transfer on-site information. 3D presentations (and especially vector interactive models) may result in large data files and require wide bands and special treatments (e.g. appropriate selections of data and generalizations)
- 3D visualization – data presentation on handheld and desktop, wired and wireless equipments.

The emergency response system this chapter discusses is a spatial decision support system that facilitates coordination and implementation of emergency response operations such as pedestrian evacuation and rescue within micro-scale urban indoor space. One important similarity of these multi-level structures is that they involve compartmentalized zones or areas connected by complex transport routes such as corridors. In addition, different levels of these structures are connected by a limited number of vertical conduits such as elevators and stairways. Many GIS-based analytical techniques can be applied for directing quick evacuation or rescue in these micro-spatial environments if their internal structure can be represented using a navigable 3D GIS data model (Lee 2001). Further, as the horizontal and vertical conduits within multi-level structures are ultimately connected to the ground transportation system, much would be gained in emergency response through establishing a real-time 3D GIS that links together the traffic systems within these structures with the ground transportation system.

To base an evacuation model on spatial analysis and modeling, a navigable data model of the building interior(s) and a dynamical geospatial database are needed. Determining the safest and most efficient way to evacuate a building can be dealt with as a transportation problem. This includes a navigable 3D GIS, a dynamic geospatial database, data positioning in real-time, analytical models to simulate possible trajectories of change and to formulate alternative decision scenarios, and a distributed information architecture.

3 REVIEWS ON 3D DATA MODELS

3D geospatial information has always been challenged due to a variety of data models, resolution and details, and ways of geometric and topological representations. Since 9/11 there has been special interest in 3D models to represent internal structures of micro-scale environments (built-in urban areas). Such models are mostly available from the design phase (as CAD models). Although design CAD models are, in most cases, too detailed for computing evacuation routes, for effective disaster management several different models have to be used. This section will review currently developed 3D data models, which are 3D geometric models, 3D topological and graph models, 3D city models and 3D CAD models.

3.1 *3D geometric models*

Practically most of the work on geometry model has been completed by the Open Geospatial Consortium Inc. (OGC, formerly the Open GIS Consortium). It is the membership of organizations developing standards for describing the real world phenomena and therefore most related to GIS. Although the initial work of the OGC has mostly been concentrated on traditional 2D GIS issues, the focus has progressed to the next stage. The OGC's current abstract model incorporates many of the geometry types as required in CAD and Architecture Engineering Construction (AEC) industry. ISO has also independently from OGC developed ISO/TC 211 19107 Spatial Schema

(Hering 2001). Currently the OGC Topic 1, Feature Geometry (of the Abstract Specifications) is identical with ISO/TC 211 19107.

The specifications aim at complete description of real phenomena. A 'feature' in OGC terms is an abstraction of a real world phenomenon, which is associated with a location relative to the Earth. In general, the feature can be described by vector or raster representations. Geometric and topological primitives (i.e. simple features and complex features in OGC terms) are used to construct geographic features and represent their relationships. Raster data is based on a complete tessellation of the space, in which each unit gets an attribute value. The ISO 19107 Spatial Schema standard deals only with vector data.

Actually the Spatial Schema treats the two models: geometry and topology. Each real world phenomena can be described by a geometric object (GM_Object) and/or a topological object (TP_Object). The Geometrical model provides the means for the quantitative description (coordinates and mathematical functions) regarding dimension, position, size, shape, and orientation. The geometry is the aspect of geographical information that depends on the geodetic reference system (particularly relevant for 2D GIS). Topology, in contracts, deals with the spatial relationships of geometric features within continuous mappings.

The geometry of spatial features is described by the basic class GM_Object, which is a combination of a geometry and a coordinate reference system. The geometry object can be a GM_Primitive, GM_Complex and GM_Aggregare. The GM_primitive is an abstract class derived from Geometric primitive (Figure 1). As it can be realized the Abstract Specifications provide a concept for representation of 3D objects as well as specific primitives such as freeform shapes (Bézier, B-spline, Cubic-spline, and Polynomial spline), spheres, ellipse, cone and triangulated surfaces.

It should be realised that the Abstract specifications provide only conceptual guidance in preparing Implementation specifications (Reed 2006). The way these can be implemented at different platforms (based on CORBA, OLE/COM and SQL) is described in three different Simple Feature Implementation Specifications (OGC, 2005). The set of primitives in the Implementation specifications is rather limited to the support of only 2D primitives, i.e. point, line and polygon (Figure 2). A real simple 3D object (tetrahedron, polyhedron, sphere, cone, etc.) is still to be included.

This chapter will further consider only the Simple Feature Implementation Specification for SQL (SFS), which provides guidance for implementing spatial data types in Database Management System (DBMS). In the last couple of years almost all mainstream DBMS (Oracle, Ingres, Informix, PostGIS, MySQL) offer support for spatial data types. Most of the DBMS are compliant with the model as described in SFS, but variations exist even in the supported data types. For example, Informix supports three basic spatial data types: point, line and polygon; Ingres supports one more type: circle; Oracle Spatial has points, lines, polygons and circles, as well as arc strings and compound polygons. It should be notices that all DBMS (except Informix) maintain, the 2D objects with their 3D coordinates. However, the spatial functions supplied with the data types are predominantly only 2D (the third coordinate is omitted from the computations), although some exceptions exist. For example, PostGIS has a number of true three-dimensional functions, e.g. length, and MinMax bounding box.

As mentioned above the data types currently in are 2D but they can be given with their 3D/4D coordinates. Practically this means that 3D data can be managed in DBMS (and eventually analysed). Using 3D polygons, 3D objects can be represented as polyhedrons in two ways: as a list of data type *polygons* or as data type *multipolygon/collection* (Stoter and Zlatanova 2003). Using the first approach, one or two more columns have to be introduced in the relational table, to be able to specify that a polygon belongs to a particular 3D objects. One 3D object is represented by several rows in the geometry table. In the second case, a 3D object is described in one row, since all the information about the polygon is decoded in the Oracle Spatial geometry type. Although the number of records is reduced, the redundancy of coordinates cannot be avoided. Each triple coordinates is repeated at least three times in the list of coordinates. An apparent advantage of the 3D multipolygon approach is the one-to-one correspondence between a record and a 3D feature. Furthermore the 3D multipolygon (compare to list of polygons) is recognized as one object by

146

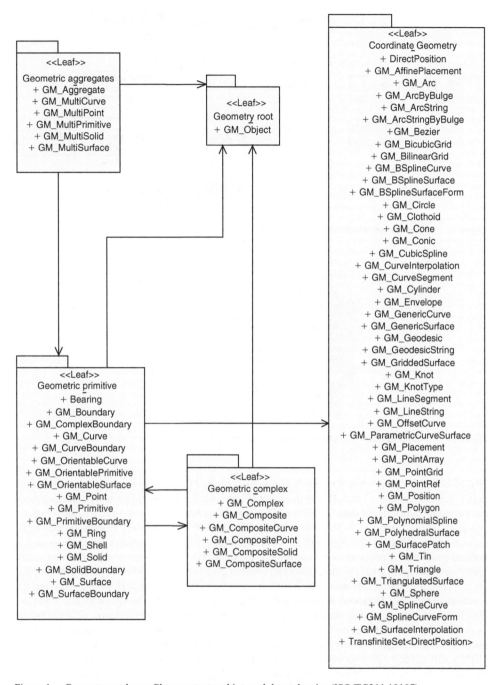

Figure 1.   Geometry package: Class content and internal dependencies (ISO/TC211 19107).

front-end (GIS/CAD) applications (Figure 3). SQL examples illustrating the creation and query of these tables can be found in Zlatanova and Storer 2006.

Spatial DBMS can play a significant role in emergency response in near future as a general data store for both operational data (*in situ*) and existing 2D/3D/4D data. Presently, 3D features stored in

147

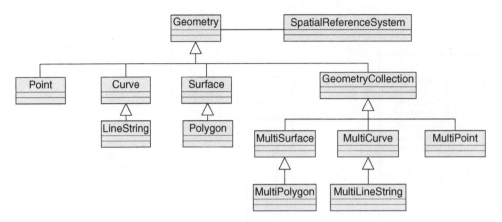

Figure 2.   Simple feature specifications for SQL (SFS).

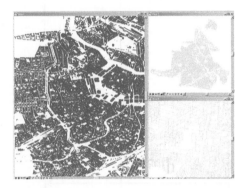

Figure 3.   3D buildings stored in DBMs and visualized in Bentley software.

DBMS can easily be combined with other features described by natively supported or user-defined data types and visualized in a 3D environment (Figure 4). Research and prototype developments have already reported new data types for a 3D polyhedron (Arens et al 2005) and freeform surfaces (Pu et al., 2006, Figure 5) as recommended by the Abstract Specification. Hopefully, the Implementation specification for SQL will be soon extended with more geometry types. It should be noticed geometric models are very convenient for rapid 3D visualisation and performing metric operations (compute area, volume, find object within given area, etc.). Most of the DBMS support several spatial indexing schemas, which further contributes to the better performance compared to topological models (Zlatanova & Stoter 2006). Presently most of the DBMS (with one exception Oracle Spatial) maintain only the geometrical model, but the most common topological operations as defined by the 9-intersection model (Egenhofer & Herring 1990) are also supported. Consequently numerous spatial analyses can be preformed as well. However, it should be realised that DBMS will never be able to support all the spatial analyses that an emergency response system may need. As discussed elsewhere (e.g. Zlatanova & Stoter 2006) DBMS would support generic functions and operations and the complex variety of spatial analyses has to be performed at a front-end application.

3.2   3D topological and graph models

The topological model is closely related to the representation of spatial relationships among objects in geographic phenomena. Over the last fifteen years, topological models for n-dimensional objects

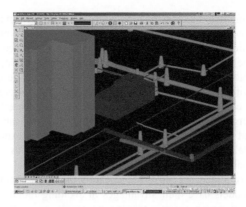

Figure 4.   3D pipes integrated with buildings.

Figure 5.   A building modeled with NURBS (CAD model and real photo).

have been developed by a number of researchers (Rikkers et al. 1994, Pigot 1992, Pigot & Hazelton 1992). However, the 3D topological models have not been implemented in the commercial 3D GIS systems (Zlatanova et al., 2002) even though models have been implemented in CAD systems such as SHAPES by XOX Inc. or GeomagicStudio by Raindrop Geomagic Inc. Likewise the geometrical model, OGC Abstract Specifications discuss 3D topological primitives, but Implementation Specifications for a topological model are not available yet.

3D entity-based data models for geospatial representation are based on the concepts used in 2D vector GISs. A number of systems have been developed to implement 3D data structures based on boundary representations (Raper 2000). B-rep has a hierarchical data structure in which an object surface is composed of four elements of predefined primitives: point, edge, face and volume (Hoffmann 1989; Li 1994). Since the 3D B-reps are extensions of representations of planar configurations in 2D B-reps, each volume in B-reps for 3D geographic entities is represented by its bounding surface (Worboys 1995). Examples of the developed data models based upon 3D B-reps are the system of 'simplicial complexes' described by Carlson (1987), the structured vector fields described by Burns (1988), 3D formal data structure (FDS) developed by Molenaar (1990) and the 'GOCAD' system developed by Mallet (1990). Tetrahedral Network (TEN) introduced by Pilouk (1996) improves 3D FDS to allow modeling of objects with indeterminate boundaries such as geological entities and pollution clouds. Zlatanova (2000) designed Simplified Spatial Model (SSS) to serve web-oriented applications with many visualization queries by simplifying 3D FDS, and Coors (2003) developed Urban Data Model (UDM), representing the geometry of a body or a surface by planar convex faces. In these models, topological relations are represented by a geometric representation of the cells and their neighborhoods defined in terms of their boundary and co-boundary cells based upon boundary representations (Corbett 1979, Pigot 1995).

Compared to the geometrical models, the topological models perform relatively bad (Zlanatova et al., 2004). First, many spatial queries are based on pure geometric properties. For example a query 'give all the features within given area' is completed on the 3D coordinates of objects. But the 3D coordinates in the topological models are stored in the node table, which requires traversal of three or two more tables (edge, face and body). In contracts, in the geometric models, 3D coordinates are organized with the features, generally in a single table. Second, the geometric model is integrated within the commercially developed DBMS allowing for efficiently, while topological models are mostly organized (with exception of Radius Topology and Oracle Spatial 10 g) in user-defined objects and tables. Lastly, DBMS maintain spatial indexing, which is not applicable for topological models. Since the tables contain only references to id's of the objects, only a general indexing is possible. As concluded in Penning 2004, a spatial index (R-three) built on a MinMax bounding box of a face, speeds up significantly the insert operation of Oracle Topology 10 g. The topological models have their advantages in avoiding redundant storage, maintaining data consistency, and performing specific topological operations such as overlap, intersections, etc. (Penninga 2004). However some of the 3D topological data models have problems maintaining efficiently local neighbourhoods, especially when the real geometry of the feature is not of importance.

In order to deal with this problem, graph models have been developed (Chalmet et al., 1982, Hoppe & Tardos 1995, Lu et al., 2003, Smith 1991, Lee 2001). Instead of representing the topological relationships between topological primitives (node, arc, face and body), the graph models present the topological relationships among 3D objects by drawing a dual graph interpreting the 'meet' relation between 3D and 3D objects as defined by the 9-intesection model (Egenhofer & Herring 1992). To plan and design an evacuation network within a building, the interior of a building is modelled as node-edge graph (Chalmet et al., 1982, Hoppe & Tardos 1995, Lu et al., 2003, Smith 1991), which is a logical network model. Similar to the node-edge graphs which use duality to represent space-activity interactions, the Combinatorial Data Model (CDM) was developed to represent more than just adjacency and connectivity relationships ($G = (V(G), E(G))$ and $H = (V(H), E(H))$, respectively), among 3D spatial objects in built environments (Lee 2001, Lee & Kwan 2005). In the CDM model, the graph H is a subgraph of the graph G because $V(H) \subseteq V(G)$ and $E(H) \subseteq E(G)$. The CDM is defined as a set of nodes (3D entities in primal space) with a set of edges (spatial relationships between 3D entities in primal space) that represent the topological relationships among entities in built environments. Both the node-edge graph and the CDM are logical network data models representing topological relations of the 3D entities (Lee 2001, Lee & Kwan 2005). As a logical data model, CDM is a pure graph representing the adjacency, connectivity and hierarchical relationships among the internal units (e.g. rooms and corridors) of a building. In order to implement network-based analysis such as shortest path algorithms in the CDM (and node-edge graphs), the logical network model needs to be complemented by a 3D geometric network model that accurately represents these geometric properties (e.g. distance or size), called Geometric Network Model (GNM). For the transformation, a node in the CDM (graph H) representing a corridor within the building is considered as a consolidated 'Master_Node'. The Master_Node is a sub-graph representing a connectivity relationship among the compartmentalized zones of the corridor generated to represent the relationships between a room and a hallway as one-to-one relations (one-to-many relations in the CDM). In other words, the node in the CDM (graph H) is converted into a linear feature in the GNM, which is a sub-graph representing a two-dimensional shape such as a hallway.

The importance of graph models is recognized by DBMS as well. For example, Oracle Spatial 10 g offers support of graph organized in two to four relational tables NODE, LINK, PathNODE and PathLINK. Additionally, it is possible to assign the real geometry to each node or link. For example if a room is associated as a node, the 3D polyhedron (or box) can be also stored together with the node. Such a structure might be quite powerful for calculations and visualization of 3D evacuation routes. Path table and path link table are optional for storage of pre-calculated paths (routes). High-level languages PL/SQL or Java API are available for building and analyzing the network.

150

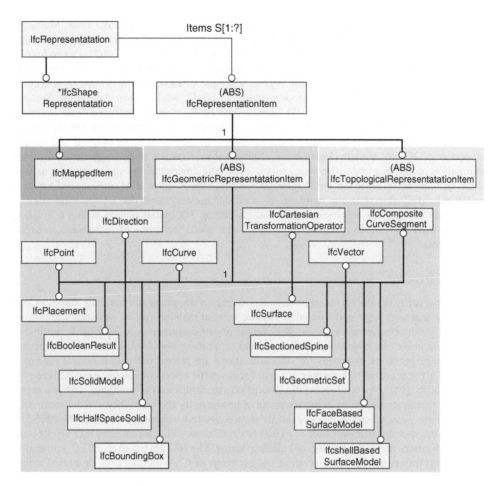

Figure 6. Shape representations included in IFC.

3.3 *3D city models*

3D City Models become a very important issue due to the increasing demand for a realistic presentation of the real world in GIS. The 3D model data are potentially of great importance to the understanding of urban structure and the mechanism of urban growth through visualizing urban and built environments. There are many applications of 3D Urban Models now available. They are based on linking visualization systems (such as CAD or net-based Virtual Reality (VR) (Faust 1995)) to data stored within a GIS. UCLA researchers have pioneered the delivery of geographic information into a 3D modeling environment (Liggett & Jepson 1995). To date, not only research centers, such as the Center for Advanced Spatial Analysis (CASA) at University College London (Smith 1998, Shiode 2001, Longley & Batty 2003) and the GIS Technology Section at Delft University of Technology (Oosterom et al., 2002), but also consulting companies, such as the Environmental Simulation Center (www.urbansimulation.com), Urban Data Solution (www.u-data.com), CommunityViz (www.communityviz.com), Miller-Hare (www.millerhare.com) have developed 3D Urban Models to support planning applications. Batty et al. (2001) listed web addresses for visualization projects in cities with a population greater than one million.

In order to represent urban objects in the systems, some applications use photo-realistic CAD-type models with Level of Details LOD technologies (Liggett & Jepson 1995, Koninger & Bartel

1998, Sugihara et al., 2000, Shiode 2001) or ESRI 3D shapefile formats (ESRI 1998, Multigen-Paradigm 2003), the others implement 3D topological data models based on B-rep (Tempfli 1998, Holtier et al., 2000, Stoter & Zlatanova 2003, Stoter & Oosterom 2002). Depending on the degree of urban environments, which is the amount of geometric content within the model, 3D visualization models deal with buildings as a simple prismatic form created by extruding the building footprint (a 2D polygon) (Jepson et al., 2001), while Urban Simulation Systems deal with buildings as a compound 3D form generated by stacking extruded floor polygons (Holtier et al., 2000). CityGML, a multi-purpose and multi-scale representation for the storage of and interoperable access to 3D city models, covers the geometrical, topological and semantic aspects of 3D city models (Kolbe et al., 2005). All systems are tied to aggregated attribute information on housing code, fire code, zoning code, and environmental code violation, as well as tax delinquencies based on a 3D floor or building object, instead of attribute information on 3D individual objects in the building. In addition, most systems don't implement suitable data models to represent topological relationships among 3D spatial objects within a building and to be used for various spatial queries to analyze spatial relations of the objects in urban environments.

3.4 3D CAD models

CAD was primarily developed for engineers responsible for designing and building 'things'. CAD was able to deal with large-scale, detailed models, without maintenance of attributes and lacking support of geodetic reference systems. The complexity of design tasks to be solved contributed to the development of a huge variety of shapes and supporting tools, which resulted in numerous data exchange problems. Attempts for standardizations are carried out within several organizations but for the scope of this chapter, we will focus on the efforts of International Alliance for Interoperability (IAI),(www.iai-international.org) on standardization of buildings. IAI is a membership of organizations aiming at improvement of productivity and efficiency in the construction in all the three design aspects, i.e. organization, process and technology. From 1999–2006 IAI developed the specifications for Industry Foundation Classes (IFC) available as ISO 16739 standard, which covers the entire process of building design. The IFC are based on STEP, yet another more general framework for representation and exchange of CAD product data, described within ISO 10303.

The variety of shapes described by IFC is much larger compared to GIS and truly three-dimensional: curves, geometry sets (consisting of points curves and surfaces as 2D and 3D elements), Surface models (which include facetted face sets and shells, always 3D), Solid Models (B-reps, Constructive Solid Geometry, Swept models), etc. (Figure 6). Some of the B-reps used in CAD systems are very similar to the geometrical or topological models widely implemented in GIS. In contract to GIS, CAD systems maintain topological models only per 3D object, i.e. spatial relationships between two distinct objects can be found only by geometrical computations. Moreover IFC has a complex thematic hierarchy, i.e. contains entities for walls (inc. curtain walls), windows and doors (Figure 7), roofs and slabs, openings, coverings and louvers, projections +/− and shading, railings, ramps and stairs. IFC contains relationships for placing items in openings, assembling and connecting elements, covering one element with another, proximity of elements (Figure 8), associating classification, documents, approvals and rules. Since the development of this standard number of tools for viewing and creating IFC are available (), but since the complexity of IFC.

Much of this information can readily be used for integration with 3D GIS and 3D city models. Apparently many of the details intended included in IFC are not needed for 3D GIS models. Therefore, similar to research in 3D City models, appropriate LOD are also investigated in AEC domain. Wix et al., 2005 report first results of integrating GIS and AEC models using GML to IFC representations. The LOD adopted by the researchers are very similar to the ones proposed within CityGML. As the developments toward 3D GIS and AEC/CAD integration progress, the interest of GIS specialists in IFC will also increase.

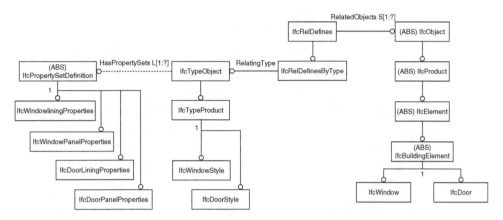

Figure 7.   Doors and windows in IFC.

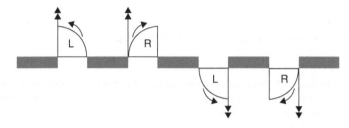

Figure 8.   Doors openings in IFC.

4   THREE COMPONENTS OF A 3D DATA MODEL FOR EMERGENCY RESPONSES

In reviewing the advantages and disadvantages of the different models in the previous section, it is a difficult task to select an appropriate data structure designed for the characteristics of the applications, for example, objects of interest, resolution, required spatial analysis, etc. (Zlatanova et al., 2004). A model designed for 3D spatial analysis may not exhibit good performance on 3D visualization and navigation. In other words, different data models might be suitable for the execution of specific tasks but not others. In order to maximize efficiency and effectiveness in the provision of operations, Oosterom et al. (2002) proposed multiple topological models maintained in one database by describing the objects, rules and constraints of each model in a metadata table. Metric and position operations such as area or volume computations are realised on the geometric model, while spatial relationship operations such as 'meet' and 'overlap' are performed on the topological model.

As mentioned before, the emergency response system is a spatial decision support system that facilitates coordination and implementation of emergency response operations such as pedestrian evacuation and rescue within micro-scale urban indoor space not only providing dynamic, specific and accurate evacuation guidance based on indoor geospatial information, but the system visualizes the information to communicate with users in macro-level (an impacted region) and in micro-level (disaster site inside buildings). In order to represent spatial objects in urban areas for ER, this chapter proposes a 3D data model, which is a hybrid data model consisting of the three models: a 3D geometric model, a 3D graph model and a 3D city model. The 3D geometric model is used for 3D geometric representation of solid features consisting of a number of 3D polygonal faces defining an enclosed boundary, while the 3D graph model is proposed to represent the topological relationships among the 3D solid features. The 3D city model can be used for visualizing the information in a 3D real view.

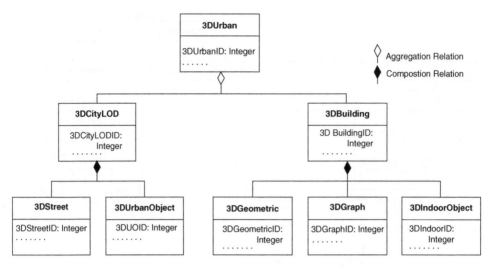

Figure 9. UML Object Diagram: Object relationships of the 3D data model.

Based on the methods of object-orientation, object-object relationships of the model are represented using an UML (Unified Modeling Language) object diagram (Zeiler 1999). Figure 9 shows the classes of the model. A 3DCityLOD and a 3DBuilding classes are associated with a 3DUrban class through an aggregation relationship that models the case where the 3DCityLOD and 3DBuilding classes are part of a 3DUrban class. The aggregation relationship is indicated by a hollow diamond headed arrow pointing from the part to the whole, and the cardinality of aggregation is indicated in the diagram. The 3DCityLOD associates with a 3DStreet and a 3DUrbanObject classes through composition relationships. The composition is a strong form of aggregation in which objects from the 3DCityLOD class control the lifetime of objects from the 'part' classes. The 3DBuilding class associates with a 3DGeometric, a 3DGraph, and a 3DIndoorObject classes through composition relationships. The 3DGeometrc represents the geometric dimension of the 3D spatial objects (such as rooms or spatial units) on a partly symbolic and simplified 3D representation in a model view (Lin & Zhu 2006), and the 3DGraph represents the connectivity relationships among the 3D spatial objects based on a geometric network representation. The reference of a 3D object in the 3DGeometric to its corresponding objects in the 3DGraph and a 3DIndoorObject should be maintained in the system. The 3DStreet is the network of ground transportation, which represents the connectivity relationships among the urban objects in a model view. The 3DIndoorObject and the 3DUrbanObject are for a virtual representation of the urban environment that enables people to explore and interact with the geospatial information about the emergency situations in the worldview giving a photorealistic 3D display.

4.1 *3D geometric representation for spatial objects in micro-scale urban areas*

As discussed in Section 3.1, a spatial object can be represented by geometry types, which are basically an ordered sequence of vertices that are connected by straight-line segments or circle arcs. The supported primitive or composed types are points and point clusters, lines, compound lines, n-point polygons, compound polygons, and circles. 3D objects can be represented using either the simple geometry type 'polygon' (with 3D coordiantes) or the geometry type 'collection' (or 'multipolygon').

This chapter proposes the 3D geometric model based on the first approach. The basic components of the model are points, polygons, and solids. To formalize the solid objects consisting of a number of polygonal faces defining an enclosed boundary, the schema of the objects is shown in Figure 10.

The primal classes of the 3D geometric model are PointZ, PolygonZ, and 3DGeometric. The PointZ consists of an identifier and position data in 3D (x, y, z-coordinates), and the PolygonZ consists of a set of Points *pt* and other attributes including an identifier, and total number of points. The PolygonZ is considered a single ring, which is a closed, non-self-intersecting loop. The 3DGeometric consists of an identifier and a list of all polygons constructing a 3D solid object representing a spatial unit (such as a room of a building) of the urban objects.

4.2 *3D topological representation among the spatial objects*

In order to represent topological relationships among 3D spatial objects in built environments (such as buildings), the 3D Geometric Network Data Model (Lee 2004a) is developed to abstract and represent the connectivity spatial relationships of the internal structure of buildings. It is derived through 3D Poincaré Duality using a graph-theoretic framework and a hierarchical representation schema, and a Straight-Medial Axis Transformation (S-MAT) modelling (Lee 2001 & 2004a). The 3D Poincaré Duality is utilized to abstract the topological relations among a set of 3D objects and to transform '3D to 2D relations' in primal space to '0D to 1D relations' in dual space. It represents connectivity relations among objects in 3D space as a dual graph, $H = (V(H), E(H))$. In order to represent the geometric properties (such as distances between nodes in the graph) of the dual graph, the S-MAT is utilized to identify linear features from a simple polygon (a hallway in this case). Each node representing subunits of a building retrieved from the graph $H = (V(H), E(H))$ is projected and connected to the medial axis to generate the graph $G = (V(G), E(G))$. The graph G is the geometric network model used to describe the connectivity relationships among 3D objects within a building. Because the 3D GNM was developed to represent connectivity relationships among the 3D objects based on a graph model, the network model can be used for emergency response systems, in order to pathfinding, allocation and tracing analyses in 3D micro-spatial environments. Such applications require a 3D network-based data model to represent the internal structures of urban-built environments and environmental factors to model pedestrian-based indoor movement, such as traffic flows, damage status, toxicity status, bottleneck locations, etc.

This section describes the detailed construction method for generating the 3D topological model for 3D objects within a building (3DGraph class) based upon the work of Lee (2006). The adjacency relationships among spatial units are the combinations of the adjacency relations in the horizontal directions and the adjacency relations in vertical directions, because of the nature of the 3D geographic entities in built-environments (Lee & Kwan 2005). Therefore, the adjacency relationships are defined by two individual procedures. The adjacency relations of 3D units in horizontal directions are derived from the topological relationships between polygons (such as floor plans). The adjacency relations in floor j of the building i can be described as the graph $Gh_{ij} = (V(Gh_{ij}), E(Gh_{ij}))$, while $j = 1$ to n in case of n story building. The adjacency relations of 3D units in vertical directions are defined by the layer-overlay functions implemented in 2D GIS. The adjacency relationships between the floor j and floor $j - 1$ of the building can be described as the graph $Gv_j = (V(Gv_j), E(Gv_j))$, while $j = 1$ to $n - 1$. The defined graphs can be combined using a UNION operation. The combined graph can describe all incidence of the topological model because the defined graphs are equivalence classes.

In order to define the adjacency relations of 3D units in horizontal directions, the hallways are transformed into linear features based on the Straight-MAT algorithm (Lee 2004a), which are a sub-network consolidated into a hallway node in the dual graph generated by the 3D Poincaré Duality. Each node representing spatial units in floor j of a building is projected and connected into the medial axis and into other nodes based on their adjacency relations. The graph $Gh_{ij} = (V(Gh_{ij}), E(Gh_{ij}))$, representing the geometric network for a floor j of a building i (Figure 12b), is combined with graph $Gv_j = (V(Gv_j), E(Gv_j))$ to produce the graph $G_i = (V(G_i), E(G_i))$ (Figure 11b), which is the topological model (adjacency) of a building i. The graph G_i is the combination of the dash and solid thick lines presented in the Figure 11b.

The connectivity relationships among 3D spatial units are defined as a subset of the adjacency relationships (Lee & Kwan 2005), as seen in Figure 11. From the property, it is known that the

```
class PointZ {                          class PolygonZ {
        Int PointZ_ID;                          Int PolygonZ_ID;
        Double x, y, z;                         Int NumPoints;
         };                                     PointZ ArrayPointZ = new PointZ[];
                                                 };


    class 3DGeometric {
        Int 3DGeometric_ID;
        Int NumPolygons;
        PolygonZ ArrayPolygonZ = new PolygonZ[];
         };
```

Figure 10.   3D Geometric: 3D Geometry model.

graph $G_i = (V(G_i), E(G_i))$, which represents adjacency relationships, is a supergraph of the graph $H_i = (V(H_i), E(H_i))$, which represents connectivity relationships among spatial units in floors of a building i, because $V(H_i) \subseteq V(G_i)$ and $E(H_i) \subseteq E(G_i)$. In this case, because $V(G_i) = V(H_i)$, graph H_i is called a spanning subgraph of graph G_i. The graph H_i representing the connectivity relationships can be generated from the graph G_i by removing edges, which are representing only adjacency relationships among the 3D spatial units (Lee & Kwan 2005). The graph H_i is presented by the solid thick lines in Figure 11b.

The 3DGraph, $G_i = (V(G_i), E(G_i))$, needs to be integrated with the network of the ground transportation system (3DStreet class), $S = (V(S), E(S))$, in order to implement multimodal transportation network analyses in urban environments using the topological models of the study area (3DUrban), $UG = (V(UG), E(UG))$. The first step of the integration is to define the connectivity relations between the building's networks (3DGraph) and the street network (3DStreet). The connectivity relations are abstracted to 'Transfer_edges', where Transfer_edges $(E(C)) = \{(n_i, n_j)/n_i \in V(G_i) \cap n_j \in V(S)\}$. The node n_i represents entrance halls of the buildings, and node n_j is defined by projection $p(n_i, E(S))$ (Lee 2004a) of node n_i onto edge $E(S)$ of the street network S. The output is represented by a transferring network, $C = (V(C), E(C))$. As the final step, the 3DUrban's network $UG = (V(UG), E(UG))$ is constructed by combining the 3DGraph $(G_i = (E(G_i), E(G_i)))$ with the 3DStreet $(S = (V(S), E(S)))$ and the transferring network $(C = (V(C), E(C)))$, because each network graph pertains to an equivalent class. The combined network, $UG = (V(UG), E(UG))$, describes a network representation of the internal structure of a building as well as among buildings within the urban area and the street network, in order to model pedestrian movements in the multimodal transportation network.

4.3   3D visualization using city models

In order to provide important information for different aspects of disaster management, 3D city models enable multi-purpose and multi-scale 3D visualization to present emergency situation information to users (rescuees and rescuers) (Kolbe et al., 2005). Although 3D visualization has to be very close to the real view, it is practically impossible to represent all the details but too few details may create unrealistic views. In contrast, high graphic density does disturb the user's understanding of the message. In order to resolve the problems, most city models support different LOD, which may arise from independent data collection processes and are used for efficient visualization and efficient data analysis. In one city model data set, the same object may be represented in different LOD simultaneously, enabling the analysis and visualization of the same object with regard to different degrees of resolution.

156

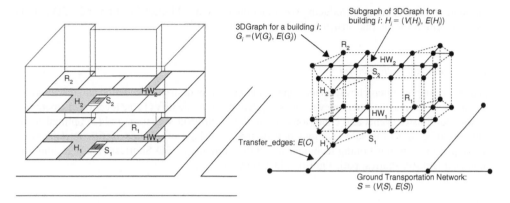

Figure 11.   An example building and 3DGraphic model.

The 3DUrban object shown in Figure 9 represents the geometrical, topological and semantic aspects of a complex 3D city model. The spatial objects represented in the model are not only buildings, but also other spatial objects such as man-made urban furniture, vegetation objects, water bodies, and transportation facilities like streets and railways. The 3DCityLOD are intended mainly for the buildings. The coarse level LOD0 is essentially a two and half dimensional Digital Elevation Model (DEM), over which an aerial image or a map may be draped. The detail LOD1 comprises detailed vegetation and transportation objects, as well as urban street furniture such as trees, light poles, traffic signals, and so on, which are represented by the 3DCityLOD objects. Buildings in LOD0 are represented by 3DGeometric, which is the block model, representing spatial properties of buildings without any roof structures or texture. LOD1 denotes architectural models with detailed wall and roof structures, balconies, bays and projections, as well as interior structures like rooms, interior doors, stairs, and furniture. The 3DIndoorObject represents the semantic properties of buildings in the detail level. The 3DCityLOD and 3DIndoorObject objects have relations to objects in other databases or data sets (Kolbe et al., 2005). For example, a building in 3DBuilding class is derived from an architectural model or 3D CAD model. The reference of a 3D object to its corresponding object in an external data set is essential. Such a reference denotes the external information system and the unique identifier of the object in this system.

5   3D TOPOLOGICAL ANALYSES FOR EMERGENCY RESPONSES

To define and develop spatial data manipulating and analytical methods to implement the planning/decision process, the emergency response system requires several important functionalities including a 3D location positioning, a network connectivity analysis, a traffic flow analysis, 3D topological analysis (3D buffer, overlay, intersect, etc.) and an indoor navigation (Dane & Rizos 1998, Miller & Shaw 2001). The 3D location positioning obtained by location-aware devices is used to identify location information of disaster sites, occupants, traffic congestion areas, and isolated zones within buildings. The network connectivity analysis is used to define isolated networks or areas, which do not have any exit node connecting to destination nodes because of being blocked by traffic congestion or disaster. The next function is an evacuation model to estimate dynamic capacities and flow rates of hallways and stairwells to update occupancy movements and traffic flow impedances in the database. The other function is to identify building evacuation bottlenecks, which are congestion locations in the network during an evacuation event. The final is an indoor navigation function to identify feasible and safe routes within a multi-level structure and to provide navigation guidance for rescue personnel. Because of the limited scope of this study, this chapter is focusing on developing or introducing two functionalities: a 3D buffer function and a 3D optimal route algorithm for internal structures of built environments.

5.1 *3D shortest route method*

In order to support emergency guidance operations such as pedestrian evacuation and rescue within urban indoor spaces, the emergency response system needs to identify (1) the location of emergency crews and disaster events and conditions within the multi-level structure, and (2) optimal routes from an affected area to safe locations outside that area. First of all, this section introduces the developed location positioning techniques, and then proposes a 3D Shortest Path algorithm, the modified Dijkstra shortest path algorithm in a 3D GIS.

5.1.1 *3D geo-location*
In the past few years, new information technology has greatly enhanced the collection of activity data. The Global Positioning System (GPS) provides location information. In addition, these locational devices are equipped with mobile GIS software (such as ArcPad) and can generate on-screen geo-referenced maps to support ufsers' operations on-site (e.g., providing navigational guidance). However, there are limitations in using GPS within a multi-level structure due to a degradation or loss of signal in certain areas of a building. Positioning techniques have been extensively investigated in Location Based Services (LBS) applications, including better localization inside buildings. Network-based or hybrid positioning technology used by most LBS providers to achieve a positional fix faster and easier than conventional GPS-based technology. Nevertheless, the problems of loss-of-fix in LBS-derived locational data still remain (Kwan & Lee 2005). Various indoor positioning technologies are described an analysed in Kolodziej and Hjelm, 2006. Dürr and Rothermel (Dürr & Rothermel 2003) proposed a fine-grained Geocast location model to determine client positions within a building based on geometric and symbolic addressing.

The global positioning techniques such as Global Navigation Satellite Systems (GNSS) to collect 3D locations have improved the quality and quantity of these data and reduced their cost. GPS receivers are currently integrated in cellular phones and PDAs, (Samet 2001). GPS devices are able to offer the easiest method and a quite accurate way of 3D positioning of the user, but only outside buildings.

The positioning within mobile networks using only the information related to the base network transmitter is a very effective method, but it is very inaccurate and practically not applicable (Zlatanova & Verbree 2003). The only advantage, compared to GPS positioning, is the possibility of working inside buildings. None of the currently experimented techniques based on mobile networks are able to obtain accuracy more than 50 meters.

Due to weak GPS signals and the limited accuracy of radio network solutions, neither technique is appropriate for indoor positioning. Wireless LANs (WLANs) are used to track mobile users in closed spaces such as buildings and tunnels (Prasithsangaree et al. 2001). The system for location positioning using WLANs runs on a standalone server and gives the x,y-position and the floor of the mobile unit. The positioning accuracy achieved by the system is up to 1 meter. Despite providing accurate positioning for indoors, however, the WLANs have problems with implementation because they require a reference database for an average signal measurements at fixed points throughout a building (Pahlavan & Li 2002, Zlatanova & Verbree 2003).

For the 3D positioning, a 3D Indoor Geo-Coding technique (Lee 2004b) has been developed to identify the location of disaster events and conditions within the multi-level structure based on the descriptive location information for the real-time emergency response decision making system. The location information like 'there is a chemical explosion at 416 McEniry at the University of North Carolina at Charlotte' obtained from 911 calls or emergency crews with wireless communication devices is transmitted back to the emergency response system in real-time. The 3D Indoor Geo-coding method translates the descriptive location information into geographic positions within the building, the x, y, z coordinates based on a given reference data set.

5.1.2 *3D shortest path algorithm*

One challenging task of 3D GIS is to support spatial analysis among different types of real 3D objects, such as a shortest path analysis in 3D space. Scott (1994) implemented a shortest path algorithm for an un-indexed three-dimensional voxel space using a cumulative distance cost approach. This approach produces a set of voxels, such that each voxel contains an attribute about the cost of traveling to that voxel from a specified start point, if there is uniform friction of movement throughout the representation. The three-dimensional shortest path algorithm moves through the 'cost volume' along the steepest cost slope from target to origin using a 3 by 3 by 3 search kernel (Raper 2000). For B-rep approaches, Kirkby et al. (1997) implemented a modified version of the Dijkstra shortest path algorithm in a 3D GIS, in which the gradient over a 2.5D surface was added into the computation. In this section, a spatial access algorithm in a 3D GIS is introduced in terms of the Dijkstra algorithm.

The network representation of the topological relationships among spatial objects in a study area is described as the graph $N = (V(N), E(N))$. Given this representation, one of the well-known algorithms for finding shortest paths in graphs is applied to the tasks of spatial access in the 3DUrbanObject, because the algorithms use logical networks containing the connectivity of the network without position. In other words, optimal path searching algorithms can be applied to the network problem in 3D space, such as spatial searching problems in the 3DUrbanObject. Dijkstra algorithm (Dijkstra 1959) (priority-first search) is implemented for this purpose. Since Dijkstra algorithm solves the single source shortest-paths problem on a weighted, directed graph $G = (V, E)$, it identifies the source from which the shortest path to all other vertices is to be found (Cormen et al. 1985, Liu 1996). The algorithm needs to be modified in order to implement spatial queries such as the shortest path from a source node to a destination node in the NRS.

Let $G = (V, E)$ be a graph with vertex set $V[G]$; w be an adjacency matrix giving the distances in $V[G]$; vertex s be a source in G; and vertex t be a destination in G. The algorithm maintains a set S of vertices whose final shortest path weights from the source s have already been determined. A priority queue Q contains all vertices in $V - S$, keyed by their d values (total distance values or traffic impedances) (Cormen et al. 1985). The edge set $EP[G]$ is the shortest path between the source vertex s and the destination vertex t in graph G. The traffic impedances (d values) are based on the environment and human factors, which are dynamic factors affecting on determining optimal evacuation route under emergency situations (Pu & Zlatanova 2005). The environmental factors include damage status, toxicity status, power status and traffic capacities on halls, hallways and stairs of the affected area. Human factors affecting people's speed of movement are population density, age and gender, level of disability, terrain effects, and so on. Based on the factors, the traffic impedance will be calculated for the study area.

The 3D shortest path algorithm (3DShortestPath (N, s, t, *AdjList*)) is as follows (Lee 2006). Initialize(G, s) initializes two attributes for each node v, $v \in V[G]$: the travel cost $d[v]$ to ∞ and the predecessor $p[v]$ to NIL, and it initializes $d[s]$ to 0. The function ExtractMin(Q) returns and removes a node u from the priority queue Q which $d[u]$ is currently minimal. While ADJ(*AdjList*, u) returns a set containing the neighbor nodes of u, $u \in V(G)$, and TrafficCost(u, v) returns a traffic cost between a node u and a node v, Traverse(v, $S(N)$) returns that vertex u, $u \in S(N)$, for which $p[v]$ is the vertex u. RearrangeQ(Q) re-organizes the priority queue Q based on the shortest travel cost from a source s to each node.

```
Procedure 3DShortestPath (3DUrbanObject N, Node s, Node t,
   LinkedList AdjList) {
 CALL Initialize(G, s)
     S(N) ← Ø
     Q ← V(N)

 WHILE Q ≠ Ø
       Do u ← CALL ExtractMin(Q)
       IF (u = t THEN
```

```
        EXIT WHILE
    ELSE
        S(N) ← S(N) ∪ {u}
    END IF
    FOR each vertex v∈ADJ(AdjList, u)
        IF d[v] > d[u]+TrafficCost(AdjList[u, v]) THEN
                    d[v] ← d[u]+TrafficCost(AdjList[u, v])
                    p[v] ← u
                    CALL RearrangeQ(Q)
        END IF
    END FOR
END WHILE
WHILE p[v] ≠ NIL
    u ← CALL Traverse(v, S(N))
    e = (v, u)
    EP(N) ← EP(N) ∪ {e}
    v ← u
END WHILE
}
```

5.2 3D buffer function

In order to identify what is near features or within a given distance, the buffer operation could be used in GIS. Suppose tourists are looking for a hotel nearby an airport. The first step for this operation can be to create a buffer object from a feature (such as an airport), and then hotels will be identified within the buffer object using an overlay operation. In 2D, the buffer object is a polygon, while the buffer object is a 3D solid object in 3D. The 3D searching operation should deal with complex geometric computational problems involved with defining topological relationships (inclusion relationships) between the 3D buffer object and well-formed 3D objects representing a micro-scale urban area (such as spatial units in a building).

In order to alleviate the problem, this chapter proposes a new approach to identifying spatial units within a specified distance. Based on the topological models of the study area (3DUrban), $UG = (V(UG), E(UG))$, the new approach utilizes an algorithm to find a minimum spanning tree (MST) in a connected and undirected graph (Kruskal 1956). A MST, one type of valued graph, is a specific network that satisfies three criteria (Chou 1997). First, the tree connects all nodes in the network with a minimal number of links. Second, the root of every tree is located at one of the nodes in the network. Thirdly, the distance between each node and the root of the tree is minimized. A minimal tree rooted at any node can be constructed for the network. Because the topological model of 3DUrban, $UG = (V(UG), E(UG))$, is a network representation having geometric properties (lengths and directions), one of the well-known algorithms for finding minimum spanning trees in graphs is applied to the tasks of 3D spatial buffer, one of topological analyses in 3D space. In other word, the algorithm can generate a minimum spanning tree from a node n_i of the UG network, and then the network segments within a specific distance (buffer's distance) from the node n_i can be identified from the MST, whose each node contains the total distance (or cost) from the rooted node n_i. From the identified network segment, the set of nodes can be determined. The nodes in the UG network represent the spatial units within the specific distance from the node, the 3D buffer object. The Prim's algorithm (1957) is implemented for this purpose. Since the Prim's algorithm identifies the minimum spanning tree from a source node to all other nodes, the algorithm needs to be modified in order to implement spatial queries such as a 3D buffer from a rooted node to other nodes within a specific distance, a buffer distance.

The following procedure explains how the nodes within a given distance from the rooted node are identified. The input data are a weighted graph $UG = (V(UG), E(UG))$ for each edge e, e ∈ $E[UG]$ having a distance value d_e for the edge. Others are the buffer distance, *b-dis*, and the rooted node *r*. The output file is a $N[\ ]$, a set of nodes.

- Step 1: to pick a node *r* as a starting node from the graph *UG*.
- Step 2: to initialize two attributes for each node *v*, *v* ∈ $V[UG]$: the distance from the rooted node *r*, $d[v]$ to ∞ and the predecessor node, $p[v]$ to NIL, as well, it initializes $d[r]$ to 0. The $N[\ ]$ is $N[\ ] + r$. Based on $d[v]$, a min-priority queue for all the nodes is generated.
- Step 3: to extract a node *u* from the min-priority queue.
- Step 4: to find the edge $e = (u, v)$, of minimum cost (for distance) extending from node *u*. Set $d[v] = d[u] + d_e$, and $p[v] = u$, if $d[v] > d[u]$. If $d[v] < b\text{-}dis$, add a node *v* to the $N[\ ]$. Update the min-priority queue based on the $d[v]$.
- Step 5: to repeat step four for all the edges extending from the node *u*.
- Step 6: to repeat step three until $d[v] > b\text{-}dis$. The nodes in the $N[\ ]$ are the list of nodes within the buffer distance, *b-dis* from the node *r*.

6 EXPERIMENTAL IMPLEMENTATIONS OF THE TOPOLOGICAL DATA MODEL

In order to evaluate the potential benefits of the 3D network-based topological data model for providing better services for emergency responses, we undertake an experimental implementation of 3D topological analyses based on the 3DUrban's network $UG = (V(UG), E(UG))$ representing topological relationships among 3D spatial objects in micro-scale built-environments. The implementations are demonstrated based on two topological analyses, 3D optimal route analysis (Kwan & Lee 2005) and 3D buffer analysis (Lee & Kwan 2005).

6.1 *3D optimal route analysis*

We describe the implementation of the 3D optimal route analysis based upon the work of Kwan & Lee (2005). The study area for this experimental case is an area in downtown Columbus, Ohio (USA), located east of Scioto River. We asfsume that a 250-pound highly explosive bomb exploded on the 42nd floor of Franklin County Municipal Building (labeled 'Disaster Site' in Figure 12), and that the shock also caused minor damage on some other floors as well as part of the stairways inside the building. Figure 12 shows the shortest routes under normal traffic conditions (in red) between the disaster building and the fire station located at 405 Oak Street. Suppose that traffic is blocked at two locations on South High Street and Mound Street (indicated by two red dots in Figure 12) nearby the disaster building. Because of these unexpected traffic blocks, the usual shortest path from the fire station to the disaster building is no longer the optimal route. Instead the route in blue in Figure 12 becomes the new shortest path (in terms of travel time). If emergency responders do not have prior knowledge about this new optimal route, they will try to access the disaster site following the usual shortest path (red route). They will then, in this scenario, need to reroute twice because of the two unexpected traffic blocks. The additional delay between the new optimal route and the hypothetical detour route represents the effect of road network uncertainty on emergency response time.

After arriving at the disaster building at Entrance A (Figure 13), emergency responders discover that this entrance is blocked by debris and cannot be used to reach the destination room (disaster site) on the 42nd floor (Figure 13). They then walk to another side of the building in order to use Entrance B (Figure 13). These responders are, however, blocked at the 28th floor as they attempt to walk up to the 42nd floor using the stairway. They then walk down to the ground level and use another stairway to go up again (Figure 14). They are blocked on the 28th floor again and have to walk down a couple of floors and walk through some corridors to go up using another stairway

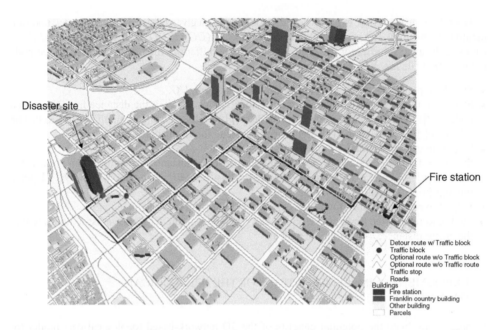

Figure 12. The shortest path between a fire station and a disaster building (from Kwan and Lee, 2005).

(Figure 14). The additional delay between the optimal route (green line) and the hypothetical detour route (red dotted line in Figure 14) represents the effect of entry point uncertainty and route uncertainty in the building on emergency response time.

In order to simulate this scenario, three travel speeds are assigned to the 3D network developed for the study: (a) 25 miles per hour for the road network; (b) 75 feet per minute for walking horizontally outside or inside the building; and (c) 40 feet per minute for going up or down vertically using the stairways inside the building. The total travel time it takes to reach the destination node c without using the system is 39.83 minutes, while it is only 24.19 minutes when the optimal route found by the system is used. This means that emergency responders can reach the destination node 15.64 minutes earlier than when such a system is not used. In the experiment, optimal routing performed using an integrated 3D network saves more than one-third of the travel time otherwise needed for reaching the disaster site. Further, the results suggest that optimal routing using only the ground transport network as in conventional 2D GIS leads to a mere 2.18 minutes saving in travel time. This means that a 3D network that integrates the street network with the building's network brings an additional saving of 13.46 minutes. This amounts to 86% of the total travel time saved due to the use of the optimal route found by using the 3D network. This experiment demonstrates that the travel time needed to reach a disaster site inside a multi-level structure can be much longer than the time needed to travel from a source node (a fire station) to the disaster building. It shows that extending conventional 2D GIS to include the internal structures of high-rise buildings can significantly improve the overall speed of rescue operations.

6.2 3D buffering analysis – spatial query based on an adjacency relationship

The implementation of the 3D buffer analysis is described based upon the work of Lee & Kwan (2005) in this section. Analysis of the urban phenomena requires those relationships to describe how the individual spatial objects interact. The topological structure can be used efficiently in a query to find neighbors – Which other 3D spatial objects are located on top or under a certain 3D object? This neighbor information can be used in environmentally oriented analyses including noise, air pollution, and emergency situations in urban environments.

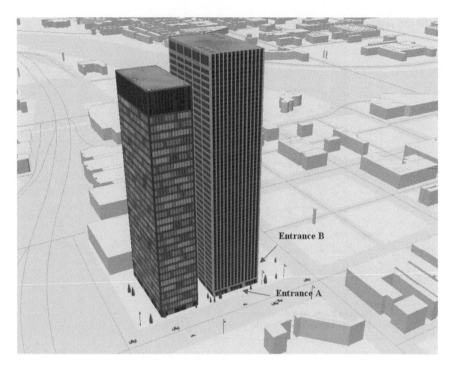

Figure 13.   Two possible entrances of the disaster building.

This example presents spatial queries based on topological relationships among the 3D object, $G_i = (V(G_i), E(G_i))$, to access the adjacency information among rooms within a building i. A click on a node in the Viewer area of 3D NRS Implementation Module runs a VB code, which delivers a Query-Result window showing a Node-ID. The Node-ID is associated with the room number. After selecting a node, the user needs to click the 'Find Adjacency Objects' button to send a request to the system. The query results are displayed in the Viewer area (Figure 15b). Figure 15b shows the result of a spatial query to retrieve adjacent rooms to TE210. Based on the sub-graph (thick lines) representing the query result, we can define that the TE210 Conference Room is adjacent to seven rooms, which are TE212 Auditorium, TE208 Computer Lab, TE209 & TE211 Research Labs, TE201 Hallway, TE110 Classroom, and TE310 Classroom. These rooms are sharing a vertical or horizontal wall with the TE210. Based on the topological information, the same result is displayed in the 3D Viewer using ArcScene of ESRI Inc. in Figure 15a. The solid object of TE210 is colored in dark green, and the solid objects of adjacent rooms to TE210 are colored in light green).

7   CONCLUSIONS

3D geospatial information has always faced challenges due to a variety of data models and no common data models. After 9/11 the interest of 3D models representing micro-spatial entities (buildings or undergrounds) in emergency responses has been progressively increasing. Although such models are mostly available from the design phase (as CAD models), design CAD models are in most cases, too detailed for computing evacuation routes. Therefore, this chapter proposed a data model to represent and to analyze 3D geospatial data in emergency management systems

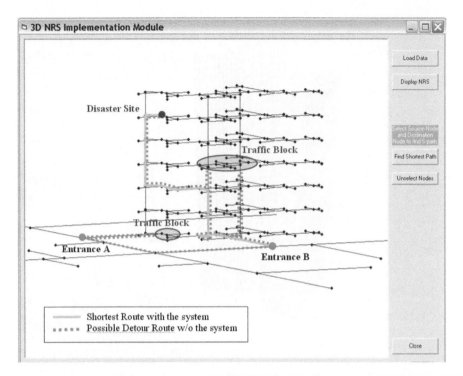

Figure 14.   The shortest path between two entrances (A and B) and a disaster site on the 42nd floor of the building (from Kwan & Lee, 2005).

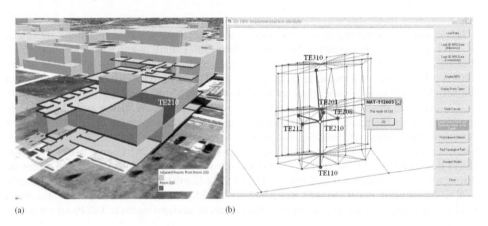

(a)   (b)

Figure 15.   Adjacency relationships from TE210 (from Lee & Kwam, 2005).

for field workers and decision makers. However, CAD models and their standard representations (e.g. IFC) have to be further studied with respect to easy elimination of unnecessary details and automatically generation of simple multi-level structures, which can be used for emergency evacuations.

This chapter described the algorithms of 3D spatial data manipulating and analytical methods to implement the planning/decision making process, which are 3D topological analysis (3D buffering) and an indoor navigation function to identify feasible and safe routes within a multi-level structure and to provide navigation guidance for rescue personnel. Also, this chapter presented the experimental implementations of 3D topological analyses, which are spatial queries based on adjacency relationships among the 3D spatial units, and 3D shortest route for evacuation from a building.

While focusing on formulating a conceptual framework of a 3D data model for ER this chapter ignores several important elements in developing a real-time emergency response system. First, successful implementation and use of the ER system depends on the availability of accurate real-time information about the emergency situation from various sources and analytical functions. The system needs to integrate with temporal databases to manage dynamic geospatial entities, which are dynamic capacities and flow rates of hallways and stairwells, in order to identify the optimal route from the source node to the destination node and the building evacuation bottlenecks within the network in real-time emergency situations. The proposed 3D Shortest Path algorithm needs to be improved in order to treat traffic cost or impedance variables as a function of all routes to predict the amount of flow per time period on the 3D network.

Additionally, several important functionalities for ER should be developed, including a 3D location positioning, a network connectivity analysis, and other topological analyses to define isolated networks or areas, which do not have any exit node connecting to destination nodes because of being blocked by traffic congestions or disaster. Lastly, the ER system should explore geo-referenced virtual environments in 3D Virtual Reality (VR) systems (Liu & Zhu 2006). Evacuation instructions sent from the emergency center to rescuers via wireless communication networks will be displayed on mobile devices with the 3D VR system. The virtual reality IBE system will be a major step in providing 3D location-based services to indoor urban areas.

REFERENCES

Arens, C., Stoter, J.E. and van Oosterom, P.J.M. 2005. Modelling 3D spatial objects in a geo-DBMS using a 3D primitive. Computers & Geosciences, 2, pp. 165–177.
Batty, M., Chapman, D., Evans, S., Haklay, M., Kueppers, S., Shiode, N., Smith, A. & Torrens, P. 2001. Visualizing the City: Communication Urban Design to Planners and Decision Makers, in R. Brail & R. Klosterman (eds), Planning Support Systems: Integrating GIS, models, and visualization tools. Redlands: ESRI Press.
Burns, K.L. 1988. Lithologic topology and structural vector fields applied to subsurface predicting in geology. Proc. of GIS/LIS 88, San Antonio, TX, USA.
Carlson, E. 1987. Three dimensional conceptual modeling of subsurface structures. Proc. of 8th International Symposium on Computer Assisted Cartography, AutoCarto 8, Baltimore, MD, 336–345.
Chalmet, L.G., Francis, R.L. & Saunders, P.B. 1982. Network Models for Building Evacuation, Management Science, 28(1): 86–105.
Chou, Y-H. 1997. Exploring Spatial Analysis in Geographic Information Systems. New York: OnWord Press.
Coors, V. 2003. 3D-GIS in networking environments. Computers, Environment and Urban Systems, 27, 345–357.
Corbett, J.P. 1979. Topological Principles in Cartography, Technical Paper 48. U.S. Department of Commerce: Bureau of the Census.
Cormen, T., Leiserson, C. & Rivest, R. 1985. Introduction to Algorithms. Cambridge, MA: The MIT Press.
Cutter, S.L. 2003. GI Science, Disasters, and Emergency Management. Transactions in GIS 7(4): 439–445.
Dane, C. & Rizos, C. 1998. Positioning Systems in Intelligent Transportation Systems. Boston: Artech House.
Dijkstra, E.W. 1959. A Note on Two Problems in Connection with Graphs. Numer. Math., Vol. 1: 269–271.
Dürr, F. & Rothermel, K. 2003. On a Location Model for Fine-Grained Geocast. In A.K. Dey, A. Schmidt & J.F. McCarthy (eds) Proc. of the Fifth International Conference on Ubiquitous Computing (UbiComp 2003), 18–35.
Egenhofer, M.J. & Herring, J.R. 1990. A mathematical framework for the definition of topological relationships. Proc. of the Fourth International Symposium on SDH, Zurich, Switzerland, 803–813.
ESRI. 1998. ArcView 3D Analyst, available at http://www.esri.com/software/ arcview/extensions/ 3dext.html
Faust, N.L. 1995. The virtual reality of GIS. Environment and Planning B: Planning and Design, 22: 257–268.
Herring, J, 2001, Topic 1 Feature Geometry (ISO/TC211 19107 Spatial Schema), OGC specifications available at www.opengeospatial.org. 184 p.

Hoffmann, C.M. 1989. Geometric and Solid Modeling: An Introduction. San Mateo, CA: Morgan Kaufmann Publishers, Inc.

Holtier, S., Steadman, J. & Smith, M. 2000. Three-dimensional representation of urban built form in a GIS. Environment and Planning B: Planning and Design, 27: 51–72.

Hoppe, B. & Tardos, E. 1995. The Quickest Transshipment Problem. Proc. of SODA: ACM-SIAM Symposium on Discrete Algorithms: 433–441.

Jepson, W., Liggett, R. & Friedman, S. 2001. An Integrated Environment for Urban Simulation. In R. Brail & R. Klosterman. Planning Support Systems: Integrating GIS, models, and visualization tool. Redlands: ESRI Press.

Köninger, A. & Bartel, S. 1998. 3D-GIS for Urban Purposes. GeoInformatica 2 (1): 79–103.

Kirkby, S., Pollitt, S. & Eklund, P. 1997. Implementing a Shortest Path Algorithm in a 3D GIS Environment. In M.J. Kraak & M. Moleanaar (eds), Advances in GIS Research II ; Proc. of the 7th International Symposium on Spatial Data Handling. London: Taylor & Francis Inc. 437–448.

Kolbe, T.H., Gröger, G. & Plümer, L. 2005. CityGML – Interoperable Access to 3D City Models. In P. van Oosterom, S. Zlatanova & E.M. Fendel (eds), Geo-information for Disaster Management; Proc. of the 1st International Symposium on Geo-information for Disaster Management', Delft, The Netherlands, March 21–23, 2005. Springer.

Kolodziej, K & J. Hjelm, 2006. Local Positioning Systems: LBS application and services, Taylor and Francis, Boca Raton, USA.

Kruskal, J.B. 1956. On the shortest spanning subtree of a graph and the travelling salesman problem. Proc. Am. Math. Soc. 7(1): 48–50.

Kwan, M-P. & Lee, J. 2005. Emergency response after 9/11: the potential of real-time 3D GIS for quick emergency response in micro-spatial environments. Computers, Environment and Urban Systems, 29: 93–113.

Lee, J. 2001. A 3DData Model for Representing Topological Relationships Between Spatial Entities in Built-Environments. Unpublished Ph.D. Dissertation, Department of Geography, The Ohio State University.

Lee, J. 2004a. A Spatial Access Oriented Implementation of a Topological Data Model for 3D Urban Entities. GeoInformatica 8:3, p. 235–262.

Lee, J. 2004b. 3-D GIS for Geo-coding Human Activity in Micro-scale Urban Environments. In Geographic Information Sciences: Springer's Lecture Notes in Computer Science Computers (LNCS 3234), eds M. Egenhofer, C. Freksa and H. Miller, 162–178. Now York: Springer.

Lee, J. 2006. A 3D Navigable Data Model to Support Emergency Responses in Micro-Spatial Built-Environments. Annals of the Association of American Geographers (Accepted).

Lee, J. & Kwan, M-P. 2005. A Combinatorial Data Model for Representing Topological Relations among 3D Geographic Features in Micro-spatial Environments. International Journal of Geographical Information Science, 19:10, p. 1039–1056.

Li, R. 1994. Data Structures and Application Issues in 3D Geographic Information Systems, GEOMATICA, 48 (3): 209–224.

Liggett, R.S. & Jepson, W.H. 1995. An integrated environment for urban simulation. Environment and Planning B: Planning and Design, 22: 291–302.

Li, H. & Zhu, Q. 2006. Virtual Geographic Environments. In S. Zlatanova & D. Prosperi (eds), Large-scale 3D Data Integration – Challenges and Opportunities: 211–232, Boca Raton: Taylor & Francis (CRCpress).

Liu, S. 1996. Object Orientation in Route Guidance Systems, Unpublished Master Thesis, The University of Calgary.

Longley, P.A. & Batty, M. (ed), 2003. Advanced Spatial Analysis: The CASA book of GIS, Redlands: ESRI Press.

Lu, Q., Hung, Y. & Shekhar, S. 2003. Evacuation Planning: A Capacity Contrained Routing Approach. In Chen, H., Zeng, D.D., Demchak, C. and Madhusudan, T. (eds) Proc. of the First NSF/NIJ Symposium on Intelligence and Security Information (ISI), Tuson, AZ, 111–125.

Meijers, M., Zlatanova, S. & Preifer, N. 2005. 3D geoinformation indoors: structuring for evacuation, In: Proceedings of Next generation 3D city models, 21–22 June, Bonn, Germany, 6 p.

Mallet, J-L. 1990. GOCAD: a computer-aided design program for geological applications. In Turner, A.K. (ed), Three-dimensional modeling with geoscientific information systems. Dordrecht: Kluwer.

Miller, H. & Shaw, S.-L. 2001. Geographic Information System for Transportation: Principles and Applications. New York: Oxford University Press.

Molenaar, M. 1990. A Formal Data Structure for 3D vector maps. In Proc. of EGIS'90, 2, Amsterdam, The Netherlands, 770–781.

MultiGen-Paradigm. 2003. SiteBuilder 3D Getting Started, Version 1.1.1.

Oosterom, P. V., Stoter, J., Quak, W. & Zlatanova, S. 2002. The balance between geometry and topology. In D. Richardson & P. Oosterom (eds), Advances in Spatial Data Handling, 10th International Symposium on Spatial Data Handling: 209–224. Berlin: Springer-Verlag.

Pahlavan, K. & Li, X. 2002. Indoor geolocation science and technology: Nextgeneration broadband wireless networks and navigation services. IEEE Communications Magazine (February): 112–118.

Penninga, F. 2004, Oracle 10g Topology; Testing Oracle 10g Topology using cadastral data GISt Report No. 26, Delft, 2004, 48 p., available at www.gdmc.nl/publications

Pigot, S. 1992. A Topological Model for a 3D Spatial Information System. In Proc. of the 5th International Symposium on Spatial Data Handling, Charleston, South Carolina: 344–359.

Pigot, S. 1995. A Topological Model for a 3-dimensional Spatial Information System, PhD thesis, University of Tasmania, Australia.

Pigot, S. & Hazelton, B. 1992. The Fundamentals of a Topological Model for a Four-Dimensional GIS. In Proc. of the 5th International Symposium on Spatial Data Handling, Charleston, South Carolina: 580–591.

Pilouk, M. 1996. Integrated modeling for 3D GIS, PhD Dissertation, ITC, The Netherlands.

Prasithsangaree, P., Krishnamurthy, P. & Chrysanthis, P.K. 2001. On indoor position location with wireless LANs. Telecommunications Program and Department of Computer Science, University of Pittsburgh. Pittsburgh, PA.

Prim, R.C. 1957. Shortest connection networks and some generalisations. Bell Systems Technical Journal (Nov): 389–1410.

Pu, S. & Zlatanova, S. 2005. Evacuation route calculation of inner buildings. In van Oosterom, P.J.M. Zlatanova, S. & Fendel, E.M. (eds), Geo-information for disaster management: 1143–1161. Heidelberg: Springer Verlag.

Pu, S., Zlatanova, S., & Bronsvord, W.F. 2006. Freeform curves and surfaces data types for integrating CAD and GIS models, IJGIS (under review),

Reed, C., 2006. Data integration and interoperability: OGC standards for geo-information, in: Zlatanova, S & Prosperi, D. (eds). 3D large-scale data integrations: challenges and opportunities, Boca Raton, Taylor & Francis (CRCpress) pp. 163–174.

Raper, J. 2000. Multidimensional Geographic Information Science. New York: Taylor & Francis.

Rikkers, R., Molenaar, M. & Stuiver, J. 1994. A query Oriented implementation of a topologic data structure for 3Dimensional vector maps. INT. J. Geographical Information System, 8 (3): 243–260.

Samet, H. 2001. Position Paper for Location-Based Services Meeting: Santa Barbara Conference on Location-Based Services, Center for Spatially Integrated Social Science, Santa Barbara, CA, available at http://www.csiss.org/events/meeting/location-based.

Scott, M.S. 1994. The development of an optimal path algorithm in three dimensional raster space,' In Proc. of GIS/LIS '94: 687–696.

Shiode, N. 2001. 3D urban models: Recent developments in the digital modeling of urban environments in three-dimensions, GeoJournal, 52: 263–269.

Smith, J.M. 1991. State Dependent Queueing Models in Emergency Evacuation Networks, Transportation Science: Part B, 25B (6): 373–389.

Smith, A. 1998. Adding 3D Visualization Capabilities to GIS, available at http://www.casa.ucl.ac.uk/venue/3d_visualisation.html

Stoter, J. & Oosterom, P. van. 2002. incorporating 3D geo-objects into a 2D Geo-DBMS, Proceedings of ACSM-ASPRS 2002 Annual Conference.

Stoter, J. & Zlatanova, S. 2003. Visualization and editing of 3D objects organized in a DBMS, Proceedings of the EuroSDR Com V. Workshop on Visualization and Rendering, 22–24 January 2003, Enschede, The Netherlands.

Sugihara, K., Hammad, A., & Hayashi, Y. 2000. GIS based System for Automatic Generation of 3-D Urban Models and its Application, Proceeding of 2000 URISA, Orlando, FL.

Tempfli, K. 1998. 3D topologic mapping for urban GIS, ITC Journal, 3(4): 181–190.

Worboys, M.F. 1995. GIS: A Computing Perspective. Bristol, PA: Taylor & Francis Inc.

Wix, J., Nisbet, N. & Liebich, T. 2005, Industry Foundation Classes: Facilitating a seamless zoning and building plan permission, IAI Government and Industry day, 31 May, Oslo, Norway, available at http://www.iai.no/2005_buildingSMART_oslo/Session%2001/20050531_IFG_IAI_Oslo_JDW.pdf

Zeiler, M. 1999. Modeling Our World: The ESRI Guide to Geodatabase Design, CA: ESRI Press.

Zlatanova, S. 2000. 3D GIS for urban development, PhD Dissertation, ITC, The Netherlands.

Zlatanova, S. & Holweg, D. 2004. 3D Geo-information in emergency response: a framework. In Proc. of the Four International Symposium on Mobile Mapping Technology (MMT'2004), March 29–31, Kunming, China, 6 p.

Zlatanova, S., Holweg, D. & Coors, V. 2004. Geometrical and topological models for real-time GIS. In Proc. of UDMS 2004, 27–29 October, Chioggia, Italy, CDROM, 10 p.

Zlatanova, S. & Stoter, J. 2006, The role of DBMS in the new generation GIS architecture. In S. Rana and J. Sharma (Eds.), Frontiers of Geographic Information Technology, pp. 155–180, Berlin: Springer–Verlag, 2006.

Zlatanova, S., Rahman, A. & Pilouk, M. 2002. Trends in 3D GIS development. Journal of Geospatial Engineering, Vol. 4 (2): 1–10.

Zlatanova, S., Rahman, A. & Shi, W. 2004. Topological models and frameworks for 3D spatial objects, Journal of Computers & Geosciences, 30 (4): 419–428.

Zlatanova, S. & Verbree, E. 2003. Technological Developments within 3D Location-based Services, In Proc. of International Symposium and Exhibition on Geoinformation, Shah Alam, Malaysia: 153–160.

Geospatial Information Technology for Emergency Response – Zlatanova & Li (eds)
© 2008 Taylor & Francis Group, London, ISBN 978-0-415-42247-5

Multidimensional and dynamic vehicle emergency routing algorithm based on 3D GIS

Q. Zhu & Y. Li
LIESMARS, Wuhan University, P.R. China

Y.K. Tor
School of Civil & Environmental Engineering, NTU, Singapore

ABSTRACT: An increasing number of disasters are frequently reported with huge loss of life and assets, both on the surface and underground in mega cities. However, there is still not a working 3D GIS-based vehicle emergency routing algorithm with due emphasis on the third dimension and the effective and comprehensive use of dynamic emergency information. In this paper, a multidimensional and dynamic vehicle emergency routing algorithm is proposed. The algorithm is based on the functional requirement analysis of 3D vehicle emergency routing. To integrate multidimensional and dynamic information into the algorithm visually, effectively and compre-hensively, this paper discusses the construction of 3D dynamic road networks, as well as the corresponding network optimization and multi-criteria evaluation methods in a 3D GIS environ-ment. The proposed algorithm and its applications to 3D emergency routing/navigation have been implemented in VGEGIS™ software, and the advantages of 3D GIS-based algorithm over traditional algorithm are also discussed through experimental results.

1 INTRODUCTION

Mega cities are continuously extending in the third dimension, both upwards and downwards. With all kinds of natural and man-made features, an increasing number of disasters – such as fire, collapse, poison gas leakage, bombing, tsunami, etc. – are frequently reported with huge loss of life and assets, both on the surface and underground, in such mega cities. Such disasters – occurring in areas of dense population and complicated urban environment – are dynamic, stochastic, and multidimensional (on the surface and underground) and the rescue work is often made more difficult due to the absence of powerful emergency response tools.

Vehicle emergency routing, developed to help rescue workers find the optimal to-and-fro routes between the disaster areas and public service centres, is crucial in emergency response. Different research groups – including transportation agency, disaster management offices, civil engineering department, remote sensing and photogrammetry community – have made great contributions to the improvement of decision making for vehicle emergency routing. Their works emphasize different aspects – such as road extraction and visualization, disaster simulation, traffic flow research, network routing, events management, etc. With the improvements in science and technology, more and more emerging techniques have the potential to support multidimensional and dynamic routing and navigation. For example, Global Positioning System (GPS) for tracking position, Remote Sensing (RS) for detecting flooding areas, Closed Circuit Television (CCTV) system for traffic flow monitoring and 3D GIS for harnessing the valuable data, models and simulation scenarios are capable of providing the third dimension. In tandem with the diverse technologies, there are large volumes of data with different formats, geo-referenced coordinate systems, temporal and spatial scales, and static and dynamic attributes, all of which present great opportunities as well as challenges to any multidimensional and dynamic emergency routing application.

However, vehicle emergency routing – regarded as a branch of Intelligent Transportation System (ITS) and primarily implemented under the framework of Geographic Information System for Transportation (GIS-T) in a 2D GIS – lacks the effective and comprehensive consideration of multidimensional and dynamic emergency information. The current 2D GIS-based emergency routing algorithm, therefore, is required to extend its capability to better suit the increasing complicated emergency situation in mega cities.

This paper is divided into six sections. Following this introduction, Section 2 reviews the state of the art of vehicle emergency routing; Section 3 discusses the functional requirements; Section 4 introduces the 3D dynamic network model and Section 5 presents the multidimensional and dynamic vehicle emergency routing algorithm and its implementation in VGEGIS™ software; and finally the concluding remarks are given in Section 6.

2 THE STATE OF THE ART OF VEHICLE EMERGENCY ROUTING

2.1 *Emergency response*

As discussed before emergency response has quite specific characteristics. First of all, the methods used in the emergency response should be direct and quick (Borkulo et al., 2005). Secondly, effective and comprehensive understanding of the overall emergency situation is significant. Thirdly, the information is heterogeneous and crossing different application domains (Gong & Batta 2005, Gong et al., 2004). The desirable effects of emergency response rely much on the use of promising technologies, data integration strategies and road network data model, and the final goal is set to provide reliable and understandable solution for assisting effective and efficient decision-making for different groups, such as governor, rescuer and victim.

2.2 *Technical support for vehicle emergency routing*

In the past decades, there has been much research into vehicle emergency routing with the support of emerging technologies and corresponding information to improve the quality of emergency response. Derekenaris et al. (2001) described a system offering a solution to the problem of ambulance management and emergency incident handling. Their work is based on the integration of the technologies of geographic information system (GIS), global positioning system (GPS) and global system for mobile communication (GSM). The route is created based on the link travel time that is calculated using the road length and traffic flow. Choi (2003) investigated the way of dispatching emergency vehicles in order to maximize the number of survivals in an emergency situation, taking into consideration the location and severity of injury of the injured parties. Mollaghasemi & Abdel-Aty (2003) presented the use of microscopic traffic simulation models to evaluate the emergency vehicle routing strategies in the case of an emergency. In their work, the roadway network and the travel demand information are the two main inputs. Gong (2004, 2005) considered a dynamic disaster environment where an earthquake had struck and thousands of casualties needed attention. The key factors affecting the dispatch of ambulances to patient locations include patient priority, cluster information and distance. It also considers the road damage information and the use of back-up routes. Capaccio & Ellis (2005) propose enhancing emergency routing applications with the capability of integrating near-real-time event information to perform accurate routing. The proposal aims to use RS technology, GIS and geo-processing techniques to regenerate an effective emergency evacuation route by integrating a flooding scenario with the road network.

Besides the widely accepted technologies such as GPS, RS, GSM, GIS and CCTV, another promising technology that can improve the dynamic understanding of emergency response and routing is simulation (Jafari 2003). Different with other technologies, simulation technology shows some strength. Firstly, it is less expensive and safer than on-line tests (Mollaghasemi &

Abdel-Aty 2003). By using the virtual environment, it produces the scenarios under the different assumptions, thus minimizing the costs and threats with a predicted manner. Secondly, in response to the case that traditional technologies record the information in a "snapshot" way and the information is temporarily unavailable in some special situation such as poor weather or technical problem, the use of simulation information will provide continuous and dynamic support. The research of simulation technology and information into emergency response is diverse and can be from many aspects, for example, disaster propagation simulation is usually yielded for warning people of dangerous areas; the simulation of human behaviour is given to model the activity pattern of people in a disaster situation; traffic flow simulation indicate the travel speed in different time of a day. There is often an increasing high public and political interest to see a simulated situation before an emergency and to set strategies to the potential threats.

Many technologies, including simulation, though can provide response timely, virtually and visually – most have developed their methods in a 2D GIS. Research into emergency response in a 3D environment has drawn increasing attention in recent years with the development of 3D GIS platform, but the emphasis is mostly on the data visualization and route navigation. Coors et al. (2005) discussed the use of 3D navigation maps in routing/wayfinding on mobile devices. Their works focus on exploring geovisualization for providing effective route instructions. Slavick et al. (2005) demonstrate the utility of 3D visualization to meet military and homeland security needs. Through their adoption of 3D program, the Tactical Operations Center (TOC) shows the importance of using 3D models, military symbols and homeland security symbols. Besides, the research of 3D pedestrian navigation model (Lee 2005, Kwan & Lee 2005) shows the intention of using 3D technology in the domain of supporting the indoor emergency response in a 3D building environment. With regard to vehicle routing in transportation system, Pollitt et al. (1995, 1996) developed a network-based vehicle routing approach in a 3D spatial information system. The approach considers static impedances (road segment length in three dimensions, gradient and different routing speeds) and dynamic impedances (lane changes) with the use of the Dijkstra algorithm. However, the approach still uses a planar transportation network – the default line was digitized from maps and was converted to 3D line by registering to digital elevation model (DEM) – and the information integrated into the routing algorithm is limited and therefore, it cannot be used in a complicated emergency routing practice in mega cities.

2.3 Road network data model

Vehicle emergency routing relies much on road network data model and emergency-related data integration strategies. Road network data model was universally represented as node-arc network model by a set of nodes and a set of arcs. According to the different manners of handling intersection node, network model is broadly classified as two types: planar and non-planar data model. Planar data model forces nodes to exist at all arc intersections and, thus, ensures topological consistency of the model. Planar network model has received widespread acceptance in 2D vehicle routing algorithm, but the planar embedding requirements will make the model difficult in representing the complex network phenomena, such as underpasses or overpasses. The drawbacks have motivated interest in non-planar network models (Fischer 2004). By comparison, non-planar network permits arcs of the network to cross without a network node being located at the intersection. The merits of non-planar network make the feature-based modeling possible and eliminate the impossible turn and nodes in real transportation system (Fohl et al., 1996). However, the current non-planar network is mainly implemented in a 2D space where nodes have been only recorded as x, y coordinate. Different with non-planar network in 2D space, non-planar network in 3D space owns z value for each node, thus allows more complicated representation and network analysis. Non-planar network in 3D space overcomes the problems of non-planar network in 2D space from at least two aspects. Firstly, it allows two nodes to have the same x, y coordinates, thus improves the ability to depict the complex transportation structure, which is very important in multi-model transportation representation. Secondly, it provides the foundation of 3D spatial analysis as multidimensional data can be integrated into the model in a direct way.

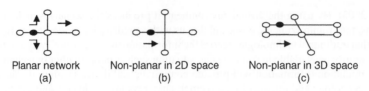

Planar network   Non-planar in 2D space   Non-planar in 3D space
(a)   (b)   (c)

Figure 1.   Node-arc network model.

In Figure 1(a), the planar network forces the crossing arcs to be fully connected and the object can move from one arc to other three. In figure 1(b), the non-planar network in 2D space allows to model bridges, flyovers, etc. Because there is no topological node located at the crossings, the object is only allowed to move forward. In figure 1(c), two arcs own the same x, y coordinates in the non-planar network in 3D space, and the object located on the top arc is constrained to move forward. Such a situation is not easy to be represented in 2D space without the support of vertical information.

Built on the basic node-arc network model, some process models, which are concerned with how transportation activities are conducted and define a process by which some transportation planning or maintenance activity can take place, are proposed (Curtin et al., 2001) – such as the linear referencing data model. The major advantage of linear referencing is its capability of locating attributes and events (such as the location of and condition of signage, the condition of pavement, and the location and severity of accident occurrences) along a linear feature with only one parameter (usually known as M) instead of two or three (such as x, y, and z). The National Cooperative Highway Research Program (NCHRP) has supported such an effort. NCHRP 20-27(2) LRS data model owns a solid foundation, welcomed by many enterprises. The main contribute of the model lies in the use of linear datum that provides the solution to the problem of multiple network and multi-level representation of geography. More recently, MDLRS data model (Multidimensional Multi-Modal Location Referencing System) – also regarded as NCHRP 20-27(3) LRS data model, is developed to integrate and effectively use data across one, two, three and four dimensions and among linear and non-linear referencing systems. MDLRS model expands the capabilities of spatio-temporal referencing and integrates the strengths of other models (Koncz 2002). However, the models are hard to implement and the maintenance of linear datum is difficult (Adams et al., 2001, Guo 2001, Koncz 2002).

Object models are those that seek to identify or enumerate as many transportation objects as possible and logically organize them in such a way that they can be most profitably used (Curtin et al., 2001). A notable effort to accomplish these goals is referred to as the Geographic Data Files (GDF) – a European standard used to describe and transfer road networks and road-related data. Related to this type of object model is the idea of an enterprise GIS-T data model. Different with NCHRP 20-27(2) model, GIS-T Enterprise LRS data model is event-data-centric (Guo 2001) instead of datum-centric architecture (Butler & Dueker 2001). Another object model that can be widely recognized is the UNETRANS (Unified network for transportation) model. In fact, UNETRANS data model can be regarded as a logical data model of GDF. This model provides templates for transportation-based entities based on ESRI's Geodatabase model to enhance project implementation and assist application development, including six sub models: reference network; routes and location referencing; asset; activities; incidents; mobile objects (Curtin et al., 2001).

However, most road network data models are derived from a 2D node-arc network. Since the newly developed MDLRS data model strengthens the connection between 2D referencing system and 3D referencing system and therefore provides the supports for multidimensional data transformation, it still lacks the ability to visualize 3D objects (such as flyover) and to support 3D operation (such as 3D distance measurement). Secondly, those data models simplify the complicated transportation system as a 2D network and even treat the flyover as a point, thus only suit for macroscopic information handling but will be problematic in complex urban transportation system. Thirdly, those models are primarily used for linear data management and dynamic segmentation but not directly support linear data integration with route planning. The unified LRS/routing model is reported to improve the reliability of path finding, such as showing the safest route for over-sized vehicles by considering the edge restriction and avoiding roads with low-capacity bridges that could collapse under the extra weight (Hardy 2003), however, there is not a working model of this kind in 3D urban environment and applied in vehicle emergency routing.

Besides the consideration of linear data integration into the widely accepted link-node centerline network models, in recent years, ITS and other transportation applications have highlighted the need for lane-based network data models (Miller & Shaw 2001). Lane-based data model specifically improves the ability to represent availability of lanes, individual lane properties, connectivity among parallel lanes and at turns, and lane movement restrictions (Malaikrisanachalee & Adams 2005). It seems that lanes could be the best choice of representing transportation system because vehicles must move on a certain lane at most travel time. However, lanes cannot limit vehicle movement as strictly as carriageways. At the same time, the current positioning technologies are not accurate enough to locate a vehicle at the level of lane and lane-based data model could add the complexity of the routing algorithm. As a result, the research of the basic modeling entities of road network data model and their implementation issues proposed a big challenge to GIS-T applications and emergency response context.

2.4 *Shortest path algorithm*

Our traditional vehicle routing algorithm has been developed from static and deterministic algorithm to dynamic and stochastic algorithm (Eklund et al., 1996, Gendreau et al., 1996, Lu 2001, Guo et al., 2002, Bent et al., 2004) with the appearance of complicated transportation phenomena (such as turn restriction, multi-lane, lane changing). However, because both 2D- and 3D-vehicle routing algorithms are all based on graph theory with the use of a node-arc model, the computational mechanism is the same. Therefore, many improved algorithm, such as Dijkstra, Floyd and A* heuristic shortest path algorithm can be incorporated into both 2D and 3D route finding. More discussion about the efficiency analysis of shortest path algorithms in transportation network can be found in Zhan & Noon (1998). In this paper, the focus of research aims at the functional requirement analysis for 3D vehicle emergency routing and the construction of 3D dynamic road network data model.

3 FUNCTIONAL REQUIREMENTS FOR 3D VEHICLE EMERGENCY ROUTING

Vehicle emergency routing is interdependent, dynamic and complicated. When disaster happens, taking earthquake as an example, it may make traffic systems (such as flyovers) collapse; destroy buildings and thus cause traffic congestion. It may also bring flooding – which will block traffic in some lowlands – as well as fire, which further threatens human life and makes people panic and flee from buildings to road surfaces. Some unpredictable events will also occur.

Functional requirement analysis is the base for system design and problem solving. The functional requirements for vehicle emergency routing can be analyzed from three aspects, namely the demand, the supply and the allocation. Their inter-relationship is to balance the demand and the supply with the use of the designed allocation model. In this paper, the term 'user' refers to 'the demand' and means 'who' concerns it, also includes 'what' kind of service and rules are desired. The 'information' refers to 'the supply'; it is the 'digital knowledge' that directs the 'user' to understand 'where' the vehicles and facilities are located, and 'what' and 'when' the event happens, etc. 'The allocation' connects 'the demand' and 'the supply'; it is a framework that contains the algorithms, data integration strategies and system platform. Two perspectives, namely user's perspective and information perspective, are chosen to demonstrate the demand and supply aspects of the functional requirements while the choice of system platform is to justify the allocation aspect.

From the user's perspective, emergency routing and navigation relate to life threatening and asset damaging events. In contrast to other traditional 'shortest' path applications (such as travelling salesman problems), emergency vehicle routing primarily calls for minimum travel time rather than minimum travel distance, or other considerations.

Table 1. Comparison of various vehicle routing optimization rules.

Rule (Mode)	Description	Key factors
Shortest Distance	Spatial distance	Geographic distance (2D/3D)
Fastest	Time-critical	Link travel time
Most economic	Costs	Road surface, fuel cost, distance, waiting time
Safest	Safe and comfort	Potential threats probability
Easiest	Spatial cognition	Familiarity degree, landmark density

Table 2. Different objects and factors in emergency routing application.

Objects	Factors	Description
Vehicle	Weight-power ratio	Physical and predefined attribute
	Weight	Dynamic location
	Height/Width	
People	Rescue workers	Cognitive and predefined attribute
	Victims	Dynamic location
	Pedestrian/Civilian	Dynamic
Road	Length	
	Slope	Physical and predefined
	Supporting capability	
	Surface material	
Event	Traffic flow	Dynamic
	Allowed travel speed	Physical and predefined attribute
	Disaster propagation	Dynamic
Facility	Service capability	Dynamic attribute
	Location/Address	Physical and predefined location

In Table 1, the comparison of various vehicle routing optimization rules is given. Because of the differences, the key factors are diverse, for example, 'Shortest Distance' mainly considers the geographic distance in 2D or 3D space, giving people the feeling of spatial measure; 'Fastest' rule concerns the link travel time, showing the temporal measure. 'Most Economic' rule is based on economic measures and 'Safest' rule asks for the consideration of probable threats. 'Easiest' rule derives from the spatial cognition of human beings, depending on the degree of familiarity and the density of landmarks in the environment.

Much attention was focused on the single optimization rule such as 'shortest' or 'fastest' in previous vehicle routing practice (Fu & Rilett 1995, Jafari et al., 2003). However, in most cases of emergency response, the decision for a sound or optimal emergency route needs much more consideration. For example, when the fire truck has a lower power-to-weight ratio (Eklund et al., 1996), it cannot follow the fastest path if, for example, the road has a steep slope. Another example is the delivery of victims. To guarantee the safety of rescued victims, consideration of potential damage along the road (such as dangerous material, unreliable infrastructure) ought to be included in the planning of the emergency evacuation routing. Besides, driving on a safe and easy-navigation route is more important for rescue workers or victims under the pressure of time and stress. As a matter of fact, it is a multi-criteria analysis process, i.e. the choice of rules and the key factors should be adaptable to the real emergency (Duckham & Kulik 2003).

As seen in Table 2, five kinds of objects need to be considered in vehicle emergency routing, namely, vehicle, people, road, event and facility. Some of them have static attribute, for example, the physical condition, while others own dynamic characteristics, for example, dynamic attributes or dynamic location. Physical or cognition conditions can be predefined and organized beforehand; dynamic conditions need updating according to a set of time interval, for instance at every five minutes.

The relationships between different objects are complicating. Attributes of vehicle such as the weight-power ratio, weight and height/width, determine which road the vehicle could use. The relationships between the vehicle and the road are simplified as one-to-one. On the other hand, event and road have many-to-many relationships, i.e. one event can relate to many road segments and one road segment can own many events. Facility and road have a many-to-one relationship. The victim and rescue worker are related to vehicle and facility and the pedestrian/civilian is assigned to event.

From the information perspective, data is the basis of spatial analysis and the source of knowledge. The choice of data is closely related to the emergency routing rules (modes) while the effectiveness and comprehensiveness of vehicle routing also depends primarily on the use of data and its integration schema with routing algorithms. Because of the diversity of data sources, data formats, referencing coordinate systems, spatial dimension and scale, the multidimensional and dynamic data integration schema challenges the transportation applications (Huang 2003, Koncz 2002). At present, more and more researchers have realized that the fundamental problem in supporting effective and comprehensive emergency routing and rapid interoperability lies in the automatic conversion of heterogeneous data with the use of formal semantics. However, the process of building such formal semantics is closely related to the classification of the domain data. According to the temporal sequence, the data have been classified into three categories: historical data, online data and predicted data. Historical data shows experience and past knowledge, usually pre-defined or time-dependent, and serves as the basis for problem solving. Online data is the most challenging but well recognized for dynamic routing. It is used for updating and calibrating the solution, and then to improve the fitness and adaptation of the results. Predicted data serves to tell people what will happen in the near future and to warn people. Predicted knowledge is an indivisible part of the knowledge system and crucial for emergency vehicle routing when potential threats exist. From the aspect of information integration, transportation information and disaster simulation information are the two main inputs for emergency routing. Traditional, transportation agencies record and report transportation information using 1D linear referencing systems (LRS) as mentioned above, for example the traffic events, road pavement and limited travel speed. By comparison, the simulation information is platform dependent and it can be 2D, 3D, or 4D (considering temporal change).

Past vehicle emergency routing approach is developed on a 2D GIS platform, and therefore the data integration and information analysis is also done in a planar environment. 2D visualization for vehicle routing navigation lacks the feeling of realism, and moving transportation network visualization from 2- to 3-dimensions has the obvious advantage of obtaining an extra degree of freedom (Chen 2002). Besides, it is possible for 3D GIS to depict the vertically disjointed structures (overpasses or underpasses) of the transportation network, terrain environment, multi-level buildings and their surrounding circumstances. With 3D GIS, it is also possible to simulate traffic flow and human behaviour in a complex urban system. These are beyond the ability of the traditional 2D GIS because 3D objects presented as 2D projections in GIS may lose some of their properties (texture, graphic, height, etc.) and their spatial relationships to other objects (Miller & Shaw 2001).

In Figure 2, a scenario of 3D GIS-based vehicle emergency routing is illustrated. In a 3D environment, the whole emergency routing comprises two parts: the vehicle routing in urban transportation network and pedestrian routing in multi-level building environment. These two parts are traditionally implemented separately in practice; however, it will make a problem that the service of rescue workers cannot directly face to the victims because there was no consideration of the entrance or exit of the multi-level infrastructure. In mega cities, the relationships between transportation systems and large-scale public buildings are especially significant in the vehicle routing practice (regarded as multi-model evacuation). This relationship can be possibly built through the logical relationship between building address point and building entrance points on the building skeleton structure. It should also notice that, with the continuous extending of urban

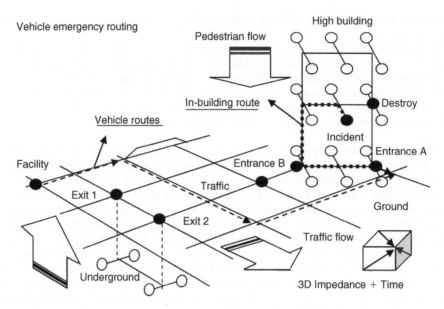

Figure 2.   Scenario of 3D GIS-based vehicle emergency routing.

space upwards and downwards, the impedance that affects the choice of emergency routes would be no longer constrained in a 2D plane. Pedestrian flows from high building and underground infrastructure, traffic jams, escape panics and some special events such as fires will synthetically affect the decision-making of vehicle routes in multidimensional and dynamic emergency situation.

A 3D GIS-based vehicle emergency routing system with due emphasis on the third dimension and the effective and comprehensive use of dynamic emergency information will improve the current vehicle emergency routing practice. Two points will be highlighted in this chapter: 3D dynamic road network and corresponding 3D dynamic routing algorithm.

4   3D DYNAMIC ROAD NETWORK

After the analysis of functional requirements, the major objects involved in the emergency routing and their interdependence are identified. Firstly, emergency vehicles are in motion and are tracked by on-board GPS (or related technologies) in real time. If the network structure is changed by any sudden events, it needs to recalculate the routes from current vehicle points (new start points) to the destination (end point). Secondly, sudden incidents may break the transportation network, and this situation should be reflected timely in an updated network, i.e. to depict the change in road connectivity in real time. Next, traffic flow and travel speed are dynamic and also change with time and even weather (sunny day, stormy day)! All these call for a dynamic road network with abilities to reflect the dynamic events and dynamic attributes. Lastly, for a quick and flexible response, the events need to be referenced directly and interactively on the road network.

As shown in Figure 3, original point, incident point, destination point and the 3D road network are linked to each other with individual time-variant attributes. The original point refers to the vehicle current location at the time when regenerating the optimal route. The current location information can be automatically determined by the use of GPS and similar technology. The destination point refers to the buildings or districts address point on the road network. The address point information can be generated by geocoding method with the use of street names and other local names, and it connects the road network data model and building evacuating data model (multi-model evacuation). The relationships between the address point and the building entrance

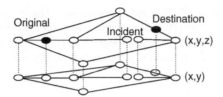

Figure 3.   3D dynamic road network.

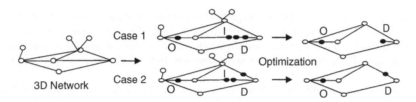

Figure 4.   Optimization of 3D dynamic road network.

points are one-to-many; they are not physically connected, but logically linked. Incident points affect the network connectivity and shows as dynamics. To enable the different dynamic incidents, different approaches can be explored, for instance, positioning by mouse cursor or map-matching by location. That is, the user is able to select a point (dynamic event) in the road structure (on-road) or a structure near the road network (off-road) interactively, or connect transportation-related data to the road network automatically to effect the change or instruction, and the algorithm will determine and display the optimal path. In the advanced 3D network analysis, incident points can be created by geo-processing methods in a 3D environment.

The proposed road network possesses 3D spatial location as well as temporal attributes, i.e., the road attributes in the road network is not static but time varying and stochastic. Therefore, the 3D dynamic road network is represented by the notation $N = \{V, E, W(t)\}$, where 'N' stands for the dynamic road network; the nodes 'V' correspond to junctions; arc 'E' connecting nodes represent road segmentation; weight 'W' shows the ability of arc connectivity, and $W(t)$ is a function of $eij \in E$ with several parameter involved, including time parameter t, connected nodes i and $j \in V$.

However, for quick response, a better way of improving routing efficiency is through optimizing the 3D dynamic road network and reducing the search space. Many heuristic strategies, such as case-based, knowledge-based methods (Ginty & Smyth 2001, Liu et al., 1994) and hierarchical representation methods (Quek & Srikanthan 2000) have been shown to have the ability to improve the efficiency of routing in urban transportation networks. However, most of the heuristic strategies deal with a static road network. In the 3D dynamic road network, original point (O) and destination point (D) add only one vertex each to the basic 3D network, and every incident point (I) adds two vertices on the network segment and breaks the segment between the two vertices. The next step, therefore, is to eliminate the dangling points (vertices) of the network until the optimization result (the degree of each vertex $d(v) \geq 2$) is achieved.

In Figure 4, there are two cases: case 1 is when incident points and destination point locate at the same road segment; case 2 is when the two locate at the different road segments. In the case 1, one more rule is added into the network optimization process, which makes the special points (such as original point O and destination point D) retain. After optimization, the urban road network is trimmed as a concise sub-graph, where impossible links and nodes are eliminated from the huge urban transportation system.

In short, the overall process of building a 3D dynamic road network can be shown briefly in three steps. Firstly, from planar or non-planar networks in a 2D space to 3D road networks, the ambiguous situation of under/overpass and network overlay in the 2D graph can be clarified. It improves the

177

abilities of visualization and the effective and comprehensive data integration. Secondly, by moving from a static network to a dynamic network, real-time information about the vehicle and other events can be directly integrated into the routing process. Lastly, after optimization, the unnecessary vertices are eliminated and the total number of vertices is greatly reduced, thus improve the ability of rapid computation.

5 3D VEHICLE EMERGENCY ROUTING ALOGRITHM

For a well-connected network, Dijkstra algorithm implementation has a lower computational complexity (Eklund et al., 1996, Liu et al., 1994, Pollitt 1995, Zhan & Noon 1998) and therefore is commonly used in many dynamic and stochastic vehicle routing cases. An important aspect of implementing the Dijkstra algorithm lies in the determination of link 'weights' based on predefined rules.

In 3D emergency routing applications, one rule is 'fastest', calculated by travel distance and travel speed. However, safety concerns and the requirements of easy direction also need to be considered as a calibration. According to real life experience, vehicles will travel slower on steeper slopes, roads with fewer lanes and a large number of traffic lights, etc. In this chapter, several steps can calculate the expected link travel time. Firstly, assign and calculate the parameters of the involved factors, such as standard travel speed, 3D road distance, slope gradient, the number of lanes of each road segment, etc. Secondly, score and weight the speed-related factors and calculate the speed weights. Thirdly, calculate the expected link travel time by the assigned formulas, as follows:

$$Speed = Standard\ speed * Weights \qquad (1)$$

$$Time = \frac{Distance}{Speed} \qquad (2)$$

$$Weights = \sum_{i=1}^{n} score_i * weight_i \qquad (3)$$

Distance: the 3D distance of road segment in real world
Speed: the vehicle speed under different travel conditions
Time: the time taken to travel between two points
Weights: the summation of the product of score and weight of the contributing criteria or factors.

Link travel distance is three dimensional and shaped by the urban landscape and the structure of transportation system. Link travel speed is changeable and is difficult to denote in a single value due to complex transportation phenomena and traffic rules. In such a situation, knowledge-based methods provide an effective way of handling this kind of decision-making problem (Liu 1996; 1997). The methodology that seemed to be well suited to support such a decision making process in selecting the 'best' alternative under the presence of multiple-choice criteria and diverse criterion priorities was the Multi-Criteria Evaluation (MCE) approach. Important features of MCE are its relative simplicity (as compared with the other decision support methodologies) and its effective abilities in handling both qualitative and quantitative data (Pettit & Pullar 1999) and therefore it is commonly used in different decision domains such as location-based services (Raubal & Rinner 2004). In this chapter, an example of using the MCE matrix to score and weight the three major physical factors, namely, slope gradient, lane number and traffic light number, is given (Table 3). For different cases, and even at different times, the choice of factors, score and weight can be different. For example, for a heavy fire truck, the factors of road grade and road bearing capability will be meaningful, but for a light ambulance, this kind of physical factors can be ignored because they affect the expected travel time slightly. Another example is the consideration of rush hours.

Table 3.　Example MCE matrix to determine speed weights.

Factors		Weights		
		Grade	Score	Weight
Physical	Slope gradient	−5% ∼ 5%	1	0.2
		>5% or < −5%	0.6	
	Lane number	1 ∼ 2	1	0.3
		>2	1.5	
	Traffic light number	1	1	0.5
		>1	0.5	
Dynamic				
Historical	Traffic volume	Attribute of the road network		
Online	Incident	Interactive input		
Predictive	Disaster simulation	Linear referencing to 3D network Geo-processing		

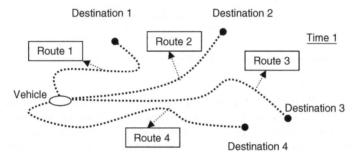

Figure 5.　Route sequences in Time 1.

In different places, the time span and distribution of rush hours is quite different because of different life style.

In Table 3, time-varying and dynamic factors comprise three parts. Traffic volume belongs to historical data and is recorded into road segment as attributes. Historical data is usually predefined and can be associated with road segments in the stage of data preparation. It determines the usual traffic condition and sets as the basis of routing. Online incident can be input interactively or using linear referencing method while predictive simulation information is incorporated into routing by geo-processing method. Both online and predictive information serve as a calibration for the routes and affect the network connectivity and impedance of each road segment.

Additionally, emergency vehicle routing shows dynamic characteristics not only in the vehicle position (original) and network status, but also in the choice of facilities (destination), such as a hospital. When considering the situation of transferring the victims to the best suitable public service centers, it is unrealistic to assume that every hospital in the emergency has the same service capacity at all times. Thus, the emergency routing algorithm should include such a requirement and the optimal route for each emergency vehicle is the result of many comparisons. As shown in Figure 5, there are four choices of destination at Time 1 moment; and route sequencing is required before the final route is set. The principle of such a choice is based on the expected travel time from one point to multiple points.

Next, several simulated cases show the results of the proposed algorithm and its comparison with the conventional algorithm. The proposed algorithm is implemented in the VGEGIS™. VGEGIS™

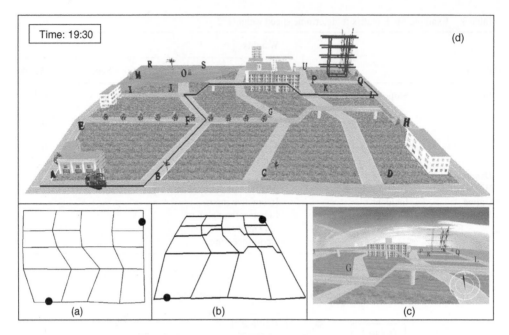

Figure 6.   Implementation results of 3D GIS-based emergency routing/navigation.

software integrates 3D GIS and VR technologies; it supports 3D modeling, 3D analysis, 3D simulation, 3D interactive visualization and navigation (Zhu & Lin 2004). The conventional algorithm is based on planar or non-planar network model in 2D space, supported by many commercial GIS software, such as ESRI's ArcGIS 9.1, Caliper's TransCAD, Geomedia's Transportation Analyst, Autodesk's Map 3D, etc.

The proposed algorithm can be operated in both 2D view (Fig. 6a) and 3D view (Fig. 6b) in the VGEGIS™ software. By interactively selecting and automatically loading dynamic events, the route will be able to avoid risky road segments (such as flooded areas and congested roads as shown in Fig. 6d) to guarantee safety. Besides, at certain moments such as 19:30 h, the route is adaptive to the current traffic conditions (traffic flow and travel speed) and possible service objects (such as fire sites) at that time. Buildings with texture, landscape, traffic signals, etc., can be freely added into the routing system in order to improve the quality of routing/navigation visualization (Fig. 6c).

The conventional 2D GIS-based routing algorithm usually implements in a 2D window (Figs 7a, b) and visualizes the results in a 3D window (Fig. 7c). In most cases, road networks are designed to represent the macroscopic transportation phenomena and therefore, the bridges and flyover are depicted as a single node. As such, the routes generated by planar network and non-planar network are quite different (Figs 7a, b). One the other hand, because the 2D road networks lack the z coordinate, the 3D route can only be achieved with the support of extra elevation information, such as DEM data, but this solution will be unacceptable when bridges are projected on the ground (Fig. 7c). It is possible for a 2D GIS to implement routing with the use of 3D geometry network (Fig. 7d); however, the spatial relationship between the resulting route and the 3D geometry network is hard to discern in a 2D window. When the resulting route is transferred into a 3D window, different cases maybe occur, for example, the route is projected into a plane (Fig. 7e), or the route is projected on the 3D geometry network but it is problematic (Fig. 7f). These phenomena may happen because of the planar embedding requirements, and the upper topological nodes may overwrite the lower ones (with same x, y coordinates) in a 2D GIS.

Besides the enhanced modeling and the vivid visualization capability, 3D GIS-based routing algorithm improves the reliability of the path finding by considering more intrinsic 3D information

180

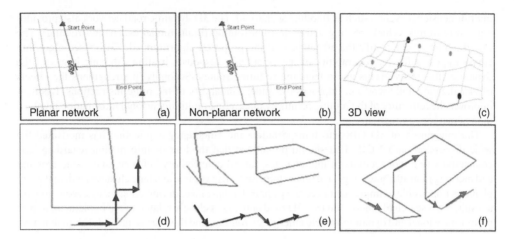

Figure 7.   Implementation results of conventional 2D GIS-based routing/navigation.

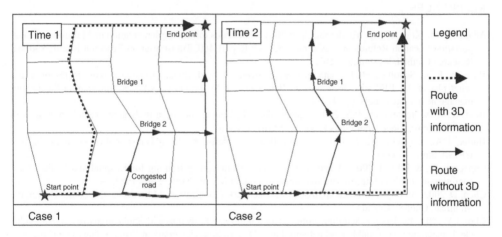

Figure 8.   Comparison of the implementation results between traditional 2D and 3D GIS-based methods.

involved in the emergency situation. In Figure 8, two comparison cases reveal the advantages of 3D GIS-based algorithm over conventional algorithm from the perspective of information integration. In the case 1, the route generated by the conventional algorithm (without considering 3D information) is turned out to be 'longer' (needs more time) because of the extra cost spent in real 3D space on steep Bridge 2. In the case 2, the route without considering 3D information is turned out to be unreliable because inadequate clearance of the Bridge 1 is not fully considered for a fire truck to pass. In addition to the static 3D information, dynamic 3D simulation information (such as flooding) challenges the traditional 2D-GIS based routing algorithm, but this situation has been well resolved in a 3D GIS (Fig. 6d).

6   CONCLUDING REMARKS

In this chapter, the importance of 3D GIS-based emergency routing algorithm has been highlighted. It states that the 3D dynamic road networks have the potential to depict the real traffic systems (over/underpasses, network overlay, etc.) in more detail and their dynamic relationships with the multidimensional events (such as flooding scenarios). The 3D dynamic road network also provides a gateway to link vehicle routing and pedestrian routing (multi-model evacuation routing). The experimental results testify the advantages of the proposed algorithm: 3D GIS-based algorithm firstly improves the interoperability of the emergency response; it allows flexible inputs of the dynamic emergency events and affects the resulting routes. Secondly, it strengthens the effects of visualization and therefore, improves the understandability of the route instructions. Lastly, it improves the reliability of the routing results by integrating multidimensional information involved in the emergency situation.

The integration of 3D GIS into transportation and emergency application has motivated the wide discussion on 3D GIS-T. It should be noticed that the loss of information regarding lane connectivity (e.g. connectivity at interchanges, lateral connectivity between lanes, lane crossing and merging) and attributes (e.g. lane types, forecast traffic volume and lane closures) will make city planners impossible to implement necessary control at lane level during emergency evacuation. As a result, the research of lane-based true 3D road network is closely related to the implementation of multidimensional and dynamic vehicle emergency routing, and this part is under development now.

REFERENCES

Adams, T. M., Koncz, N., & Vonderohe, A. P. 2001. Guidelines for the Implementation of Multimodal Transportation Location Referencing Systems. NCHRP Report 460, Transportation Research Board, National Research Council, Washington, DC.
Borkulo, E. v., Scholten, H. J., Zlatanova, S., & Brink, A. v. d. 2005. Decision making in response and relief phases. In: PJM van Oosterom, S. Zlatanova & E.M. Fendel (Eds.), Geo-information for disaster management – late papers, pp. 47–54.
Bent, R. W. & Hentenryck, P. V. 2004. Scenario-based planning for partially dynamic vehicle routing with stochastic customers. Operations Research 52 (6): 977–987.
Butler, J. A., & Dueker, K. J. 2001. Implementing the Enterprise GIS in Transportation Database Design. URISA Journal 13 (1).
Capaccio, R., & Ellis, J. 2005. A Proposal to Enhance Emergency Routing for Hospital and Critical Care Facilities, available at http://www.directionsmag.com/article.php?article_id=1983.
Chen, D. 2002. An object oriented approach to 3D network visualization. M.Sc thesis. Concordia University. Montreal, Quebec, Canada.
Choi, J. Y. 2003. Stochastic scheduling problems for minimizing Tardy jobs with application to emergency vehicle dispatching on unreliable road network Ph.D. Dissertation. University at Buffalo, The State University of New York.
Coors, V., Elting, C., Kray, C., & Laakso, K. 2005. Presenting Route Instructions on Mobile Devices – From Textual Directions to 3D Visualization In: Dykes, J., A. MacEachren and M.-J. Kraak (Eds.) Exploring Geovisualization, Amsterdam: Elsevier, 2005, pp. 529–550.
Curtin, K., Noronha, V., Goodchild, M., & Grise, S. 2001. ArcGIS Transportation Data Model. Environment systems research institute, Redlands, CA.
Darken, R. P., & Peterson, B. 2001. Spatial Orientation, Wayfinding, and Representation: Handbook of Virtual Environment Technology. Stanney, K. Ed.
Derekenaris, G., Garofalakis, J., Makris, C., Prentzas, J., Sioutas, S., & Tsakalidis, A. 2001. Integrating GIS, GPS and GSM technologies for the effective management of ambulances. Computers, Environment and Urban Systems 25: 267–278.
Duckham, M., & Kulik, L. 2003. 'simplest' paths: automated route selection for navigation. COSIT'03, Lecture Notes in Computer Science. Springer-Verlag.
Eklund, P. W., Kirkby, S., & Pollitt, S. 1996. A dynamic multi-source Dijkstra's Algorithms for vehicle routing. Australian and New Zealand Conference on Intelligent Information Systems, IEEE Press, 329–333.
Fan, W. C., & Su, G. F. 2005. Progress in emergency management research for a mega-city in China. Proceedings of the fourth international symposium on new technologies for urban safety of mega cities in Asia (USMCA 2005).
Fischer, M. M. 2004. GIS and Network Analysis. In Hensher, D., Button, K., Haynes, K. & Stopher, P. (Eds.): Handbook of Transport Geography and Spatial Systems, pp. 391–408. Elsevier, Amsterdam, New York, Oxford.

Fu, L., & Rilett, L. R. 1995. Estimation of Expected Minimum Paths in Dynamic and Stochastic Traffic Networks. Transportation Research – Part B 32: 499–514.

Gendreau, M., Laporte, G., & Seguin, R. 1996. Stochastic Vehicle Routing. European Journal of Operational Research 88: 3–12.

Ginty, L. M., & Smyth, B. 2001. Collaborative Case-based Reasoning application in personalised route planning. the Fourth International Conference on Case-Based Reasoning (ICCBR): 362–376.

Gong, Q., & Batta, R. 2005. Allocation of Ambulances to Casualty Clusters in a Disaster Relief Operation, available at http://www.acsu.buffalo.edu/~batta/papers/Qiang-IIE%20submission.pdf.

Gong, Q., Jotshi, A., & Batta, R. 2004. Dispatching/Routing of Emergency Vehicles in a Disaster Environment using Data Fusion Concepts. Paper presented at the Proceedings of the 7th International Conference on Information Fusion, Stockholm, Sweden.

Guo, B. 2001. A feature-based linear data model supported by temporal dynamic segmentation. Ph.D., University of Kansa.

Hampe, M., Elias, B., & Hanover. 2004. Integrating topographic information and landmarks for mobile navigation. Paper presented at the Symposium 2004, Geowissenschaftliche Mitteilungen, Wien, Austria.

Hardy, P. 2003. Unifying LRS and Routing Networks, available at http://www.gis-t.org/yr2003/gist2003sessions/Session613.ppt.

Huang, Z. 2003. Data integration for urban transportation planning. ITC Dissertation.

Jafari, M., Bakhadyrov, I., & Maher, A. 2003. Technological Advances in Evacuation Planning and Emergency Management: Current State of the Art, available at http://www.cait.rutgers.edu/finalreports/EVAC-RU4474.pdf.

Jain, S., & McLean, C. 2003. A framework for modeling and simulation for emergency response. Paper presented at the Proceedings of the 2003 Winter Simulation Conference.

Koncz, N. A. 2002. Integrating time, space, movement and geographic information systems: Development of a multi-dimensional location referencing system data model for transportation systems. Ph.D. dissertation. Wisconsin: University of Wisconsin Madison.

Kwan, M.-P., & Lee, J. 2005. Emergency response after 9/11: the potential of real-time 3D GIS for quick emergency response in micro-spatial environments. Computers, Environment and Urban Systems 29: 93–113.

Lee, J. 2005. 3D GIS in Support of Disaster Management in Urban Areas. Directions Magazine, available at http://www.directionsmag.com/article.php?article_id=2049.

Liang, K., Li, Z., Zhang, Y., & Song, J. 2002. Towards GIS-T Information Fusion. IEEE 5th International Conference on Intelligent Transportation Systems, Singapore.

Liu, B. 1996. Intelligent Route Finding: Combining Knowledge, Cases and an Efficient Search Algorithm. Paper presented at the Proc. 12th Int. Conf. Artificial Intelligence, Cases and an Efficient Search Algorithm.

Liu, B. 1997. Route Finding by Using Knowledge about the Road Network. IEEE Transactions on Systems, Man, and Cybernetics Part A: Systems and Humans, 27 (4).

Liu, B., Choo, S.-H., Lok, S.-L., Long, S.-M., Lee, S.-C., Poon, F.-P., et al., 1994. Integrating Case-Based Reasoning, Knowledge-Based Approach and Dijkstra Algorithm for Route Finding. Paper presented at the Proceedings of the Conference on Artificial Intelligence Applications: 149–155.

Lu, X. 2001. Dynamic and stochastic routing optimization: algorithm development and analysis. Ph.D., Civil Engineering, 2001.

Malaikrisanachalee, S. & Adams, T.M. 2005, Lane-based Network for Transportation Network Flow Analysis and Inventory Management. In TRB 2005 Annual Meeting CD-ROM, available at http://www.topslab.wisc.edu/publications/adams_2005_0800.pdf.

Miller, H. J., & Shaw, S.-L. 2001. GIS-T Data Models. In Geographic Information Systems for Transportation: Principles and Applications (480 p.): Oxford University Press.

Mollaghasemi, M., & Abdel-Aty, M. 2003. Post-Disaster Dynamic Routing of Emergency Vehicles, available at http://catss.engr.ucf.edu/projects/documents/abstracts/pdf/project_41.pdf.

Pettit, C., & Pullar, D. 1999. An integrated planning tool based upon multiple criteria evaluation of spatial information. Computers, Environment and Urban Systems, 23, 339–357.

Pollitt, S. 1995. A 3-D spatial information system for emergency routing in Okayama city. BSc Thesis. Department of Computer Science, University of Adelaide, Australia.

Quek, K. H., & Srikanthan, T. 2000. A hierarchical representation of roadway networks. 7th World Congress on Intelligent Transportation Systems.

Raubal, M., & Rinner, C. 2004. Multi-Criteria decision analysis for location based services. Proc. 12th Int. Conf. on Geoinformatics-Geospatial Information Research: Bridging the Pacific and Atlantic, University of Gävle, Sweden, 7–9 June 2004.

Scarponcini, P. 1999. Generalized LRS approach for interoperability of As-Is Legacy data. Proceedings, Geographic Information Systems for Transportation Symposium, San Diego, CA, 1999.

Slavick, D., Christopher, & Moore. 2005. 3D Visualization Bolsters Homeland Security, available at http://www.geointelmag.com/geointelligence/article/articleDetail.jsp?id=162588.

Zhan, F. B., & Noon, C. E. 1998. Shortest Path Algorithms: An Evaluation using Real Road Networks. Transportation Science 32 (1): 65–73.

Zhu, Q., & Lin, H. 2004. CyberCity GIS: 3D city models for virtual city environment Wuhan University Press (in Chinese).

Zlatanova, S., Holweg, D., & Coors, V. 2004. Geometrical and topological models for real-time GIS. Paper presented at the Proceedings of UDMS 2004, 27–29 October, Chioggia, Italy, CDROM, 10 p.

Part 4
Positioning, virtual reality and simulation

Geospatial Information Technology for Emergency Response – Zlatanova & Li (eds)
© 2008 Taylor & Francis Group, London, ISBN 978-0-415-42247-5

3D positioning systems for emergency response

K. Kolodziej
IndoorLBS.com, USA

ABSTRACT: The chapter explores the different types of indoor, urban, and seamless indoor-outdoor location-aware applications, their requirements in terms of the infrastructure needed to support them, and the current limitations. The chapter gives detailed coverage on the most promising technologies and indoor positioning. The TV-GPS positioning technology that is featured in the chapter has the promise for enabling seamless indoor-outdoor positioning. The chapter also describes the design and implementation of several positioning systems and real-world applications and show how these tools are being used to solve problems that can be related to the reader's own applications.

1 LOCATION DETERMINATION: INDOOR AND LOCAL POSITIONING

Today, there is a vast array of so-called location technologies that are involved in the calculation of a user's or object's position in a space or grid, based on some mathematical model. Positioning here means allowing a mobile device to be aware of it's location with different degrees of precision and accuracy. The technology required for provision of automated location information to mobile devices has been in continual development for several decades. While the majority has its roots in military (e.g., GPS), modern consumer technology is also raising to meet the challenges, specifically in metropolitan areas. Telecommunication initiatives like the US FCC's E911 and Europe's E112 have generated a lot of interest in applications and services that are a function of a user's or an object's location, referred to as location-based services (LBS).

Unfortunately, millions of square meters of indoor space and urban areas are out of reach of GPS systems. Conventional GPS receivers do not work inside buildings due to absence of line of sight to satellites, while cellular positioning methods generally fail to provide a satisfactory degree of accuracy. The delivered position fixes cannot even be used for determining whether a target person stays inside or outside a certain building, not to mention that it is by no means possible to locate it with the granularity of rooms or floors.

Fortunately, over the past decade, advances in location positioning technology have made it possible to locate users and objects indoors. These alternative technologies are now being introduced to the market enabling indoor (in-building) positioning. Different technologies will demand different capabilities from the device, while they'll bring various constraints. Outside the remit of 2G, 2.5G, 3G, and 4G cellular networks, exist other families of positioning technologies that are often referred to as 'local positioning,' which make use of short range networks such as 802.11 (Wi-Fi), Bluetooth, radio-frequency identification (RFID, ultrasound, Ultra Wide Band (UWB), infrared data association (IrDA) wireless specifications, or television (TV) radio signals.

There are different types of indoor, urban, and seamless indoor-outdoor location-aware applications, their requirements in terms of the infrastructure needed to support them, and the current limitations. The chapter gives a detailed coverage on the most promising technologies, which are WLAN fingerprinting, RFID positioning, and indoor positioning with non radiolocation positioning with infrared and ultrasound. Also, the chapter addresses the problem of absence of a common

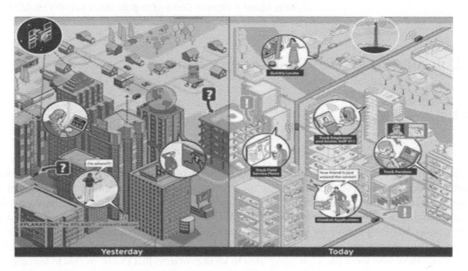

Figure 1.   Today's possibilities with positioning technologies (Source: Rosum).

integrated approach for universal positioning technology. This results partly in a demand for stand-alone indoor solutions. The TV-GPS positioning technology that is featured in the chapter has the promise for enabling seamless indoor-outdoor positioning. The chapters describe the design and implementation of several positioning systems and real-world applications and show how these tools are being used to solve problems that they can relate to their own applications. There are books specifically about this (e.g., Kolodziej and Hjelm, 2006).

1.1   *The need and utility of advanced 3D locator systems*

There is a need to be able to accurately locate and track incident responders in situations such as: inside of threatened buildings, collapsed buildings, and subterranean facilities or underground. Accurate location and tracking is necessary in order to allow emergency managers, including fire chiefs and other incident commanders, to rapidly and effectively deploy and re-deploy their forces or understand and respond to the consequences of potential threats to their forces. The systems have to be fast, have to be able to find information with respect to the context and have to be able to integrate different data (existing or coming from the field) for further analysis and decision-making.

Consider the Advanced 3-D Locator System (HSARPA, 2005) under development by the Department of Homeland Security. The system needs to provide timely operational support for all-discipline, all-hazards scenarios in a broad range of environmental conditions and terrain.

Users of this systems are the following:

- Department of Homeland Security (DHS) Emergency Preparedness and Response
- Federal, State, Local and Tribal incident responders and managers
- DHS/Emergency Preparedness/Federal Emergency Management Agency (FEMA)
- All lead and supporting Federal agencies of the National Response Plan
- Law Enforcement agencies
- Fire Departments.

There is a huge utility to DHS of such a 3D locator system. There are over 2 million emergency responders (ERs) in the US with the mission to save lives, while staying alive themselves. Emerging technologies are critical for meeting the U.S. Fire Administration's goal: reduce firefighter

Figure 2. Need for a 3D positioning inside buildings (Source: Rosum).

fatalities by 25% in the next five years. Existing technology allows first responders to monitor their own safety, e.g., Personal Alert Safety System (PASS) and Heads-Up-Display (HUD) units, and to send and receive messages from incident command (IC), however reoccurring failures and user error have led to fatalities. A lack of status, condition, location, task, resource, threat, and exit accessibility information, or inability to convey this information among emergency responders (ERs), Strike Teams, and Incident Command (IC) can result in civilian and responder casualties.

The following are some of the required performance measures for this locator system:

– Locator must send information including location-related information.
– Locator must wirelessly transmit inside or outside of structures and through rubble to an off-site incident command post, on-site incident command posts, emergency responders, and/or other authorized parties including within teams of responders.
– Locator must be self-initializing, self-calibrating, self-adjusting and must have self-diagnostic capabilities to ensure speed and reliability.
– Locator must operate outside all buildings and inside of almost all buildings, no matter their structural state and environmental conditions.
– Primary incident command posts should be able to monitor the status of the locator and its host from a radial distance from 30 meters to 100 meters (per relay).
– Locator must be able to specify the location of its host in three dimensions within 6 meters (3 meters desired).
– The base station software must be able to display location and identification of personnel.
– The base station must be able to display general-to-specific information (the ability to drill down from an overall scene to a specific individual) about an operation/incident and its emergency responder participants.
– The base station must include visualization tools that:
  • Allow incident commanders and site personnel to easily interpret incoming displayed information.
    ○ Display the location of an emergency responder in easy to understand coordinates. (One form of display must be a wire-frame like view of the building structure with the position of each responder indicated. The wire-frame view must include a scale showing grid spaces of approximately 10 feet in every direction.)
    ○ Allow the user to identify, group, and categorize responders as desired.
– Optionally, the coordinates returned by the locator can be input to a Geographic Information System (GIS) system (including a building map or equivalent for underground structures).

189

Unfortunately, an integrated system that allows bidirectional mission-critical information transmission including *3D location* of ERs, does not yet exist. Moreover, no *single* product accurately tracks ERs outside of, into, and within:

- intact or partially-collapsed single or multi-story buildings,
- intact or partially-collapsed underground environments (e.g., facility sub-levels, subways, utility tunnel systems, mines, caves, etc.), and
- extensive outdoor, industrial, or urban incident scenes.

A functional, user-friendly, and affordable system providing only 3D location information will save many lives. However, a system that integrates 3D with existing and key emerging technologies could be the greatest life saving tool since the self-contained breathing apparatus was introduced 40 years ago. Such a system should incorporate industry specifications, lessons learned, and communication standards within a framework that seamlessly interoperates between multiple agencies, levels of government, and overlapping first responder jurisdictions; it should be scalable and extensible to allow new data feeds, revised standards, and modular to integrate updated components and technologies.

Building a positioning system that works well indoors is a challenge, because signals reflected off walls, floors and ceilings tend to confuse sensors, and often there are obstructions between sensors and objects being tracked. GPS and cellular-network-based positioning aren't appropriate for indoor use due to loss of line-of-sight as well as signal blockage, fading and shadowing.

Nevertheless, the indoor world can provide a more controlled environment, and several positioning methods can be used alone or combined:

- Triangulation
  - Lateration: Time of flight, attenuation (e.g., MIT Cricket)
  - Angulation: Arrival angle of a signal against a baseline can be measured using signal strength or time difference of arrival (e.g., UbiScene's UbiTags)
- Scene analysis
- Use of a feature as a reference point (e.g., Microsoft RADAR)
- Proximity
  - Physical contact through pressure sensors (e.g., SmartFloor)
  - Monitoring (e.g., Active Badge)
  - Observing (e.g., automatic ID systems).

2   MICRO-GEOGRAPHY SCALE

Of primary interest is that one point in space in the "Outdoor World" (e.g., one address or set of coordinates) potentially represents entire sets of points in the "Indoor World." For example, a "rookie" first responder arriving at an emergency site (a specified address or position) in an unfamiliar part of the city might be surprised on arrival to find quite a large building or even a campus of interconnected buildings sharing the same address; he may have been prepared to search a single, small building or home. Realistically, a single building may be expanded into a collection of floors, and then onward to halls, rooms, etc (logical divisions of indoor space).

Within a typical urban area, E911-allowed margins-of-error of up to 100 meters usually include a range of possible addresses (e.g., "The 400 block of Main Street," which is a collection of indoor spaces). To achieve maximum precision for emergency positioning service in urban areas, the location of a phone may be interpolated as set of geographic coordinates (as in GPS-assisted determination of Latitude/Longitude). In rural settings, often lacking well-named and sufficiently dense street-address networks, representing the position of an emergency caller using coordinates is a highly useful level of accuracy and detail.

A major reason that geographic coordinates alone are not optimal in denser urban settings is that emergency response teams usually travel over the road network to reach an emergency (i.e., they do not travel in a straight line as the bird, or helicopter, flies). Typical street-address structures of a city can be navigated intuitively, and it is a common practice to "geocode" or cross-reference geographic coordinates within a reference model of the land-based road network.

There are common data standards for performing geocoding or address matching operations in the outdoor world. Network (vector) data models such as TIGER have achieved high levels of efficiency, descriptiveness, and functionality, such as containing street-line level attributes for much of the US Census Bureau (Drummond, 1995).

One premise in this chapter is that the spatial efficiency demonstrated by outdoor-oriented geocoding models extends to the indoor world, where they facilitate location-based service authentication and delivery (Beal, 2003;Vittorini & Robins, 2003). Specifically, it is proposed that the process of using any type of position or translating absolute position (represented by coordinates) into relative or "contextually symbolic" ones (such as, "in Room 101") is a critical enabling step for delivering "seamless" indoor LBS, which by definition makes information available in all forms, whether being specified by an address, xy coordinate, or "in Room 101" (Kolodziej, 2003). The terms "micro-geography" and "micro-LBS" can be used to describe the spatial nature and geolocation methodologies of indoor environments for a variety of applications.

Within the indoor world, different positioning types can co-exist not only on different floors, but also on the same floor. Consider also the fact that, indoors, even a 3 meter potential for error could mean a difference of two or three floors in vertical space and a number of rooms in any horizontal direction. Plus, public and private facilities can be intermingled in wireless space, which doesn't obey the barriers of walls and floors within buildings (see Hassan-Ali & Pahlavan, 2002; Kuikka, 1999). Part of the indoor LBS challenge is to design solutions for a world of heterogeneous, overlapping and interrelated spatial zones, reflecting the various ways that people and property tend to move and congregate indoors.

Moreover, new indoor-specific spatial data models are being proposed (Kolodziej, 2003), and these alternative technologies enable many kinds of indoor LBS applications that require infrastructure and investment. Indoor location-aware applications require micro-detailed geo-referencing to satisfy users' growing needs. It's not enough to geo-reference a building if the position of users and other objects inside the building also is relevant. Objects are used as landmarks, and relationships among the objects are crucial for symbolic representation of the whole system.

3 COMMUNICATION INFRASTRUCTURE

3.1 *The network component*

Deploying a positioning system that works well indoors is a challenge, because signals are reflected off walls, floors, and ceilings, which tend to confuse sensors. In addition, the coverage has to be complete in the areas where positioning is needed. Where existing networks can be leveraged, this will speed up deployment and simplify the creation of services. However, it is possible that the coverage of these networks needs to be complemented with additional base stations to give full coverage with full positioning detail in all the areas where positioning is required.

The proliferation of lightweight, portable computing devices and high-speed wireless local-area networks has enabled users to remain connected while moving about inside buildings. Because indoor settings have the disadvantage of absorbing and diffusing the radio frequencies of GPS and cellular network systems ((i.e., Global Systems for Mobiles (GSM), Code Division Multiple Access (CDMA)/CDMA2000, and Universal Mobile Telecommunications System (UMTS)), their positioning mechanisms (i.e., Time-of-Flight (TOF), Angle-of-Arrival (AOA), etc.) are not appropriate to provide the location of a user inside buildings, at least not in cases where indoor base stations are not deployed. Nor is the resolution typically good enough from these systems. Usually, it is in the range of 10 m (25to 30 ft).

Accuracy requirements for location-based applications are much higher. To pinpoint a shelf inside where a product is located, accuracy to within 30 cm (a foot or so) is required, specifically as (x, y) coordinates (absolute positioning). To achieve this level of accuracy, either a dedicated positioning system, such as the MIT Cricket, or a location function in a local-area network can be deployed to achieve this accuracy level.

The role of the network can be to provide the sensor system, such as where base stations are used to triangulate the terminal. But the network is also the communications interface between terminals and servers. In many environments of interest, like shopping malls, schools, convention centers, hospitals, etc., the communication infrastructure already exists to provide data-networking capability to mobile terminals.

Many such locations are equipped with 802.11 standards-based wireless LANs. However, it is not necessarily sufficient just to have a hot spot in an area to deploy location-based services. The system requires a number of additions. These include, for instance, functions to compute positions based on the information from the network (either cell information, radio fingerprint, or triangulation in the same way as with wide-area networks). It is also likely that coverage has to be increased to cover all places where location-based services will be required. After the deployment of the other infrastructure types that are essential for indoor location-based services, such services will complement this data-networking capability of, for instance, wireless LANs (802.11 family of standards). This, in turn, will add value to such a network. Due to its wide deployment, WLAN is a strong candidate to handle the positioning centrally in an indoor system.

There are two main scenarios for indoor location-based services: use only within the network, or use with the wide-area network as well. In terms of seamless communication infrastructure, roaming between access points is already supported, and Wi-Fi networks, for example, can be extended to create "clouds of connectivity" inside the so-called hot spot (i.e., locations with high connection frequency, such as an office building). How to connect the hot spots to the wide-area network is not yet clear, however. Standardization is ongoing in 3GPP, the standards body that works on these issues, and it is likely that this will be solved during 2005, with products emerging in 2006.

There are two main philosophies of organizing the sensor network communication: either over a separate network or over the existing air and backbone network. For practical and economic reasons, it is likely that the second case will be the primarily deployed case. And for most purposes, this implies an 802.11-type network. This has implications on deployment, which we will go into in a later section. But it means that as a developer, you must be able to at least roughly model the traffic, to make sure that the network does not get congested just at the moment the information is needed.

Finally, there are two main ways to build a location system: passive or active. They can be combined, e.g., as in assisted GPS, but here we will look at the differences between them.

In passive systems, the location data are publicly offered and the terminal takes care of the processing to interpolate location. GPS is an example of a passive system. Satellites provide a timing signal and a known position; the terminal has to calculate its own position. Architectural challenges associated with passive systems include optimum formats for data, synchronization and broadcast intervals, and logical signal differentiation. From the perspective of limited devices there is also the problem of antenna effect and signal frequency, which will affect battery life. A passive positioning system provides an open infrastructure, anonymously broadcasting location information. Any privacy issues will be dependent on the terminal. A passive system, consisting, e.g., of RFID tags, does not communicate actively with the access points. Instead, it merely transmits back information when requested (through a radio trigger).

The active system uses terminals that actively communicate with the access points. This requires power, which has to be available for communication at the intervals prescribed by the system. This communication may need to occur anyway (as in the case of mobile telephone systems).

Location system implementations generally use one or more techniques to locate objects, people, or both. Most of the methods used to define a position are based on geometry computations such as triangulation (by measuring the bearings of an object from fixed points) and trilateration (by measuring the distance). In addition, scene analysis, proximity (detecting physical proximity), monitoring wireless cellular access points, and observing automatic ID systems can also be applied.

3.2 The sensor component

The sensors are the focus for the next two sections, but in general, the sensors may be a feature of the network (such as in mobile location determination systems based on GSM, UMTS,etc.), or it may use separate sensors (beacons). There may be hybrids, which use, e.g., an RFID tag to determine the area of the absolute position, and then a relative position to the base stations.

It is important to realize that different positioning systems express location in different ways – different measurements (geometric vs. symbolic), different spatial frame of reference, or different uncertainty. As a result, it is important to first determine which type of positioning (absolute vs. relative) the location-based service will require.

It is a challenging problem (to say the least) to build an ideal indoor positioning system that provides accurate and precise location at a high update rate. Different underlying sensors will give different results of accuracy and precision. Accuracy requirements for indoor location-based applications will vary significantly. For example, RFID tags are considered to be proximity sensors. These can detect a tag when it passes within a relatively short distance of a sensor (usually a few centimeters (1 foot or less). Other technologies can sense when a tag is within a room (relative positioning), but cannot identify its absolute location within the room.

Much depends on the requirements of the application. What is actually the need of the system when it comes to information accuracy? Is it sufficient to know you are in the right building, or do you have to know which room you are in? Which shelf in that room? This will determine how the sensor system should be organized to serve the application in the best way.

Most available local positioning systems are network based rather than terminal based (sensors are integrated into the mobile device), meaning that the network calculates the position of the mobile device, as opposed to a receiver inside the device calculating its own position and using the network to notify others of its location. Network-based positioning is especially common for local positioning systems that are typically based on small radio frequency or infrared cells, since the processing capacity in the terminals is too small for any calculations within any reasonable time. In fact, all of the local positioning systems that were found as part of the detailed survey done for this section were networked based.

4 THE POSITIONING INFRASTRUCTURE

There are, in the main, three ways to create a positioning system for indoor location: deploying a set of sensors or beacons independent of other networks, using existing infrastructure such as Wi-Fi networks, or using the existing wide-area networks, with their positioning capabilities.

Each of these approaches has its advantages and problems. In practice, since they will tend to overlap, there will be a need for the software system to manage handover of the location information.

4.1 Positioning systems and algorithms

The core of any positioning system is the means for measuring the whereabouts of a terminal, and the algorithm to compute those whereabouts in relation to some known location (whether absolute or relative). This can be done in the terminal or in the network. The nodes can be active, transmitting a signal themselves, or passive, just receiving the signal (leaving it to some other entity to compute the position). These two dimensions are complementary, not contradictory.

In addition, active systems transmit a signal themselves; passive systems just receive a signal. Most global positioning systems (GPS) are passive, simply receiving the signal (since it would not make much sense to communicate with the satellites anyway); most mobile phone systems (when used for positioning) are active, since the phone is required to send out a "keep alive" signal to the base stations now and then, to make sure the system does not remove it from the list of connected terminals. This also implies that in entirely passive systems, there is no way for the sender to know who uses the positioning signal.

In either case, you can further make the choice of basing the system on the existing communications network or a dedicated network that is only used to transmit positioning signals. If you base it on an existing network, there is the risk that the positioning signal crowds out other (useful) signals; if you base it on a dedicated network, you have to provide additional receivers in the terminals, which may bring an extra cost.

Moreover, active systems incorporate signal emitters, sensors, and landmarks placed in prepared and calibrated environments. Passive systems are completely self-contained, registering naturally occurring signals or physical phenomena. Examples include compasses sensing the Earth's magnetic field, inertial sensors measuring linear acceleration and angular motion, and vision systems sensing natural scene features. Most of the outdoor tracking is based on a sort of passive-target systems utilizing vision (Azuma, 1997). Vision methods can estimate camera position directly from the same imagery observed by the user. But vision systems suffer from a lack of robustness and high computational expense. Unfortunately, all tracking sensors used in passive systems have limitations. For example, poor lighting disturbs vision systems, close distance to ferrous material distorts magnetic measurements, and inertial sensors have noise and calibration error, resulting in a position and orientation drift. Hybrid systems attempt to compensate for the shortcomings of a single technology by using multiple sensor types. Among all other approaches, the most common is passive magnetic combined with a vision system because inertial gyroscope data can increase the computing efficiency of a vision system by providing a relative frame-to-frame estimate of camera orientation, and a vision system can correct for the accumulated drift of an inertial system.

4.2   *Classifying positioning systems*

There are a multitude of methods for determining the traveler's current location. These vary in the extent to which they require sensing of the environment or reception of signals provided by external positioning systems. At one extreme, there is inertial navigation, which requires no external sensing. At the other extreme are methods involving the matching of perspective video images of the environment to three dimensional models stored in computer memory. In between are methods employing dead reckoning and a variety of local and positioning systems in which the navigator determines current position using signals from transmitters at known locations.

Because of the high degree of accuracy of its position fixes, some of the local positioning systems (e.g., Ekahau) are the preferred choice for environments where obstructions and multipath distortion are accounted for. For environments in which indoor signals are only intermittently available, the positioning system of choice needs to be supplemented by inertial navigation or dead reckoning. When indoor signals are unavailable, a network of location identifiers will be needed to assist the user with navigation.

Figure 3 portrays a taxonomy of geo-location methods for local positioning systems with external sensing. Location-sensing techniques can be divided into three general categories: location fingerprinting (scene analysis), triangulation, and proximity. (All these are based on external sensing, in contrast to inertial navigation, which requires no external sensing.) These various techniques can be discussed in terms of being physical or symbolic and relative or absolute.

A physical technique generally results in a set of coordinates, such as longitude and latitude coordinates, whereas symbolic techniques provide more of an abstract description, such as location in terms of which building an object is in. The difference between absolute and relative systems is the frame of reference. In an absolute system, a frame of reference is used. A relative system, however, uses local references where physical and symbolic techniques may be combined. Signal strength is often used to determine proximity or sometimes range (through an attenuation model).

As a proximity measurement, if a signal is received at several known locations, it is possible to intersect the coverage areas of that signal to determine a "containing" location area. If one knows the angle of bearing (relative to a sphere) and range (distance) from a known point to the target device, then the target location is precisely known in three dimensions. Similarly, if one knows the angle of bearing to a target from two known locations, then it is possible to triangulate to determine the (x, y) coordinates of the target. This can be extended to multiangulation for higher dimensions.

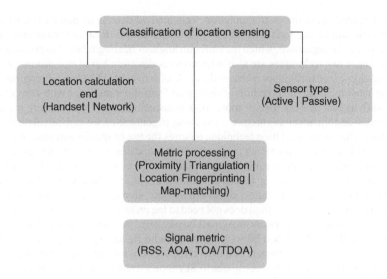

Figure 3.   Geo-location methods.

In a similar fashion, if one knows the range from three known positions to a target, then using multilateration it is possible to determine the target location. Video cameras can be used to describe spatial relationships in scenes using image processing techniques, and thereby determine position. Proximity schemes are often enhanced by using received signal strength (RSS) to provide additional indications of the distance from the source. These schemes rely on propagation loss estimates often derived from a careful mapping of the environment characteristics.

With the proximity technique, in wireless local-area network (WLAN) location sensing we can measure the signal strength at an MC (receiver) and the signal strength at the transmitter. We could then use a radio wave propagation model to calculate the distance the signal has traveled. This approach provides the location of an MC relative to the known location of an access point (AP). Using only one measurement would allow us to place an MC in part of a building, as opposed to another part, but it would not be able to easily provide us with a coordinate for the location. This approach depends on the propagation model used and is hampered by interference caused by walls and furniture. An alternative proximity approach involves creating a table of measured signal strengths and then comparing the signal strength of an MC to the values in the table. Sampling and recording more values in the table would increase the resolution of the system. This approach is not as susceptible to error due to interference; however, it involves more work to set up the location system. An example of proximity-based localization is radio frequency identification (RFID) proximity cards (e.g., Bewator Cotag, Wavetrend). RFID proximity cards are in widespread use. They are active or passive devices that activate by proximity to a fixed reader. They are often used in access control systems. Location can be deduced by considering the last reader to see the card.

The location fingerprinting (scene analysis) technique examines a scene such as a room from a fixed vantage point. One such approach is adopted by the Microsoft Research group in the radar implementation. Radar measures signal strengths of mobile devices from the vantage point of a fixed base station. These measurements are then used to calculate the position of the devices on a two-dimensional coordinate system local to the building. Indoor positioning systems utilizing the location fingerprinting method require a user to record signal strengths of all APs in range from a point in a table. This is done for various points in the WLAN area. To find out the location of a mobile client, the signal strengths from all APs are measured and then the values are compared to the entries in the table. The closest entry in the table is the probable location of the user. The more points a user calibrates, the better the resolution of the software.

The third general technique is triangulation. This method derives its name from trigonometric calculations and can be done via lateration, which uses multiple distance measurements between known points, or via angulation, which measures an angle or bearing relative to points with known separation. These two techniques are also referred to as direction-based and distance-based techniques. Direction-based techniques measure the angle of arrival (AOA). Using directional antennas, the receiver must measure the direction of the signal from the transmitter with respect to a fixed direction, such as east or west. Two or more AOA measurements can tell us where paths between the transmitters and the receiver intersect, and using trigonometry we can calculate the location. Because this particular triangulation technique requires the use of special antennas, it would not be suitable for a WLAN location-sensing application that mandates the use of standard components. Distance-based techniques involve measurement and calculation of the distance between a receiver and one or more transmitters whose locations are known. These techniques involve using one or a more of the following signal attributes: signal arrival time, signal strength, and signal phase.

The map-based positioning method does not need to log on to any of the Wi-Fi networks to figure his position, and some of the networks may well be password protected. But the Wi-Fi base station does send out an ID number when hit with a blast of radio frequency energy from the Wi-Fi card inside a laptop or handheld computer. And even if the signal is too weak for a laptop to connect to, it can still get the ID, which it then compares to Skyhook's database and plots on a map.

4.3 *Classification based on where location estimation takes place*

Positioning systems can be terminal based or network based. In the first case, the terminal receives the information from the sensors or sensors (which may be a part of the network, as is the case when positioning is done using some 802.11-based technologies, or built into the terminal) and computes its own position. It then has to communicate that position to other entities that want to use it (for instance, servers that have location-based services). As wireless devices, the mobile devices are configured to communicate with a network through a wireless interface.

In the second case, no computation is done in the terminal, but the position is computed in the network and communicated to the terminal if required. Mostly, the location-based services are produced in the network as well, and the service access is transmitted to the terminal (for instance, through a web page or another type of client). Of course, hybrids exist (such as assisted GPS), but these are the two main types. Each has a different set of constraints and a different set of advantages.

Preferably, the mobile device needs no special client-side configuration, modules, or programs to be detected and tracked, since detection and tracking are preformed on the network side of the interface. The availability of applications and access to data may be selectively provided or inhibited as a function of the location of the mobile device and an identity of the mobile device or its user, or both.

Although network-based sensing is a compelling business case for operators, many systems have chosen to compute location on the handset because of the lack of standard methods for querying location from the network, and because this solution provides for simpler regulatory compliance in jurisdictions with strong location privacy legislation (i.e., private use of others' personal information is exempt from most data protection provisions). Computing location on the mobile device also respects the user's control for disclosure. A disadvantage of placing a network of location identifiers within the environment is the cost of installing and maintaining the network relative to the coverage achieved. One alternative is to use computer technology to locate the traveler and then make use of a spatial database of the environment to display to the traveler his or her location relative to the environment.

In some cases, such as a high-end portable laptop, it could be possible to compute position information, whereas on a portable handheld such as an iPaq, computing the position information could be cumbersome and utilize too much computational effort. In such cases it could suffice to report just the position with the highest accuracy, to just merge the two most accurate positions, or to use a less accurate merging technique. It is therefore important that the platform can be optimized for the underlying host characteristics and needs of the programs utilizing it. As well as merging positioning information from local devices, the position can also be augmented with information from other non-local position platforms via limited range *ad hoc* networks such as a WaveLAN segment or Bluetooth networks.

Overall, sensor measurements are attained and then these measurements are collated to acquire the user's position. In most indoor systems the user carries either a transmitter or a receiver. If the user carries a transmitter that broadcasts signals, then sensors fitted in the room are used to detect the emitted signal. These signals may be radio frequency signals, infrared signals, or ultrasonic pulses. The measurements that are used to pinpoint the user may be either the signal strength of the signal detected by the sensor or the time-of-flight of the signal from the transmitter to the sensor. For example, the BAT system, which is one of AT&T's indoor locationbased systems, can find the three-dimensional position of a user when given three or more time-of-flight measurements. All systems that exploit transmitters and receivers have their own pros and cons, but the fact that the users must carry either a transmitter or receiver is a universal problem for such systems.

Location determination methods are also classified according to the point of computation: server based or client based. The location methods can be further differentiated using the architecture of the implementation to emphasize autonomy. The architecture can vary from infrastructure dominated to infrastructure-less schemes. Infrastructure schemes can be active or passive. In active schemes, there are sensor servers that provide measurement and timing signals that are received by the clients (equipments). The location computations are performed by the clients.

Sensors may be requested or they may be constantly generated. In passive infrastructure schemes, the sensors (or measurement signals) are generated by the clients and the locations are computed by one or more servers, as in cellular base stations performing pattern matching or angle-of-arrival schemes. The infrastructure-less class of systems, where location is performed by clients only, are either independent or cooperative. Independent schemes involve a single client operating from a known position with self-tracking capability, such as an inertial navigation (or variants such as star tracking, compass, etc.). Infrastructure-less, cooperative schemes, or *ad hoc* location-sensing schemes, involve measurements made by multiple clients that must be communicated among clients to determine position. The cooperative schemes cannot determine location individually and must self-organize to perform the communication necessary to share the measurement information.

4.4   *Classification based on signal metrics*

The basic function of a wireless positioning system is to gather particular information about the position of a mobile device and process that position into a location estimate. The particular information could be one of the classical geolocation metrics for estimating the position:

– Received signal strength (RSS)
– Angle of arrival (AOA)
– Time of arrival (TOA)/time difference of arrival (TDOA)

Calculating direction using AOA requires additional antenna hardware that needs to be precisely calibrated; the systems based on the time of flight of electro magnetical waves require very accurate timing measurements, and thus high synchronicity.

In a networked environment of many nodes, cell site density, the distribution of nodes, the ability to share information, terrain, and physical obstruction play major roles in the accuracy derived by the system.

A positioning system may or may not be able to communicate and thus share data used to solve for the position. The systems that cannot communicate have to rely solely on their ability to detect an electromagnetic signal. They can employ methods based on signal strength, Doppler shift, and AOA measurements. If communication is available, data can be modulated onto the navigation signal or be available on a secondary communication link. Combinations of modulated data and secondary links, for example, differential GPS, are possible too. The data source of the modulated signal assists in deriving the GPS solution, the secondary communication link in correcting and refining the solution. TDOA and TOA measurements require a minimum of transmitted data consisting of

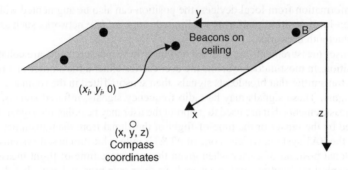

Figure 4.   The coordinate system used in MIT cricket.

at least a pulse shape, used for synchronization and thus measuring the time of flight of a signal, to be exchanged.

Depending on the signal and the range anticipated, larger amounts of data can be transmitted. Hitachi released TDOA-based location technology in October 2002. This system uses two types of access points, a Master AP and a Slave AP. Slave APs synchronize their clocks with that of a Master AP. Slave APs measure the arrival time from a mobile terminal, and the Master AP determines the location of the mobile terminal by using the time differences of arrival between the signal reception times at multiple Slave APs. Triangulation methods can be used in conjunction with angle and range data to solve for remote positions.

Range vector data from range measurements such as TDOA, TOA, and received signal strength indicator (RSSI) or similar are considered the input data. The absolute position of a device/user position can be computed using range measurements (triangulation) to known positions. Multiple range measurements (TOA, TDOA, or RSSI) or angles (AOA) are necessary to determine the exact three-dimensional position on a remote navigation receiver.

5   EXISTING POSITIONING SOLUTIONS

MIT Cricket localization motes developed by MIT use radio frequency (RF) and ultrasound measurements to determine a time-of-flight signal that determines distance between listener motes and previously emplaced beacons. Significant limitations of Cricket include the necessity of emplacing the beacon motes at specific distances from partitions, configuring beacons for broadcasting locations, and requiring beacons to be in place before the localization network is functional. Moreover, system performance in noisy environmental may be questionable.

MIT Cricket uses proximity lateration and time of flight (TOF) (distance readings), as explained in more detail later. To enable its beacons (sensors) to gather the distance information (ratio of height to distance), Cricket implemented a local coordinate system using four active beacons instrumented with known positions in space. The beacons are configured with their (x, y, 0) coordinates (as shown in Figure 2.1.1) and broadcast this information on the RF channel, which is sensed by the receiver on the compass. In this sense, the MIT Cricket system behaves as a form of indoor pseudo-GPS.

MIT Crickets is an example of not only a position-sensing system, as we will see in the next section. It is also an example of a service discovery system, with its own directory system, maintained by the MIT Facitilites group, which is the building maintenance office of the university.

MIT Cricket also has its own mapping software application, Floorplan ActiveMap, for a mobile device. As the user moves in a building, the navigation software running on the mobile device uses the listener API to update its current position. Then, by sending this information securely to a map server, it can obtain updates to the map displayed to the user.

In addition, mobile devices learn about services in their vicinity via the floor plan application that is sent from a map server application, and interact with services by constructing queries for services at a required location. Services appear as icons on the map that are a function of the user's current location. The services themselves learn their location information using their own listener devices, avoiding the need for any per-node configuration. MIT Facilities recognized that, for example, a common coordinate system is essential for doing facility management, especially in an environment such as MIT's, where one building is actually a series of 14 separate interconnected CAD plans that do not necessarily know where they lie on a local/global spatial reference system. For this reason, MIT Facilities invested in figuring out what the correct orientation is for each floor plan (i.e., a world file) to allow the construction of a room location model based on real-world coordinates. This is very important because MIT Facilities can now do horizontal and vertical adjacency studies not allowed with the CAD files as they were, and MIT Facilities can now overlay seemingly dissimilar data sets and perform analyses not possible before. A simple example for this would be to know where the closest fire hydrant is to a particular lab with a particular use. This is querying architectural information with something that is typically a civil/survey question. Even though this is not directly related to indoor location-based applications, MIT Facilities is aware that it does open the floodgates to such applications like indoor location services (i.e., way finding, emergency management, etc.).

Another system of significance most easily employable in crisis response is the Aether Wire & Location system, which uses ultra-wideband transceivers, called Localizers, for precise position location and low data rate communication. (As opposed to MIT Cricket, this system is commercial.) Localizers determine location by sharing range information within a network of units distributed in the environment. The range between pairs of Localizers is determined by cooperatively exchanging ultra-wideband signals consisting of coded sequences of impulses. As Localizers are activated, each acquires as many contacts as possible. As local groups of nodes form into clusters, nodes in one cluster link with one or more nodes in other clusters, forming bridges between the clusters. Range information is constantly shared so that all Localizers are aware of all other Localizers in the network. Using precise timing techniques, the Localizers are able to establish these ranges to an accuracy of about a centimeter. The advantages of Aether Wire's CMOS Ultra-Wideband (UWB) Localizer technology are:

– Accurate position (1–5 cm);
– Low cost, low power units, potential for single-chip CMOS system;
– Minimal interference with other communications systems due to UWB technology and low energy;
– With their wideband spread spectrum nature, the units are more tolerant to interference and multipath, and offer inherent privacy;
– UWB penetrates most substances, and can operate effectively in RF-hostile environments;
– The network approach allows units to be great distances apart without requiring the power to directly reach that distance.

The position calculation is network based. However, all members of the network can be mobile (i.e., it is not necessary to make any member be stationary unless the application wants that). A good example of a moving network is a squad of rescue personnel who desire to track each other as they move in an emergency rescue operation.

Another system of note comes from the Rosum corporation. The system provides 30–50 m accuracy indoors (and 5 m accuracy outdoors.) Potential applications for its technology to include E911 position fixing and asset and people tracking. Rosum's technology is based on the triangulation of TV signals. Rosum's TV-GPS positioning system utilizes unmodified broadcast TV signals for position location. There are 2,800 TV transmitters and 4,500 TV transmitters in the United States alone, and they are well-correlated with population centers.

TV broadcast signals, which were designed for indoor use, are about ten thousand times more powerful than signals emanating from GPS satellites. TV signals are also in a lower frequency

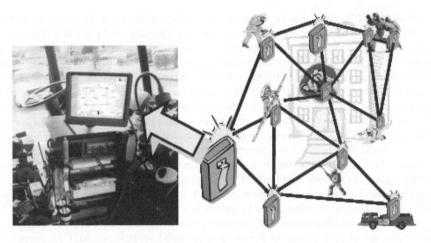

Figure 5. Use of Aether Wire & Location system in a fire emergency rescue operation.

range than GPS signals, which means they do a better job of penetrating walls. For even greater accuracy, it's possible to install "Pseudo-TV" transmitters around the outside perimeter of a building to provide position fixing within areas as small as a single room. This could be useful for tracking people inside burning buildings. Firefighters could temporarily set down a number of Pseudo-TV transmitters around a burning structure and see where the people trapped inside are (provided they are carrying a compatible device). The Rosum TV Measurement Module (RTMM) receives local TV and GPS signals, measures their timing, computes the pseudoranges and sends that information to the Location Server (LS). The LS computes the position of the RTMM and sends that location back to the RTMM or to the tracking application server. In addition to the wide-area positioning system, Rosum also develops a limited-area, 3D positioning system. This system is used by first responders in emergency situations. Rescue personnel can be tracked from a field command center, reducing precious time spent giving location updates and eliminating blind searches in man-down situations.

Other systems include the Multi Channel-Ultrawideband (MC-UWB), which is a single signal technique exploiting Orthogonal Frequency Division Multiplexing. The signal characteristics permit resolution of multipath effects in difficult RF environments. No time synchronization between locators is required, important for ad-hoc, ephemeral networks characterizing incident response situations. Initial testing indicates centimeter scale accuracy.

The "Walking GPS" system developed at the University of Virginia explores localization using a two component hybrid system where a GPS receiver is deployed and the sensor nodes infer their position with respect to the GPS-enabled node. The sensor mote uses RSSI to infer distance from GPS mote. Testing with statically emplaced motes indicate 1–3 meter accuracy.

6 CONCLUSION

The positioning technology that is most easily to employ in crisis response is the technology that is scalable and not impacted by a fire or an earthquake. Systems from Rosum and Aether Wire & Location fall under this description; Rosum's technology is deployed on a metro scale (TV towers), hence if a building is on fire, the positioning infrastructure will not be destroyed. Aether Wire's system is *ad hoc*, self-configuring, so it can be deployed in any situation. Wi-Fi positioning system are not the best solutions in disaster management situations. The disadvantage of Wi-Fi based location is that the infrastructure is not dependable in fires. Whichever system wins out, the potential demand for the capabilities promised by these technologies is still unknown. The success or failure of Rosum or any other location based services infrastructure provider will rest on the desirability of the applications that will be developed to make use of them.

Indoor location-aware applications require micro-detailed geo-referencing to satisfy users' growing needs. It's not enough to geo-reference a building if the position of users and other objects inside the building also is relevant. Objects are used as landmarks, and relationships among the objects are crucial for symbolic representation of the whole system.

It's important to realize that symbolic doesn't need to be less accurate/preferred than geometric, because the real issue is whether the underlying data model and services are based on symbolic (adjacency/topology) or geometric reasoning. For example, a developer can use location to "snap" to a room or floor to enable services based on vicinity or containment vs. using geometry and Euclidian distance as the basis for judging what's close or far.

Whether a position is measured in absolute (i.e., x, y, z coordinates) or relative terms (i.e., "in room 101"), it alone almost certainly isn't the ultimate solution. Any robust and scalable location-sensing architecture needs to develop a hybrid data model to characterize location as a heterogeneous mix of sensors. This way, the LBS can understand the environment and provide users with the location of objects or places they're interested in finding. Scales and the sensors used to assess our environment are critical considerations when moving between models.

BIBLIOGRAPHY

Bahl, P. & Padmanabhan, V. N. (2000). RADAR: An in-building RF-based user location and tracking system. Proceedings of INFOCOM 2000.
Baus, J., Kray, C., Krüger, A., & Wahlster, W. (2001). A resource-adaptive mobile navigation system. Proceedings of the International Workshop on Information Presentation and Natural Multimodal Dialog. Verona, Italy 14–15 December.
Beal, J. (2003). Contextual Geolocation: A Specialized Application for Improving Indoor Location Awareness in Wireless Local Area Networks. 2003 Midwest Instructional Computing Symposium, University of St. Scholastica.
CIO Magazine & Synchrologic (2001). CIO outlook 2001: Architecting mobility. Ten critical steps in supporting mobile enterprise computing. Alpharetta, GA: Synchrologic, Inc, available at www.synchrologic.com
Dey, A., & Abowd, G. (1999). Towards a better understanding of context and context-awareness. GVU Technical Report GIT-GVU-99-22. College of Computing, Georgia Institute of Technology, available at ftp://ftp.cc.gatech.edu/pub/gvu/tr/1999/99-22.pdf
Drummond, W. J. (1995, Spring). Address Matching: GIS Technology for Mapping Human Activity Patterns. APA Journal, 240–251.
HSARPA, 2005. Rapid Technology Application Program Broad Agency Announcement 05–10 (BAA 05-10) Department of Homeland Security. Homeland Security Advanced Research Projects Agency (HSARPA)
Kishan, A., Michael, M., Rihan, S. & Biswas, R. (2001). Halibut: An infrastructure for wireless LAN location-based services. Stanford, CA: Stanford University.
Hassan-Ali, M & Pahlavan, K. (2002). A new statistical model for site-specific indoor radio propagation prediction based on geometric optics and geometric probability. IEEE Transactions on Wireless Communications, 1(1), 112–124.
Kindberg, T., Barton, J., Morgan, J., Becker, G., Caswell, D., Debaty, P., Gopal, G., Frid, M., Krishnan, V., Morris, H., Schettino, J., & Serra, B. (2000). People, Places, Things: Web Presence for the Real World. HP Cooltown, available at http://www.cooltown.hp.com/dev/wpapers/webpres/WebPresence.asp
Kolodziej, K.W. & J. Hjelm (2006) Local Positioning System: LBS Applications and Services for indoor us, CRCpress Taylor & Francis, Boca Raton, USA.
Kolodziej, K., (2003). Open LS for Indoor Positioning: Strategies for Standardizing Location Based Services for Indoor Use. MIT thesis, Massachusetts Institute of Technology, Cambridge, MA.
Kuikka, P., (1999). Wideband radio propagation modeling for indoor geolocation applications. Seminar on telecommunications technology. Helsinki, Finland: University of Helsinki, Department of Computer Science.
Lee, J., (2001). A Spatial access oriented implementation of a topological data model for 3D urban entities. The Ohio State University, Department of Geography Levenshtein, L. I., (1966). Binary codes capable of correcting deletions, insertions, and reversals, Soviet Physics–Doklady, 10 (8) 707–710.

Lo, C.P., Albert K. & Yeung, W. (2002). The concepts and techniques of geographic information systems. Upper Saddle River, NJ: Prentice-Hall, Inc.

Nicklas, D., & Mitschang, B. (2001). The NEXUS Augmented World Model: An extensible approach for mobile, spatially aware applications. 7th International Conference on Object-Oriented Information Systems.

Orr, R. & Abowd, G. (2000). The Smart Floor: A mechanism for natural user identification and tracking. Proceedings of the 2000 Conference on Human Factors in Computing Systems. The Hague, Netherlands: ACM Press.

Pahlavan, K., Latfa-aho, M & Ylianttilia, M. (2000). Comparison of indoor geolocation methods in DSSS and OFDM wireless LAN systems. IEEE VTC 2000.

Pahlavan, K. & Li, X. (2002, February). Indoor geolocation science and technology: Next-generation broadband wireless networks and navigation services. IEEE Communications Magazine, 112–118.

Prasithsangaree, P., Krishnamurthy, P. & Chrysanthis, P.K. (2001). On indoor position location with wireless LANs. Telecommunications Program & Dept. of Computer Science. University of Pittsburgh.

Priyantha, N., Chakraborty, A., Balakrishnan, H. (2000). The Cricket Location-Support System. Proceedings of 6th Int'l Conference on Mobile Computing and Networking. ACM Press, New York. pp. 32–43.

Schiller, J. (2000). Mobile Communications. London: Pearson Education.

Smailagic, A, Siewiorek, D., Anhald, J., Kogan, D. & Wang, Y. (2000). Location sensing and privacy in a context aware computing environment. Institute for Complex Engineered Systems. Carnegie Mellon University.

Vittorini, L.D., & Robins, B. (2003, November). Optimizing Indoor GPS Performance. GPS World, 40–48.

Wallbaum, M. (2002). Wheremops: An indoor geolocation system. (Department of Computer Science, Aachen University of Technology.) IEEE PIMRC.

Want, R., Hopper, A., Falcão, V. & Gibbons, J. (1992). The Active Badge location System. ACM Transactions on Information Systems (TOIS), Volume 10, Issue 1, pp. 91–102.

Geospatial Information Technology for Emergency Response – Zlatanova & Li (eds)
© 2008 Taylor & Francis Group, London, ISBN 978-0-415-42247-5

Virtual Reality for training and collaboration in emergency management

E. Kjems & L. Bodum
Centre for 3D GeoInformation, Aalborg University, Denmark

ABSTRACT: Rescue training has been carried out for centuries but new advanced technology involving Virtual Reality (VR) unveils new opportunities. They are still very limited but do give the possibility of rescue training in e.g. buildings or planes on fire without anybody getting hurt. Properly training programmes in surroundings and under conditions similar to the real "thing" are important parts of a virtual training facility. Such a facility can provide the right environment for an emergency rescue scenario that can be lived through over and over again without having the costs for training rising up through the sky. Different aspects of ER can be involved in the virtual training environment and be practiced one by one or all together. Among them are e.g. communication and cooperation between the different rescue groups. These aspects are very important in order to minimize the number of mistakes and delays during an operation. Even if working ER systems using VR are very sparse this article focuses on systems solely based on immersive VR technologies. For that, some sort of display system must be present and provide an experience of "being there" which cannot be obtained through an ordinary computer monitor. This also means that systems based on picture based platforms like QTVR (QuickTime Virtual Reality), game-like platforms or decision support systems will not be discussed in this article. The article begins with an introduction to VR and gives a few examples of use and outlines the potential of VR within ER since the real killer application yet has to appear. The article rounds up with a more general view on VR for those of you who are interested in the technology and might be interested to invest in the equipment used.

1 VIRTUAL REALITY FROM THE START

The VR technology has been on its way since the fifties when the first ideas for a virtual computer generated environment were born. But not until the eighties it became a real possibility to demonstrate the technology and its ideas. The technology was then, as it still is perceived today, very advanced and very demanding in relation to the capacity of the computer (Rheingold 1991). In the beginning of the nineties it was still necessary to have a large number of calculating units either running in a network or combined in one machine like the work stations from Silicon Graphics (SGI) with its RealityEngine technology. SGI kept on developing the line of heavy graphical work stations and main frames through the nineties and hold a dominant position within the VR market during that period. Every institution or company that seriously aimed at this technology used large sums of money to obtain these very expensive VR-systems. Nonetheless, this was the only way to gain some first experience with real VR. The systems also unveiled their fantastic possibilities and potential but also their weaknesses and needs for further development.

The situation is somehow different today. Regular PC's with mainstream hardware placed in a cluster can perform similar to large Unix based workstations though the system architecture is in favor of the expensive workstations. This results from the need of connecting peripheral devices for head-tracking, navigation etc. Nevertheless most computers have become quite fast and provide reasonable results generating lifelike 3-D scenarios. Another reason for that is the

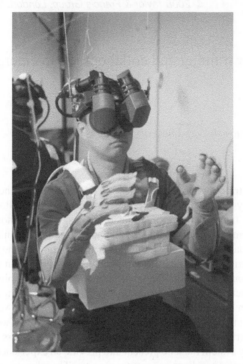

Figure 1.   Spacewalk simulator at the VR Laboratory, Johnson Space Center, July 2002 (JSC) (with courtesy to NASA).

growing improvement of 3-D modelling software and in particular the improvement and growing supply of tools to present 3-D models in real time (VR).

The computer technologies used for visualizing high-end 3-D graphics are in most cases closely linked with the computer games industry. Actually, the applications developed for both the professional and the entertainment segments are closely connected. Especially the output capacity of computer graphics had a huge boost in the beginning of this millennium. This was caused by the competition on the market for graphics boards for gaming which at that point had a tremendous growth-rate, because PC's were sold for reasonable prices and feasible for everyone. The same hardware components can be used for a professional VE.

However, there is still a need for more computer power; better software; and first of all better interaction to achieve the very high demands for a really intuitive VR application. Many systems seem to be working awkwardly especially when looked upon from the outside. The technology is giving the premises and users must still deal with odd equipment, cables etc as shown in this spacewalk simulator at NASA (see figure 1). This is very much the case for virtual ER systems and may beside the costs be one of the biggest impediments to the dissemination of this technology.

The term VR is used very broadly but this article solely focuses on immersive VR. VR in general is characterized by a real-time interactive computer-simulated environment. The interactivity has a twofold task. The important one is the possibility to calculate the graphical output in a way that the head-motion and thereby the perspective from the viewer (or user) is correct at all time and position. Secondly the interactivity provides the user with an immediate control of the action like in games.

The expression Virtual Environment (VE) is often used instead of VR which merely is the technological part and VE consist of the whole range of hardware and software used for a particular application. Perhaps going from VR to VE is an expression of going from technology to a medium. Using the expression medium in connection with a VE is not accepted in general but developments and the utilization of VE in a great number of scientific and commercial domains have shown that the technology now, in 2006, is so mature that it can be seen as a real medium for presentation and communication. In this chapter VR is the technical description of the solutions and VE covers the field in a more widely manner and includes context and principles for the modelling parts (Sherman and Craig 2002).

Media such as radio, television, and Internet used for entertainment, education, and more general information spreading, must be capable of passing on accurate metadata in order for the public to make a fast and safe judgment and take their precautions related to a given situation. These demands will have its influence on a VE so that it is possible, at all times, to interpret a situation based on what is experienced through the senses; what is felt, seen, and heard in a virtual environment. Through this, the risk of mistakes is minimized and the possibility of making better decisions in a crisis-situation is optimized.

VE has the same dilemma as especially television (TV). When a medium is used merely for entertainment it is labelled as a toy rather than as an item for serious use. Even though the television medium with its audiovisual interface has proven itself to be good for educational purposes, it has been used very slightly for that compared to the increasingly time spent in front of a TV during the last 50 years. VR has an even bigger potential for educational purposes than TV because of its truly interactive interface. But the situation for hardware and software, which was very expensive and complex in handling only a few years ago, made it almost impossible to make serious use of educational applications. A new domain called edutainment is evolving heavily and it looks very promising to the educational use of VE and other media.

One of the domains that receive great attention but only counts a few applications is ER. There are a few good examples where VE has been used within ER. These systems are often targeted at specific applications such as fire fighting in buildings (Freund, Rossmann et al., 2003) (Gamberini, Cottone et al., 2003) or responses after earthquakes .(Li, Zhang et al., 2005) (Yi-li, Yang et al., 2002). Usually these systems cannot be used in general but have been developed in connection within a specific project.

The following section will present some of these examples as well as outline the potential of VE's. Unfortunately immersive VE's have only been used in a few cases more or less experimental while commercial VE systems are not immersive at all but based on game engine like platforms controlled in front of a monitor.

2 APPLICATION AREAS FOR VR/VE

The three most obvious areas for VR/VE applications used within ER are:

- Training
- Collaboration
- Remote Control

Training addresses the rescue team in action and focuses primarily on the single rescue worker dealing with certain situations and also debriefing. Collaboration is about the interoperability between the single rescue workers and the head of operations primarily in relation to audible and visual communication. With remote control means a system that is providing the head of operations with an overview over the situation and the possibility to navigate the rescue worker on a remote-controlled basis. This involves 3-D models of the building in question and some sort of positioning device.

2.1 *Training*

One of the most obvious areas for VR applications is connected to training where very dangerous situations are trained over and over without anyone getting hurt. Usually this is a simulation known from flight simulators where pilots repeatedly are exposed to technical problems in a plane and must learn to react correctly according to the actual situation. Flight simulators have been developed and constantly improved for many years and even simulators for the mass marked, which are sold as computer games, are so realistic that they do not lack anything compared to the more commercial and advanced flight simulators from the aircraft plants as Airbus or Boeing. The big difference lies in the available hardware, where the flight enthusiast has to settle with a desktop computer, and where the pilot at Airbus can settle into a complete cockpit in order to increase the realism (McCarty, Sheasby et al., 1994). An alternative to a cockpit modelled in 3-D computer graphics is a high-resolution image where information of even the slightest details can be identified. The images can be recorded with a special camera, and besides the image itself, one can refer to other information through hyperlinks. This kind of simulator has been used due to the fact that the United States Air Force wanted to train its pilots and other employees in interiors and instruments, which they did not know or were not used to from the normal everyday life. In a catastrophic situation it can be necessary for pilots to fly other kinds of aircrafts and another type of personnel to be stand-ins if for some reason or another, the pilots are unable to continue flying (Rosenberg 2003).

One of the main reasons that training using VR/VE technology within ER has not yet been disseminated is the lack of realism. It is simply very difficult to simulate e.g. a fire fighters entry into a burning house without real physical obstacles, heat and smoke formation, and water. Nevertheless a training simulation primarily focuses on the visual elements, which is why one will choose to train orientation and passable ways in connection with a fire. Another thing is that one's senses can be cheated by visual or audio-visual means. Thus, one can simulate different situations only with the help of audio-visual elements but still obtain the feeling of hitting obstacles (by sound for instance); meet other kind of challenges like holes in floors or even get stressed by exploding noise.

In traffic accidents and other disasters where many victims have been involved, the preparedness needs to function with great flexibility and with full control in order to rescue as many persons as possible. Traditionally, this has been trained through preparedness exercises in realistic environments and with figurants. This acquires long time planning and at the same time involvement of more parties as daily routines and systems have to be stopped. The Department of Civil and Environmental Engineering and Geodetic Science at the Ohio State University in cooperation with the Ohio State Highway Patrol has developed a VR/VE system, which simulates traffic accidents at the highway. i.e. that the system is used to training the police in handling everything from the arrival at the crash scene, where the situation is being assessed, to the decision making of which vehicles have to be removed first and how to control the traffic during the clean-up (Hadipriono, Duane et al., 2003; Hadipriono, Duane et al., 2003).

In addition, when it comes to procedures and working operations in hospitals it is possible to use VR/VE for training of personnel. At large accidents in some countries, it is necessary to involve voluntary personnel to assist the trained personnel in the work. The Department of Communication, Computer and Systems Sciences, University of Genoa, Italy in cooperation with the San Martino Hospital, Genoa, Italy has built a VE training environment where the voluntary personnel, which might get involved in a large accident, can test various treatment scenarios by using VR/VE (Leo, Ponder et al., 2003). Using the VE system, a trainee should recognize and assess a medical emergency situation correctly, show level of competence and appropriate knowledge, select and apply appropriate medical procedures, select and use specific medical equipment, act within tight time constraints and under high pressure, overcome psychological barriers, and handle life-like events that are usually not part of the "theory".

When only a few immersive systems for ER exist, how do we know that a VE can actually enhance the skills of a rescue worker? There have been carried out a lot of "transfer of training" studies due to simple motor functions or more complex situations like way finding in buildings with immersive and non-immersive VR systems. All studies show a positive result, which means that virtual training can be as good as training in real facilities. Of course vision fidelity, haptics and other kind of interaction influences the outcome but in general the results point in the same direction all together. (Uhr 2004).

Several non-immersive VR systems have reached to a reasonable level and are commercially available for ER. Systems like DiaboloVR (http://www.e-semple.com) or ADMS (http://www.admstraining.com/) have proven their value and may be a good start for a virtual training system. (Louka and Balducelli 2001). But the degree of immersiveness has a direct influence on the single rescue workers benefit of the training. While non-immersive systems concentrate on cooperation and the operation as such, immersive systems concentrate on the single worker. The whole situation is seen from the rescue worker which is the basic nature of immersive VR. Immersive VR needs a point of view due to the perspective orientation which again makes it obvious to place that point at the rescue worker in a real immersive system. (Psotka 1995). Thus collaborative immersive VR systems need these systems connected somehow in a network.

2.2 *Collaboration*

The cooperation between rescue workers and rescue teams is normally carried out through radio communication. All information available by all teams is passed on to the other team-members and head of operation, helping to provide a permanent survey over the situation and allow to giving precise orders following up the situation.

Working collaborative within a VE will demand a system that enables the user to interact with another or several others. Interaction here means that the participants are represented somehow in a VE and can be depicted by each other through a simple avatar (3-D model), a picture (billboard), name tag (sign) or other kind of ID. In such a scenario several VE systems must be connected to each other typically over the Internet or on a training-site with a common server. In each VE one will find a rescue worker immersed into the scene standing at a certain position and acting together with the other participants. Due to the immersiveness and its demands for the right perspective each worker will have to act independently and the collaboration will be close to a realistic situation except from the eye contact which might be missed. (Louka and Balducelli 2001). Even CAVE like VE systems that normally are suitable for one single person only, gives one the possibility to deal with a whole team of rescue personnel working together in a virtual environment when combing them in a network (Steed and Frécon 2005).

Using collaboration for ER makes a lot of sense also during training exercises since supervisors can put themselves into a workers viewpoint following exactly what the trainee is seeing. Recording the whole sequence gives a very good basis for debriefing and training evaluation afterwards since views and gestures can be reproduced easily even with immersion.

A non-immersive example of a multi-modal and collaborative system that uses a rather simple user-interface in two dimensions but has the geospatial information as its main platform is DAVE_G. This acronym stands for "Dialogue Assisted Visual Environment for GeoInformation", which describes a system that can be very useful in ER situations (http://www.geovista.psu.edu/grants/nsf-itr/feature.html). (Rauschert, Agrawal et al., 2002; MacEachren 2005)

There are several other examples of systems that can handle collaboration also with regards to an immersive VE. One of the perhaps less known applications within the area of collaboration is COVISE (Collaborative Visualization and Simulation Environment) which originally was developed at Stuttgart University at HLRS (http://www.hlrs.de). COVISE is an excellent platform to connect different virtual systems with each other. In that way one can communicate and act simultaneously in the same model. Since COVISE also can handle particles and flow calculations (CFD) it is an excellent platform for visualizing fire and smoke and other dynamic phenomena, which all together are valuable features for an ER system providing an immersive VR based user-interface (Wössner, Schulze et al., 2002).

There are several vendors of VR related platforms who will claim to be collaborative and supporting true immersive VR functionality, but very often these platforms are immature or rather low level programming libraries like for instances VR Juggler. VR Juggler is a set of software libraries for developing cluster based collaborative VR systems. VR Juggler is very much scaleable and can be used with a broad range of applications and VE's. One would use VR Juggler if one wants to develop own dedicated VE. VR Juggler has its roots at the VRAC at the Iowa State University (http://www.vrjuggler.org). (Cruz-Neira, Bierbaum et al., 2002)

Another more complete and fully commercial VE package is Virtools (http//:www.virtools.com). Though Virtools is rather expensive it does provide one with a whole range of professional VR functionality among them also collaboration and immersive projection. Though most commercial VR systems that are dedicated for ER primarily work as non-immersive applications they provide sensible graphics and quite good collaboration features. Making these systems available as immersive collaborative VE's is not a difficult task and should be considered seriously but of course these developments are controlled entirely by the market forces and the demand of the customers. Unfortunately at the moment the immersive part of an ER is conceived as too expensive in relation to what one gains from it, and that might even be true. There has not been made any proper cost-benefit analysis on these applications with regards to the implication of immersiveness.

2.3 Remote control

Watching movies like "Enemy of the State" make most people have the understanding that technology today can provide one with tiny little positioning devices that can retrieve one's exact position somehow even within an arbitrary building and send this information directly to the other end of the world. Well, not yet. But we are getting there. With relation to ER this technology would provide the head of operations with crucial positioning information of e.g. rescue workers. These positions can be shown in a 3-D model of the building in question, and help to coordinate and direct the rescue workers around the building. Even if the building is dark and filled with smoke the rescue worker has a "third eye". The immersion part here would be used by the head of operation, since it is easier to explain the correct moving direction actually experiencing the surroundings by one-self.

These indoor systems are mostly seen in movies where the obvious advantages of having such a system are shown. There are quite a few technologic challenges such as the positioning of the rescue team inside buildings and having 3-D models of the buildings at hand. Furthermore, the technology is very expensive and a rescue team might prefer having some of the worn-out equipment replaced instead of using high-tech they cannot really trust. Nonetheless, the development is evolving as discussed in the previous chapter.

Tests have shown that one could be positioned indoor in large buildings with up to 1-meter accuracy, though the solution required expensive positioning devices within the building. One might suggest that these devices should be standard equipment of large buildings in the future as the indoor positioning can be used for other things as well (tracking staff, redirecting phones automatically or locating important equipment). However, it will probably not be a requirement when building multi-family housing, which very often is the primary target of rescue operations. (Nakanishi, Koizumi et al., 2005). The new Galileo satellite system for positioning is promising a certain degree of indoor positioning, so perhaps we can expect some help here.

For the remote control system we also needed a complete 3-D model of the building. From a technological site this part is easy because there are many tools to provide 3-D models. Unfortunately these models are only relevant to a few and very time-consuming and therefore expensive to produce. A lot of cities are getting their buildings modelled in so called 3-D city-models (see the next section of the book). But usually these models are only shells of the surface and the outer limits of the building and cannot be used for inside operations. There is some light at the horizon because the building industry in more and more countries like in Scandinavia and the Far East like China have come to an agreement of exchanging all building data electronically using the IFC standard. This means that every building detail is provided as an object and if available contains a geometrical description of its content. So all new buildings should be collected some how in a database and made available for rescue operations among others.

3 VE SYSTEMS AND IMMERSIVENESS

The following section presents a short technical introduction of the various parts of a VE system. The basic idea with this section is to guide a novice within this field into a deeper understanding of the terminology already used in this chapter and the different types of VR. This section will only focus on the important parts and will not be a complete review of available components used in a VE. The three main areas of an immersive VE are the following:

- *Immersiveness* in connection with the display technology and the way the virtual world is experienced.
- *Tracking* technology for head-tracking and all kind of devices used for navigation and interaction with the virtual world.
- *The computer-generated model* by virtue of the model fidelity, modelling tools, and visual effects and events.

This means the degree of immersiveness combined with a tracking device and some 3-D modelling for the purpose of VR constitute a VE. Note, no word spent on computer-power or on graphics boards. One should simply aim at equipment with a good price/performance ratio. In that case it is easier to get some new hardware within a reasonable time. The only remark is about the graphics board. Be sure to buy a board with appropriate abilities in relation to stereo and cluster operation, see later on in this section.

What is immersiveness? This term has been used several times and has been treated as a very important parameter for a VE but has not yet been addressed properly in this chapter. Immersiveness describes the level of realism given by the VE system. As more immersiveness the system provides as more the user gets the feeling of being there and being part of the virtual world or being present. Mel Slater says: "Presence is a human reaction to immersion" (http://presence.cs.ucl.ac.uk/presenceconnect/articles/Jan2003). Our experience though not scientific at the VR Media Lab at Aalborg University is that it is easier for children to be immersed in a virtual world. With their unreserved attitude to the "new" technology and its often awkward interaction as well as children's ability to immerse into a story with the use of one's imagination, children are capable of abstracting themselves from unaccustomed technological surroundings and very quickly become a part of a virtual universe. Older persons seem to have more difficulties in adapting to an artificial world, as they tend to think whether they act correctly according to the technology and thus loose focus from the content. This observation has also been described as mental immersion versus physical immersion. (Sherman and Craig 2002) So, children seem to immersive much faster mentally than elder people who keep on wondering about the physical immersion and have difficulties to get immersed into the story.

A few rules about immersiveness are commonly known. The human field of view (FOV) has to be as enclosed in the VE as possible though there will be an individual variation from person to person which of course will result in different experiences. A rule of thumb tells that the FOV should exceed 60 degrees to provide an immersive feeling. Head-tracking as mentioned earlier is of great importance because the perspective view has a great influence when perceiving depth in a view. This indicates that we need some kind of stereo image because the human ability of experiencing depth is given by the fact that we have two eyes available. Creating stereo images demands a great deal from a VE and may be the main reason why this technology still not is more commonly used. Finally an annoying thing about immersiveness is that motion sickness is increasing as more the user gets immersed but the degree of motion sickness diverge quite a lot among the users. (Psotka 1995)

3.1 *Stereo*

The stereo effect is used to give the left and right eye an image each, so that one feels that objects in the image have different distances to the viewer and thus give a distinct feeling of presence. Especially, stereo makes sense where the viewer is relatively close to objects; i.e. within 10 metres. If the objects are further away than 10 metres, the stereo effect fades away. Thus, it is obvious within ER to use stereo as this provides a more realistic feeling of the spatial surroundings in e.g. buildings. However, the stereo effect is also one of the elements in a VE system that contains a number of technologic challenges and makes the difference compared to ordinary display systems.

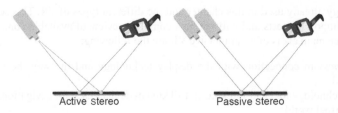

Active stereo   Passive stereo

Figure 2. Principle for passive and active stereo projection.

There is a distinction between active and passive stereo. In short, passive stereo is the less expensive solution where light is polarized differently for the left and the right eye, see figure 2.

The polarization can be made in various ways. The most known is the colour polarization which was used in cinemas in the 1950s for the first time. The most used polarization today within a VE is linear polarization or circular polarization, respectively. The latter has more degrees of freedom when moving one's head in relation to the image. Passive stereo only needs a pair of low-cost LCD-projectors, a pair of light cardboard glasses, both with built-in polarization filters and a good reflecting screen that makes sure that the light beam not is reflecting omni-directional, but straight back to the viewer. The expensive active stereo requires that the glasses synchronizes with the projectors so that the glasses are blacked out alternately in the left and right glass at the same rate as the image for the left and the right eye is projected on the screen. With an image frequency of up to 120 images per second, each eye gets 60 images per second. As the "spectacle glass" is closed a little bit more than half the time and the projector has to divide the light between two eyes, one can see that the glasses feel like very good sunglasses where the light reaching the eye is reduced by more than 80%. Something similar happens at passive stereo, however the reduction of light is caused by the polarization and as each eye has its own projector the reduction will just be a little bit more than 50%. These indications vary from product to product. Thus, active stereo has a disadvantage regarding light intensity and requiring rather expensive glasses. The light intensity is no longer a problem, as DLP projectors have a very good light intensity and therefore compensating for this loss, but it still requires the glasses.

One of the things that makes active stereo somewhat more expensive than passive stereo is the requirement for 120 Hz image frequencies, because ordinary projectors are not built for this frequency. The LCD technology cannot at the moment at least reach this frequency because of the afterglow in the crystals. The analogue CRT technology can provide both the frequency and the correction of the image geometry; on the other hand, it has some problems with the light intensity. The new digital DLP technology can provide the image frequency with the expensive models and supplies enough light in even the smallest models. A correction of the image geometry e.g. to a curved screen requires good software and is used in the high end of the model range.

Colorcode 3D (http://www.colorcode3d.com) or Infitec (http://www.infitec.net) are other systems on the market that try to achieve the stereo effect and new approaches are on their way, however some of these have other goals than just showing 3-D graphics in stereo and other drawbacks like e.g. Colorcode 3D is not that suitable for people with distinct eye-dominance either on the left or the right eye and Infitec uses very much light compared with other systems.

3.2 *Display systems*

The stereo effect is a very important part of the presence one's experiences moving around inside a virtual building. Another part is the image one looks at. This issue is not that simple since there are many ways of representing a 3-D model on a display.

The light beam creating the image has to hit an obstacle in order to present the image. The light beam has to reflect on a defined surface. Traditionally this surface is e.g. a white screen or a wall, but in principle, there is nothing standing in the way for using a surface consisting of e.g. water,

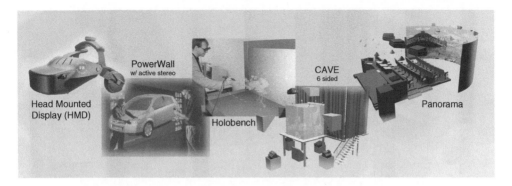

Figure 3.   Iconic overview over different immersive display-systems.

which for instance is used by the American military in target practice. Something less fierce than water can also be used, e.g. the Fogscreen where pictures are projected onto a neutral type of gas (http://www.fogscreen.com).

When the special display types are left out, the different VE setups can be organized into these typical groups:

– An ordinary computer monitor where sizes like 30″ is common. Several screens can be combined and form a large surface.
– Single screen solutions known from meeting rooms. These screens have grown quite a lot size wise and become increasingly suited for presenting VR/VE.
– Multiple screen solutions or large screen solutions where a large number of projectors are combined in order to create a large combined image on the screen. Here, the screen can be curved in order to strengthen the effect of the immersiveness.
– A special edition of the multiple screen (multi pipe) solution where the screens are placed perpendicularly to each other and form a little room is called a CAVE. This setup is the ideal immersive solution providing a virtual scenario for only one person. A CAVE can consist of four, five or six sides. (Cruz-Neira, Sandin et al., 1993)
– A non-immersive device looks more like a table where the screen is mounted on the table and on the table's edge perpendicularly to the table surface. This solution is often called a holobench or workbench.
– A Head Mounted Display (HMD) is a display consisting of two small monitors placed right in front of one's head and provides a screen size of e.g. 50″. These displays are to most people synonymous with VR and cyber-space since they were used in the early VR days. There are different types available. The newest HMD's use technology like Microvisions virtual retinal display technology (http://www.microvision.com) where laser light draws the image directly on the retina, which is placed on the backside of one's eye.

The pixel resolution in the images can vary a lot, but should not be less than 1000×1000 pixels per projection surface or "approximately 1,00,000 pixel pr. square meter" and should be "approximately 150.000 pixels pr. square meter". Fortunately it can be observed that the manufactures of graphics boards and display-systems are increasing the resolution consistently.

Another technical feature related to the display-technology which should be mentioned is the front-projection versus the back-projection technology. The two systems are illustrated in figure 4. At the front projected display the image is projected on a screen and reflected back to the viewer while the back projected image is coming from behind the screen and beamed through the screen onto the viewer.

Both have advantages and disadvantages. The back projection has the great advantage that the viewer can stand close to the screen as the body will not make any shadow in front of the light beam

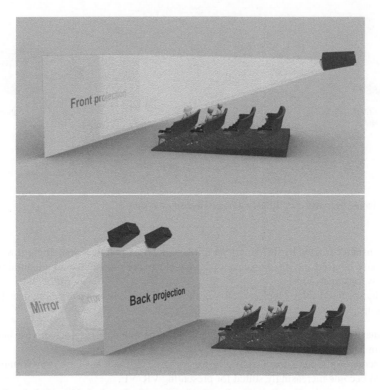

Figure 4.   Room-design with front-projection versus back-projection.

which is usually seen from an ordinary front projection or from old-day-cinemas. On the opposite the back projection requires somewhat more space behind the screen and the intense light-beam is more clearly spotted in the center-area even though good lenses and an appropriate arrangement can eliminate this disadvantage. Both active and passive stereo can be used with both systems.

The above list is not in any way complete, especially not as several new systems are on their way, but it does present some principles among available display-systems in use. In the following, this list will be commented and a kind of recommendation will be given in relation to what kind of screen technology can be or should be used to the various applications.

3.3   *Which display to use*

What degree of immersion do I want? – Is the first question one should ask. The degree of immersiveness is increasing the more the peripherical view is enclosed cf. figure 5. This means the less one gets the real world involved in the virtual one, the better the immersiveness. Well, this is not entirely true since HMD's e.g. are not having the real world interfering but have a massive dark frame around the view which of course reduces the FOV and also the immersiveness. Nevertheless HMD's are the first step to immersive VE, but are not really cheap. Three levels of immersiveness can be obtained respectively from a CAVE (fully immersed), to a Panorama (partly immersed) to a Powerwall (slightly immersed). The price difference between a Panorama and a Powerwall is not that significant but when it comes to immersiveness and the first person perspective the CAVE and even the HMD are preferable. The Panorama is an excellent facility for presenting building, city- and landscape models in a sensible scale for a medium sized group of 10–20 people while the Powerwall can handle much larger groups but lacks of real immersion. Therefore these two facilities

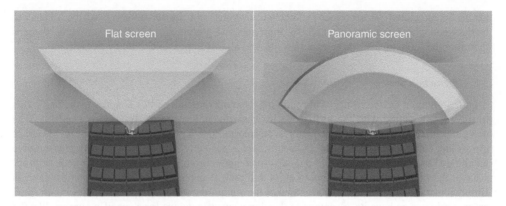

Figure 5.   Comparison regarding screen shape and field of view.

primarily can be used for debriefing or for the head of operations during training procedures or even for following life action scenarios.

Without any doubt, ordinary screen solutions will take one far, especially at this point as they have obtained a significant size. However, the conditions of presence are simply not given. One cannot feel presence in a virtual world if the immersiveness up to a certain extent is not present.

It is a general opinion that HMD's provides a good immersive experience though many people get annoyed of the heavy equipment and feel silly wearing it. The advantage of good HMD's lies in the built-in tracking, as the point of view is a well-defined point from the viewer. Thus, the image will always obtain a correctly calculated perspective in accordance with the position of one's head. Though some calibrating here is quite often necessary and this task is not always as easy as it could be.

Actually, there are no VE systems that provide the optimal solution. The closest one comes to a good presence experience is in a CAVE where only the surroundings limit free mobility and the active stereo glasses limit the field of view, however the CAVE does provide the possibility to create a virtual universe in all view directions including floor and ceiling and gives one the idea of being moved physically to another place. Eventually, a plug-in for the brain has to be invented in order to remove the last distinction between the real and the virtual world.

There are several "of the shelf" VE systems on the market that are more or less suited for ER. As more complex one chooses his solution as more maintenance one will have to take into account. But the maintenance will not be one's only concern. Running a VE with all its hardware and software is still a bit more complicated than running a desktop computer. That's why most of the VR centres have staff people who are specialized in this kind of systems. Especially if the system is run on a daily basis it is advisable to employ dedicated personnel for the handling and maintenance of the VE.

3.4   Tracking

As earlier mentioned, tracking is as important as the display type for creating real immersiveness. With tracking means retrieval of certain points within a CAVE or in front of a screen. These points are the point of view and view-direction of the observer called head tracking and possibly the position and orientation of the interaction device. Tracking are usually (there are others) obtained either optical, with sound or with electro magnetic fields. All technologies have their advantages and disadvantages due to the surrounding and interference of obstacles and certain materials, clumsiness due to wearing, accuracy etc. The tracking technology should be discussed in detail with the vendor since finding the right solution often is a question of finding the best compromise.

Tracking though is not to be excluded in a VE even if it makes everything about the VE just so much more complicated. Tracking requires extra peripherical units; the geometry of the model has to be calculated for a correct perspective viewing; and the surroundings of the VE has to be designed with respect to the choice of materials and placement of cables or emitters. Perhaps therefore, building up a VE is an experts work and almost impossible to do by oneself.

3.5 Interaction devices

A quite important device, which will round off the hardware part of this chapter, is the one wants to use for interaction with the virtual world. Unfortunately none really good is available. The computer mouse has more or less been transferred to the VE with a doubtful success. There are gloves and cubic mice (Fröhlich et al., 2000), stylus pens etc. but they all lack of really intuitive handling. The main problem is that one has to navigate through a virtual space with an arbitrary scale at the same time as one wants to point and pick at things in a spatial environment. This is simply unnatural why this combination makes it even more unnatural to put into a device. In some cases it might even have its advantage to develop a specific device for a specific purpose like the cubic mouse is designed for the exploration of a CAD-model or geospatial model. In the area of ER it makes sense to develop interaction devices specific for the different parts of action during a training session. E.g. tracking of tools and dealing with collision detection during an operation in a building. Perhaps the navigational part has to be thought through thoroughly due to some special movements or devices that are used in special cases or situations. Virtual Reality is characterized by its artificial environment where primarily the fantasy draws the limits.

As important as these devices are, this chapter will not discuss this matter any further but urge the reader to discuss this matter with the vendor of the VE system, as the devices should be chosen very close in connection within the scope of the VE, the tracking system, the display system and so on.

4 CREATING THE MODEL

One major part of using a VE system is the content. One tends to focus a lot on the hardware but forget to think about the software and the content of the system. Nevertheless it is getting easier and easier to find some major software solutions helping one to build up the virtual environment. Unfortunately these very flexible systems can be quite expensive. The price is tight connected to the flexibility. The flexibility is due to scalability and functionality of the VE. This means that systems that can handle different setups including multi-pipe VE systems like the CAVE or Panorama are more expensive than systems, which primarily are envisaged for the PC with a single or dual monitor. There is simply a huge difference between being able to handle multi-pipe systems with tracking functionality and just making a game like application for mouse and monitor.

On the other hand the difference on the content site is much less pronounced. Building up the model is quite similar whether one chooses a real VE or a game like system. The way of interaction might give some difference and more possibilities in the VE system, but preparing for it is more or less the same workload.

There is a large amount of software that can be used for creating a 3-D model of the situation one wishes to exercise even for one who is not familiar with 3-D modelling. Whether one wants to create one precise model of a single building for the purpose of ER operations in a building on fire, or one wants to build up a whole city or landscape area for the purpose of disaster management, the workload will have a direct ratio with the numbers of polygons one needs in the model. Only a few years ago the VE system had a kind of limit at about 1,00,000 polygons. Models above that size, tended to be very slow within a VE. Today models easily exceed 1 mio. polygons and are still presented with a decent performance.

It can sometimes be hard to understand why 3-D games are capable of producing very high frame-rates with nice detailed models and beautiful textured surfaces full of structure together with complex light and shadows. Games have the advantage that they do not need to deal with the real world pursuant to structures like buildings or vegetation or for that matter correct lights and shadows. For instance, a city model tends to have a very high demand on correct texturing of the facades as one wants to recognize the buildings, whereas games reuse the same texture over and over again and save a lot of computational power that way.

Modelling a building can be quite time consuming especially if one wants to put some effort into important details. For larger areas like cities it is more and more common simply to order those from specialized companies who use laser scanning techniques or other instruments to build up areas in a semiautomatic way. The biggest issue though is to model people and natural movements for instance in a crisis situation. Getting people looking natural or injured and act as human beings is very difficult. For special situations like pedestrians in city areas or travelers at airports one can find different simulation systems that deal with this kind of rather simple human action. But in crisis situation where some people panic, some act rational while others again don't move at all it is much more difficult to simulate this kind of behavior of a virtual person who at the same time should look as real as possible.

There is a lot of room for improvements and the ER domain should play an active role with its experience and knowledge. The technology is quite new and immature so on one hand developments can be pushed in the direction needed for ER on the other no one should expect to simply buy an of the shelf VE, that considerers all needs of an ER-VE.

5 CONCLUSION

Even though VR applications are few and specially designed VE systems for the purpose of ER training in a virtual environment are even fewer still, there are three major conclusions to draw.

First of all; some applications have been developed by now and some first experiences have emerged. Unfortunately the demands are high and the system developments for ER are not really driven by the forces of the commercial market but by governmental budgets. One way to bypass this bottleneck if lack of economical resources can be called that is to do some partnering with the military if possible. A lot of basic demands and applications are the same but the time being the financial aspects can probably be covered more easily by the military (allegation).

Secondly; the hardware is evolving day by day. Hardware gets better and cheaper as time goes, but one should also understand that even if things are getting better and more sophisticated the gear that can be bought today is quite capable of coping with the virtual environments in a way that one can benefit from it. The initial costs are high and even the maintenance costs should not be underestimated but the results are considerable. And lifes saved by well trained personnel are priceless.

Thirdly; the applications presented here are not really representative of the potential within the ER domain in virtual environments but rather an indication of a new media which only has been used sparely. This is probably caused by the lack of knowledge about the possibilities and about the requirements to get started but also lack of economic resources.

This article was meant to help clear up the uncertainty about some of these issues, but can not give any suggestions on where or what to buy. One thing, which has not really been touched until now, is the cost of it. It is almost impossible to give a price which can be justified. Nevertheless if one wants to work with real immersive solutions one can start up with a passive solution on one screen for less than €20,000. If one wants a three pipe solution to build up a Panorama or three sided CAVE with active stereo one needs a couple of hundred of thousand Euro(€). But then again, this is still just the price for the hardware.

REFERENCES

Cruz-Neira, C. and A. Bierbaum, et al. (2002). VR Juggler: An Open Source Platform for Virtual Reality Applications. 41st Aerospace Sciences Meeting and Exhibit, Reno, NV, AIAA.
Cruz-Neira, C. and D.J. Sandin, et al. (1993). Surround-Screen Projection-Based Virtual Reality: The Design and Implementation of the CAVE. SIGGRAPH 93, Annual Conference Series, Anaheim, CA, ACM Press.

Freund, E. and J. Rossmann, et al. (2003). Enhancing a Robot-centric Virtual Reality System Towards the Simulation of Fire. Intl. Conference on Intelligent Robots and Systems, Las Vegas, Nevada, IEEE/RSJ.

Gamberini, L. and P. Cottone, et al. (2003). Responding to a Fire Emergency in a Virtual Environment: Different Patterns of Action for Different Situations. Ergonomics 46(8): 842–858.

Hadipriono, F.C. and J.W. Duane, et al. (2003). Implementation of a Virtual Environment for Traffic Accident Simulation; Part 1: The 3D Models and Control Panels. Journal of Intelligent & Fuzzy Systems 14: 191–202.

Hadipriono, F.C. and J.W. Duane, et al. (2003). Implementation of a Virtual Environment for Traffic Accident Simulation; Part 2: Developing the Virtual Environment. Journal of Intelligent & Fuzzy Systems 14: 203–214.

Leo, G.D. and M. Ponder, et al. (2003). A Virtual Reality System for the Training of Volunteers Involved in Health Emergency Situations. CyberPsychology & Behavior 6(3): 267–274.

Li, L. and M. Zhang, et al. (2005). ERT-VR: an Immersive Virtual Reality System for Emergency Rescue Training. Virtual Reality 8(3): 194–197.

Louka, M.N. and C. Balducelli (2001). Virtual Reality Tools for Emergency Operation Support and Training. In Proceedings of TIEMS 2001 (The International Emergency Management Society), Oslo.

MacEachren, A. (2005). Moving Geovisualization toward Support for Group Work. Exploring Geovisualization. J. Dykes, A. M. MacEachren and M.-J. Kraak, Elsevier Science Ltd.

McCarty, W.D. and S. Sheasby, et al. (1994). A Virtual Cockpit for a Distributed Interactive Simulation. IEEE Computer Graphics and Applications 14(1): 49–54.

Nakanishi, H. and S. Koizumi, et al. (2005). Virtual Cities for Real-World Crisis Management. Digital Cities 2003. P. van den Besselaar and S. Koizumi. Heidelberg, Springer-Verlag.

Psotka, J. (1995). Immersive training systems: Virtual reality and education and training. Instructional Science 23(5–6): 405–431.

Rauschert, I. and P. Agrawal, et al. (2002). Designing a human-centered, multimodal GIS interface to support emergency management. Proceedings of the ACM Workshop on Advances in Geographic Information Systems: 119–124.

Rheingold, H. (1991). Virtual Reality. London, Mandarin Paperbacks.

Rosenberg, B. (2003). USAF Employs Super-High-Res Camera for Disaster Training. Aviation Week & Space Technology. 159: 90.

Sherman, W.B. and A.B. Craig (2002). Understanding Virtual Reality – Interface, Application, and Design. San Francisco, Morgan Kaufmann Publishers.

Steed, A. and E. Frécon (2005). Construction of Collaborative Virtual Environments. Developing Future Interactive systems. M.-I. Sanchez-Segura. Hershey, PA, Idea Group.

Uhr, M. (2004). Transfer of Training from Simulation to Reality, Investigations in the field of driving simulators. Zürich, ETH.

Wössner, U. and J.P. Schulze, et al. (2002). Evaluation of a collaborative volume rendering application in a distributed virtual environment. Proceedings of the workshop on Virtual environments 2002, Barcelona, Spain, Eurographics Association.

Yi-li, S. and C. Yang, et al. (2002). The exploiture of earthquake surroundings VR system. Journal of System Simulation 14(11): 1509–1512.

Geospatial Information Technology for Emergency Response – Zlatanova & Li (eds)
© 2008 Taylor & Francis Group, London, ISBN 978-0-415-42247-5

Visual analytics in flood forecasting

M. Jern

ITN, Campus Norrköping, Linköping University, Norrköping, Sweden

ABSTRACT: The ever-increasing massive amounts of multisensor, multidimensional and time-varying digital information streams represent a major challenge for emergency operations. A solution to this challenge could be found in the emerging Visual Analytics (VA), the science of analytical reasoning facilitated by interactive visual interfaces and creative visualization. This chapter will address the use of VA technology in emergency response based on flood forecasting as a means of risk communication as well as risk exploration. A flood forecasting tool "FloodViewer", developed with the spirit of VA, is demonstrated and evaluated based on a real world case study scenario in Italy during the near-flooding event that occurred on November 20–21, 2000 along the Arno River that floats through Florence. This prototype highlights the different approaches that can be used in explorative visualization and illustrates the use of dynamic and animated time geovisualization that targets improved flood forecast insights such as how flood level could rise based on precipitation data from gauge stations and radar. The proposed visual methods could encourage the development of new tools and improve the flood forecasting reliability and precision and hopefully shorten the time required to detect events that lead to catastrophic flood events.

1 INTRODUCTION

The problem of flooding is as old as time. However, while natural flooding of large areas did not create situations more dangerous than others in a prehistoric world, the expansion of human activity and cities has made preventing damage caused by floods or harnessing over-bank flows for one's own purposes as in ancient Egypt a necessity that remains vital to this day. Given the large number of high risk situations as well as the extremely high costs in terms of casualties and damages involved, it can be reasonably estimated that a flood forecasting decision management system, which combines advanced visualisation, interaction and GIS capabilities with modelling for the planning and the preparation of risk maps, emergency plans and the real time analysis of possible interventions, has a large potential market. But without also strengthening the dissemination link in the chain, the performance of the whole system is degraded. The advancement of the accuracy of flood forecasting, and better targeting of the dissemination of the resulting message, is not only for the benefit of the forecasters but also of the public. In this chapter we also present an innovative visual user interface that will enable the users to take a more active role in the process of investigating flood forecasting and illustrate it with an example of monitoring a day of risk in the Arno basin area of Tuscany.

Norrköping Visualization and Interaction Studio (NVIS) in Sweden which, since its start in 2000, has grown rapidly and today constitutes one of the strongest focal points for advanced visualization in the Nordic region. NVIS now has a staff of approximately 40 with a mix of PhD students, senior researchers and technicians. Our recent applied Visual Analytics (VA) research involves adopting information and geographical visualization in emergency response. We collaborate with Swedish Meteorological Service SMHI and an EC-funded consortium for a flood forecasting system, developed in collaboration between University of Bologna, ET&P Italy, CNR Italy, University of Newcastle, Gematronik Germany and several European Water Management institutes.

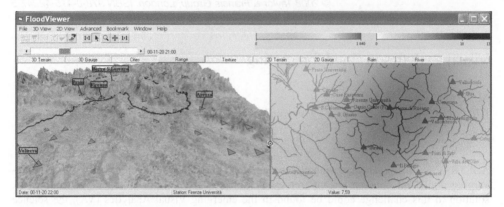

Figure 1.   Applying Visual Analytics (VA) methods to flood forecasting enables meteorologists and hydrologists to be more interactive and efficient in the process of visualizing and investigating flood scenarios, providing better understanding of the data and uncertainty behind the forecasting system. The right view shows hourly dynamic animated precipitation visualizations of radar rainfall data for a 24 hours time period, superimposed on the river net and gauge stations using opacity blue colour contouring. Left view represents a 3D landscape overview of the potential flooding area in Toscana Italy, highlighting the Florence University gauge station.

The basic role of any real-time quantitative precipitation and flood forecasting system lies in its capability, within the forecasting horizon, of assessing and reducing the uncertainty in forecasts of future events in order to allow improved warnings and operational decisions for the reduction of flood risk. A key objective here is seeking to improve the communication and the dissemination of results to the authorities involved in real-time flood forecasting and management (Todini, Catelli, Pani, 2004).

As well as warning about an impending flood, civil protection agencies need to know what is happening during a flood and what is going to happen. New visualization tools allow the representation of the observational data, the forecast and the uncertainty in a clear, intelligible manner via either the Internet or visual mobile communications. Such tools may be fine tuned to particular end user groups such as flood forecasting specialists, the director of civil protection, controllers of emergency services and the concerned public. An operational flood management system is considered as a sequence of five activities where visualization has a crucial task:

- **Detection** of the likelihood of a flood forming (hydro-meteorology);
- **Forecasting** of potential river flow conditions for the hydro-meteorological observations;
- **Warning** issued to the appropriate authorities and the public, severity and timing of the flood;
- **Response** by the public and the authorities;
- **Review** of the experience from the post-flood management activities to improve the process and planning for future events.

1.1   *VIS Arno River November 20–21, 2000*

Our flood forecasting model is tested with data from the Arno River during the near-flooding event that occurred on November 20–21, 2000. It is necessary to identify not only the immediate factors that caused the alarm, but also the underlying causes of the phenomenon. Indeed, the declared seriousness of the event was not only due to the quantity of rainfall, but even more so, due to the occurrence of what meteorologists call a "stationary rainfall phenomenon." Arno runs along the edge of the Apennines of Tuscany and Romagna and these mountains act as a barrier to the passage of storm clouds in the southwest direction, thereby blocking the precipitation so that it remains stationary over a limited area – the basin of the Arno River – whose soil also happens to be prevalently impermeable.

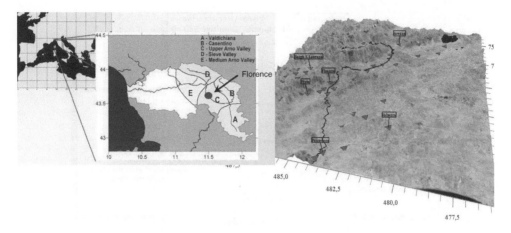

Figure 2. Location of the Arno River Basin and a 3D visual representation.

2 VISUAL ANALYTICS

Visual Analytics (VA) is an emerging and interdisciplinary frontier defined in the book "Illuminating the Path" (Thomas & Cook 2005) as the science of analytical reasoning facilitated by interactive visual interfaces. VA is seen as the logical next step in the long development of computer-based visualization methods which have progressed, as computational technology has advanced, from fundamental computer graphics used to display features present in scientific and other data, through more sophisticated visual representations of complex information to the current sophisticated data management and multimodal visualization tools which are becoming available now. A flood operator must make quick decisions about water level rising above a critical level. Immediate response to any visual query, often in time-critical situations, demands ever increasingly sophisticated tools that aid the user to manage process and interact with these massive data streams. The essence of this challenge can be captured in the following strategic statement about visual information fusion:

• Computer storage capacity and the rate of new information flow currently outdistance our ability to manage or analyze them;
• To respond, a new generation of information management and visualization tools must provide new abilities to efficiently locate and analyze information;
• User-friendly visual analytics tools for manipulating, data mining, and visualizing highly multi-dimensional data will arm a generation of analysts, just as spreadsheets did a generation earlier.

VA aims at bridging the gap between the machine and the human mind, and takes advantage of human perception capabilities and can be described as "find patterns in known and unknown large dataset via visual interaction and thinking". Several new trends are emerging from VA and among the most important one is the fusion of visualization techniques with other areas such as data mining, databases and geographic visualization to promote broad-based advances. Another trend, which has often not been well met to date by visualization researchers, is the realization that algorithmic and other technical development should be closely coupled with usability studies to assure that techniques and systems are well designed and that their value is verified and quantified.

Decision makers, analysts, and engineers are often confronted with vast amounts of complex, seemingly inconsistent or incomplete data, which is derived from a number of heterogeneous sources (MacEachren 2001). The goal of analyses is transparent knowledge, which can be applied in the context of future decisions. To provide solutions for managing information overload we need to build a bridge between the advantages of both human perception and computer science technologies.

VA needs interdisciplinary science beyond traditional scientific and information visualization to include statistical analysis, mathematics, knowledge representation, management and discovery

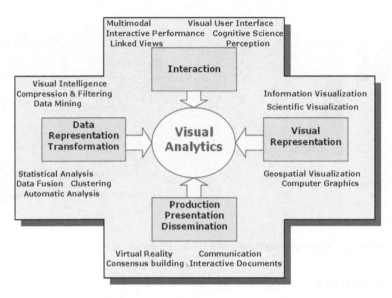

Figure 3.   Four research areas constitute the foundation for Visual Analytics.

technologies, cognitive and perceptual sciences, decision sciences, and more. VA will arise from a combination of the following four research areas (Figure 3):

Visual representations: drawing on disciplines such as scientific-, information- and geospatial visualization as well as on fundamental computer graphics.

Interaction: allowing the user to explore and understand large amounts of complex data by identifying structure, patterns, anomalies, trends, and relationships based on cognitive and perceptual principles.

Data representations and transformations: when mined, converted and fused into suitable data formats these provide for richer and more informative visualization.

Rendering, presentation and dissemination: communicate information in the suitable context to a variety audience taking advantage of state-of-the-art rendering techniques and Virtual Reality.

Combined types of innovative VA techniques will be used to provide timely, defensible, and understandable assessments and communicate knowledge effectively for action. VA offers the potential to provide managers, analysts and experts in academia, industry and public services with competitive decision-making tools to:

- Keep pace with increasing model complexity – see essential information more quickly.
- Uncover opportunities, risk and trends which would have gone unnoticed before.
- To develop a pro-active approach to decision making and understand the reasoning and validity behind it.

3   ARCHITECTURE OF THE FLOODING SYSTEM

The technology task of a flood management system has been described in detail by (Todini, Catelli, Pani, 2004) and should include:

- A set of *relational and mathematical models*;
- A *knowledge-based system* which manages data flows to and from the database; executes the mathematical models in a supervised sequence; organizes and compares the performed scenarios; guides the user in making decisions regarding a particular issue or problem;

220

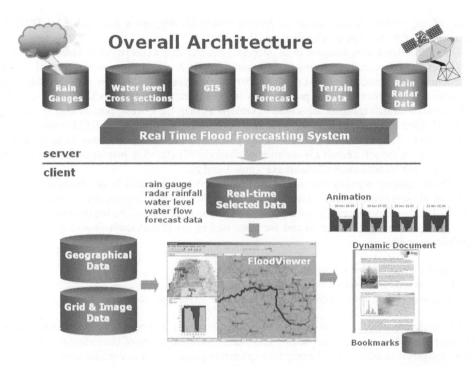

Figure 4. The overall architecture of the FloodViewer system. Requested real-time flood forecasting data (radar rainfall, rain gauge, water level, forecast data) is transferred across the Internet from the server-based database to the visual analytics client application "FloodViewer". Geographical and grid data for the selected case study area (Arno River) is stored in a local database file and provides highly explorative and dynamic visualization. Communication and presentations are generated as animations, dynamic documents and bookmarks (snapshots).

- A *database management system*, includes a series of primary and secondary data treatment procedures, designed to optimally manage the historical series of climatic measures (rainfall, temperature, etc.), information concerning watershed entities (rivers, sub-basins, etc.) and the environmental data;
- *GIS tools* which provide support for management of geo-referenced data;
- A *visual analytics component* based on advanced visualization and interaction and at the same time, easy-to-use and support qualitative analysis and interpretation of results.

The capability to fuse the relevant information from massive amounts of divergent multi-dimensional, multi-source, time-varying digital information streams represent a major challenge for emergency operations. Figure 4 shows the overall architecture of a real-time flood forecasting system. Data (radar and satellite-based rainfall observation, rain gauge measurements, water levels, forecast data in terms of probability and reliability) is collected and combined in the real-time flood forecasting system. Rainfall fields derived from satellite, radar or rain gauges measurements, all have different error structures which need to be recognised when they are combined. Block Kriging is used to create a rainfall field from rain gauge data combined with existing radar and satellite estimations to obtain a minimum variance estimate.

The flood forecasting system architecture was designed and implemented by Italian company Environmental Technologies and Products Srl (ET&P) and process spatio-temporal radar rainfall, rain gauge, water level and Block Kriging information stream in real-time to the client-side visual analytics application component "FloodViewer". Other data considered in the flood forecasting system includes: time series of hydro-meteorological quantities, structural maps, and specific series of maps for the description of the precipitation fields originating from radar, METEOSAT over regular, although non-overlapping, grid domains.

The system database is designed in an integrated manner with a GIS structure based on GRASS for direct geo-referenced management of structural data; A combination of a relational (RDB) and map geo-referenced (GeoDB) database. Structural maps as DEM, river network, sub-basin boundaries, time series of maps, precipitation field estimates from radar, METEOSAT, and any other kind of rainfall estimate are stored in GeoDB as map files. An important step in creating practical flood forecast visualization is to obtain elevation data through a DEM (raster GRID) of the area to be analysed. Elevation data was calculated for 500-square-meter tiles. A satellite raster image of the selected area is superimposed on the 3D elevation map to enhance the visual experience and identification of the area (Figures 1 and 2).

A flood planning, warning and control decision support system must be developed to meet the needs of all users in the system, which include: meteorologists/hydrologists, operational managers and civil response managers. The visual communications links that will be required between these three groups have been examined and are addressed in the FloodViewer project through three application scenarios:

(1) *Analyse and Explore* – Standalone single user version.
(2) *Collaborative Analysis* – Collaborative multi-users version.
(3) *Communication, Presentation and Dissemination* – Dynamic document.

The analyst first explores the data in a FloodViewer session and prepares the results for review by other users. This step includes setting bookmarks and annotations of interesting discoveries to share insights with other users. During the exploration process the user can also start a collaborative FloodViewer session with other experts who are not available in the same office and visually collaborate with the same dataset remotely. This cyber-space group can visually discern and concretise already discovered patterns and/or new, earlier overlooked ones enabling more accurate interpretations of flood forecasting issues. In a final step, the result of the exploration process can be presented to public through a Microsoft Office Word or PowerPoint presentation integrating a FloodViewer application component together with bookmarks and annotation data. Scenarios (2) and (3) are described in chapter 5.

3.1 *Visual representation to support data fusion*

A challenging task in VA is the capability to fuse the relevant information from divergent multi-source, multi-dimensional, time-variant information streams. New visual metaphors and data structures are required for specific data types or information streams that allow the relevant information to be combined into a single information space. For example, flood information involves heterogeneous data that must be integrated, synthesized, and viewed at multiple layers (Figure 5). Current visual representations often focus on a single attribute and do not enable analysts to understand the richness in complex heterogeneous information spaces. In real-time flood emergency management, analysts need to integrate information involving geospatial locations, rainfall data from gauge stations, weather, and rain radar sensor into a single application environment. Visual representations and interactions are needed that can represent relationships without oversimplifying information and help the analysts explore and understand the complexity of their information.

FloodViewer is fuelled with geospatial data from many sources (Figure 5) including temporal components (actual and forecast data). These huge data must be represented, combined, and transformed to enable users to detect and discover potential risk for flooding. The volume and complexity of flood data combined with their dynamic nature provide a challenge for visual and quantitative analysis of such data. The client database incorporates support for both raster and vector type representations and some data have temporal attributes such as rain gauges, radar and water level. The two basic representation schemes (raster grids and vector) used for geospatial data operates as independent and distinct representations.

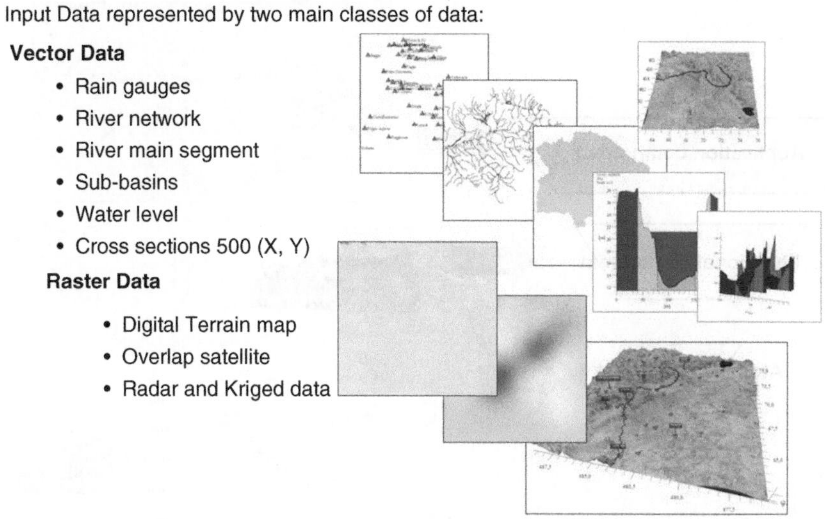

Input Data represented by two main classes of data:

Vector Data

- Rain gauges
- River network
- River main segment
- Sub-basins
- Water level
- Cross sections 500 (X, Y)

Raster Data

- Digital Terrain map
- Overlap satellite
- Radar and Kriged data

Figure 5. The FloodViewer visual flood forecasting system is fueled with geospatial data from many sources including spatio-temporal components (actual and forecast data). These huge data must be fused, represented, and transformed to enable users to detect and discover potential risk for flooding.

FloodViewer provides visual analytics support for fusion of multiple data sources based on:

- Methods for visual information fusion producing visually consistent and meaningful views of data from several sources including dynamic linking and coordination of data.
- Definitions for consistent interfaces for visual query specification and visualization of results extracted from distributed data sources.

3.2 Component architecture – reusing and distributing results

Our prototyping and application integration is based on individual encapsulated objects and customization. Customizable and scalable high-level application and functional components are designed and developed with layered component architecture (Figure 6). Generic low-level basic and functional components, each one performing a specific task in the overall VA process, are put together into high-level application components. This component approach, as implemented in Microsoft .NET, enables broad applicability, customization, application scalability, reusability of components and shortens the development time.

Interoperability is an idea that is core to the component architecture and this has been invaluable to our development of the FloodViewer. Different developers, working almost entirely independently, can contribute software components to a common, quality-assured prototype. VA components obtained from this collection can then later be combined into larger assemblies using a variety of interconnection mechanisms and are readily transferable to most risk analysis and decision support intensive sectors including emergency response. Example of visual representation components: 2D animated image, 2D contour, 3D and 4D surface, vector data (river, basin), time bar chart, axis, colour legend, 2D and 3D water cross section.

Our application development is based on the rapid prototyping approach: iterative development cycles ensuring constant improvement of the system and so reaching the specified goals. Practical testing in close collaboration with the end user partners proves the validity, usability and attractiveness of the applications produced. Intuitive interfaces, fast and individual configurability to the varying demands of different application types, and the combination of modern communication technologies will guarantee natural man-machine interaction.

This approach to system architecture could, for the first time provide the infrastructure for true customisable climate and flood forecasting system interpretation facilities. The open component

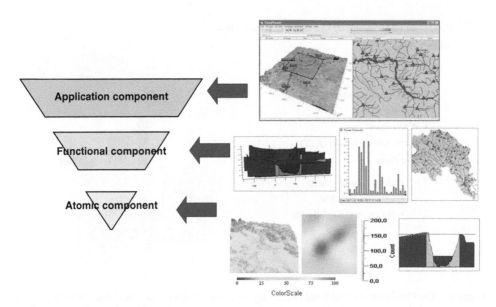

Figure 6.   Layered component architecture is the foundation for FloodViewer. Low-level VA components are combined into functional components and application components.

architecture allows easier integration with geo-computational methods such as spatial statistics, classification and databases.

4   FLOODVIEWER

Three areas are essential for understanding our flood viewer model: visual representation, interaction, and integration. Representation refers to the visual form in which the model appears. A good representation displays the model in terms of visual components that are already familiar to the user. Interaction refers to the ability to see the model in action in real time, to let the user play with the model as if it were a machine. Integration refers to the ability to display relationships between the model and alternate views of the data on which it is based and provides the user with context. All three of these techniques help the user understand how the model relates to the original data and provide an external context for the model and helps to establish semantic validity.

We have set the following generic requirements for the flood visualization tools:

- Exploratory spatial and temporal data analysis;
- Easy access to database and use;
- Support for uncertain and missing data;
- Interactive performance that gives the user a sense of immediacy and speed-of-thought;
- Tailor-made application with individual encapsulated objects;
- Shorten development time by utilising already developed and assessed components;
- Dynamic resizable linked views that maximize screen area;
- Design based on cognitive and perceptual principles;
- Application and data accessible through the Internet.

4.1   *Visual user interface*

We propose dynamic and interactive visual representations for visualizing risk and uncertainty. This is particularly true when exploring real-time forecast data from a flooding model. A static bar chart of standard deviation values would be of little use to decision makers. Visually exploring the

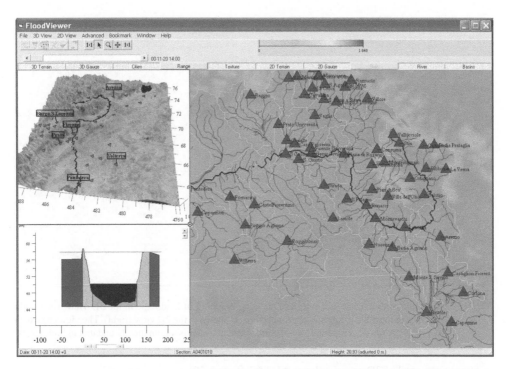

Figure 7.   Dynamic flood forecast mapping – FloodViewer visual user interface. Flood forecast data from the entire Arno River region (Toscana, Italy) is showed in resizable and coordinated 2D and 3D views. The 3D digital terrain map with overlaid satellite image of the region provides full viewing control and "focus-and-context" selection. The main or 2D working view (right) shows river network, terrain, gauge stations, basin boundaries for a selected view interactively defined in the 3D map. The attribute view (lower left) displays rainfall and water level time-variant data and links with the working view. The 3D View helps with orientation and navigation. The user can zoom in-and-out or scroll the 2D or 3D views.

implications of variance is far more important. This has direct bearings to risk visualization where one static display of a certain set of parameters does not give a full picture of the variability in the data. Exploring different combinations of risk related to the confidence of the derived risk may give a different view than the risk value alone. For example, if the risk is high, but its confidence is low, then is the risk actually high? It is the responsibility of the risk assessor to use whatever information us available to obtain a risk approximation with as much precision as possible together with an estimate of the uncertainty.

Of central importance to the FloodViewer project is therefore the Visual User Interface (VUI) that enables the user to directly manipulate graphical objects, which respond interactively and immediately to the user's input actions (Hochheiser & Shneiderman 2004). The user interacts with both 2D and 3D objects with an immediate graphical feedback response, without the need for moving to a secondary GUI menu window. The 2D view "area-of-interest" is selected in the 3D landscape view using a movable orthoplane with 8 handles that control the size and location of the coordinated 2D view. The 2D view can also be further zoomed and panned with simple mouse moves and the selected working area is automatically updated in the 3D view always updating the user of the current position in the landscape (Figure 8). This feature is supported through two-way coordinated pipeline architecture. Interesting areas of the display can be selected and zoomed in on, or flown through for closer investigation. Example of additional direct manipulation features:

• Select gauge station and view bar chart displaying detailed hourly rainfall (Figure 10)
• Analyse dynamic time animation of radar rain data

225

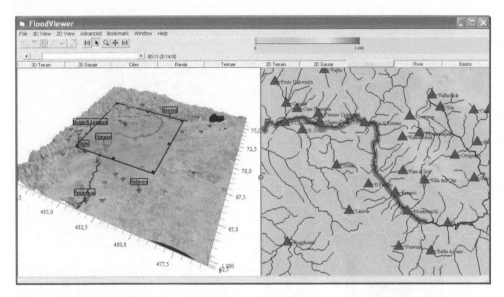

Figure 8. Example of the visual user interface showing cropping of landscape in the 3D view using eight interactive movable handles. The cropped area is shown in the 2D view (right). The views based on two-way coordinated pipeline architecture – selected area is automatically updated in both views.

- Dynamic control of 2D opacity planes "Z-thru" (Figure 10)
- Select water level station and view 2D or 3D cross-section
- Select a river segment for dynamic time animated 2D or 3D cross-section.

4.2 Multiple linked views

Flood forecasting data are not best analysed through the use of a single type of graph (Roberts 2004). In order to detect complex patterns within the data it is necessary to view it through a number of different visualisation methods, each of which is best suited to highlight different patterns and features and reveal different aspects of the data. Linking and relating the information in one view to that of other views assists the user in the exploration process and may provide additional insight into the underlying information.

Coordinated and linked views of the same data are important features in most geovisualization task and enable users to rapidly and dynamically compare visualizations and data (North & Shneiderman 2000 and Andrienko & Andrienko 2003). The user can analyse spatial and temporal flood forecasting data and derive a deeper understanding of compound properties through data correlation and multiple display methods of the same data. A common approach is to display each view in a separate GUI window and allow the user to arbitrary arrange the windows. The notion of linking and arranging views is generic and is available in some visualization systems. Tools for arranging views and controlling the size of individual views are indeed not simple programming tasks. We therefore propose using Microsoft's Visual Studio's. NET hierarchical layout management, with dynamic embedded resizable views in a single coherent GUI window that will maximize the use of the screen area.

Ensuring that a single, dynamic multiple-view window does not become a mess to look at, if the user decides to resize or close one of the views is far from trivial. Creating a single, dynamic multi-view GUI is a rather practical and easy task in Visual Studio .NET due to the hierarchical layout management features. View controls are arranged in a hierarchy where each child view is only aware of the size of the parent view (Figures 7, 8 and 9). An "anchor" property is used to specify how the control behaves when a user resizes the window. You can specify if the control

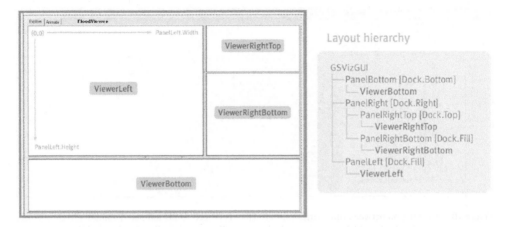

Figure 9.   Constructing a visual exploration GUI employing multiple views becomes easy with Visual Studio
.NET and without the need for programming: The anchor and dock properties, the layout hierarchy
of the FloodViewer application. Views can easily be resized or completely hidden.

should resize itself, anchoring itself in proportion to its own edges, or stay the same size, anchoring its position relative to the window's edges. The "dock" property is related to the anchor property and is used to specify that a control should dock to an edge of its container.

Each view in FloodViewer is connected to a specific visualization object. In Figure 7, the upper left view is connected to the 3D terrain map and the right view to the corresponding 2D map, while the lower left view is reserved for three chart types (2D bar, 2D and 3D water level). Using the splitter controls a user can hide or just change the size of entire panels. The viewers, being children of the panels, are changed accordingly, that is, either hidden or resized.

It's also necessary to achieve balance between 2D and 3D functionality. Some problems are best solved with one or the other, however many require the services of both technologies. In the FloodViewer, each 3D visualisation tool can have a 2D counterpart and the object-oriented nature of the technology ensures that most functionality is shared between the two. By using the two together, the power of each is amplified (Figures 7 and 8).

4.3   Analysis of spatial-temporal flood data

Visual analytics of time-series data is an important task in emergency response. Existing tools for analyzing observations over time and geography are based on traditional techniques and approaches from disciplines such as GIS, geographical, scientific and information visualization (Muller & Schumann 2003 and Andrienko & Andrienko 2004). The FloodViewer enhances these traditional techniques with more dynamic and interactive tools that are needed for decision support and risk analysis in flood forecasting. The presence of a temporal component changes the types of data analysis that may be done and consequently affect the required visual representations as well. Our research includes methods to explore data, "what-if analysis", having both spatial and temporal reference using linked, highly interactive 2D and 3D views. Rain information is stored in a structure that preserves attributes about the temporal characteristics of the data.

The Arno River case study is based on real-time radar and gauge station rainfall data for hourly time steps between 20 November 2000 14:00 and the 21 November 2000 14:00. Time-series water level data is provided for the same time period but the flood forecasting system has also estimated forecast data for additional 12 hours. FloodViewer can spatially explore these values of numeric attributes referring to different moments in time and locations in space (Figures 10, 11 and 12).

Time-controlled contour maps represent rainfall at a particular time moment based on radar data measured at 500-meter grid intervals. Time steps can be interactively changed using a GUI

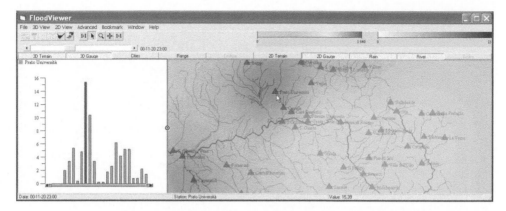

Figure 10. Correlation between time-series rainfall data visualized as 2D animated contours (right picture, time is 23.00 on 11 November) and a bar chart displaying detailed hourly rainfall for the whole 24 hours period. The figure identifies how data and visualization are linked and relate to the each other. The user selects a gauge station "Prato Universita" in the 2D contour map to see detailed rainfall data in the selected gauge bar chart. Data, analysis and visualisation "flow" together in a seamless process of discovery.

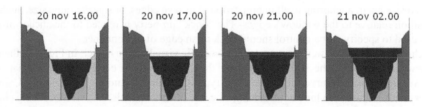

Figure 11. Water level animation allows the user to observe how the water level changes over time and at a particular time moment. The last image (right) shows forecasted data.

time slide control, which results in each time step being immediately redrawn. Map animation is achieved by moving the time slider along the time axes. The radar maps shows precipitation in mm in the previous defined region for a time moment (Figure 13). The interactive GUI events for providing the radar time animation in FloodViewer:

1. Select region-of-interest in the 3D terrain view (upper left)
2. Fine-tune the defined region in the 2D image view (right)
3. Select radar rain data visual representation (rain icon)
4. Move dynamic time slider (Figure 14) along the time axes.

4.4 *Explore water level data*

Event-based visualization with spatial context plays an important role in flood forecasting. Events represents the point in time when a change is about to happen with regard to the incoming data. FloodViewer alerts the analyst when the water level is rising above a critical level (an event) by changing the colour for that river segment from blue to red. The analyst can dynamically with a slide controller (Figure 14F) find out the potential risk for flooding in the river if the water level increases by X meter. We demonstrate interactive and dynamic 2D and 3D visualization methods for exploring time-variant water level data (Figure 14):

- 2D and 3D cross section charts – move along river
- Select river segment to be viewed

228

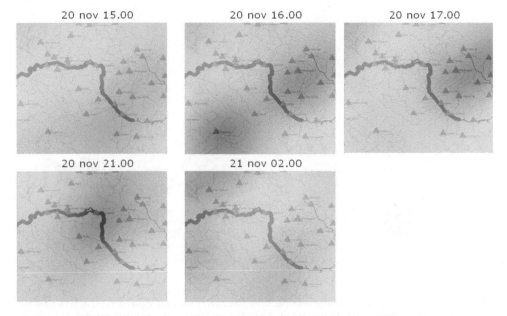

20 nov 15.00   20 nov 16.00   20 nov 17.00

20 nov 21.00   21 nov 02.00

Figure 12.   Dynamic time animation of hourly rainfall data from radar for the period 20–21 November superimposed on the river network and gauge stations using opacity blue colour contouring.

Figure 13.   Two time setting scrollbars control the display time moment for dynamic map animation. The left
time slider controls the 24 hours steps for calculated radar precipitation and water levels. The
right scrollbar controls prediction values for water levels for 0 to +9 hours. See also figure 8 for
a complete GUI overview.

- Alerts for high water levels in red
- Set water level for a What-if "additional 1, 2 or 3 meter higher water" Scenario
- Range of time includes both real-time and predicted 9 hours in advance

4.5   *Bookmarks*

Emergency risk management is a highly collaborative activity often conducted under extreme
time pressure. Robustly visual analytics systems are needed that allow team members to share
their understandings of unfolding events and see what their colleagues are thinking via shareable
dynamic visualizations that enable easy-to-use and efficient collaboration. The technology must
support re-use of previous analytic processes and gained insight. The system should also be able to
produce interactive electronic documents for communicating analytical assessments to colleagues
and operation management.

    In an emergency, if only the analyst knows the result of an assessment, that assessment has no
effect on the emergency response. On the other hand, if the analyst also must manage all aspects
of communicating the assessment, the analyst will have less time for the concentration needed to
conduct a complex analysis. Visual analytics tools must therefore have a straightforward means of

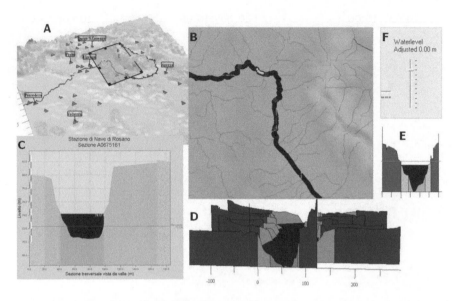

Figure 14.   Water level prediction for Stazione di Nave Rosano (C). Selected region defined in 3D terrain map (A). The 2D map (B) displays "alerts for high water level" in red colour. Cross-water sections can be viewed as animated 2D images (E) or as a series of 3D cross section charts (D). The white dots along the river segment in the upper left 2D map view represent the location of the corresponding ten 3D cross sections (right picture). The slider (F) is used to increase the water level by X meter and analyse how such scenario affects the overall flooding in the river.

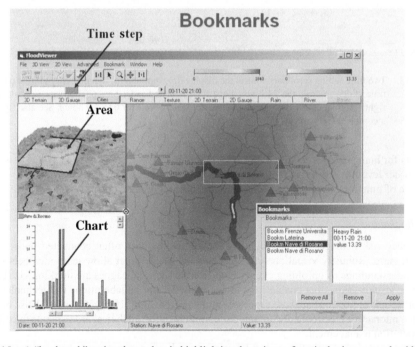

Figure 15.   A "bookmark" assists the analyst in highlighting data views of particular interest and guide other users to this discovery. The interactive image above highlights the rainfall data at a selected gauge station "Nave di Rosane" for a certain time step. Four bookmarks describing four different scenarios are stored in the repository.

saving and packaging the results of an analysis in a format that can be unwrapped for just-in-time by other members of the emergency team.

NVIS has developed a system that allows the analysts to set snapshots or "bookmarks" during the visual analytics process such as the FloodViewer. We believe that our bookmarks can help the analyst to highlight data views of particular interest and guide other analysts to important discoveries of an emergency situation. Colleagues can use these descriptive snapshots to quickly locate key information in the system by simply selecting the snapshot view they need. The Bookmark Manager (Figure 15) remembers and records the status (attributes) of a data navigation experience.

The FloodViewer analyst has selected suitable data dimensions, display properties, filtered data with the slide rangers focusing on the data-of-interest and finally highlighted the "discovery" from a certain angle (viewing properties) and can now save this status "bookmark" in an external file. A bookmark includes information about visualization attributes such as the area-of-interest, time step, viewing matrix, color scale, selected gauge or water level station, etc. When a new user initiates the FloodViewer, it will then start the VUI based on any analyst's previous bookmarks. The visualization will revert to exactly the same status as defined by the previous analyst. Users can, step-by-step, follow the analyst's way of work and understand how the results were achieved in a FloodViewer process by initiating selected bookmarks.

5 DISSEMINATION AND COLLABORATIVE VISUALIZATION TOOLS

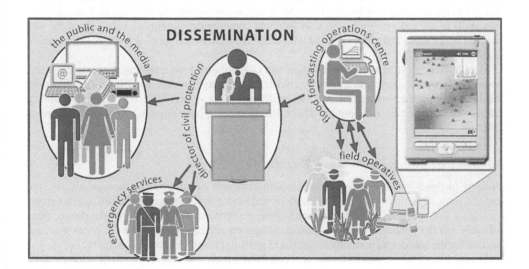

5.1 *The flood warning dissemination problem*

As well as warning about a coming flood, civil protection agencies need to know what is happening during a flood and what is going to happen. VA-based visualization tools allow the representation of the observational data, the forecast and the uncertainty in a clear, intelligible manner via either desktop or visual mobile communications. Such tools should be fine tuned to particular end user groups such as: flood forecasting specialists, the director of civil protection, controllers of emergency services and the concerned public.

At the present time, there is ample evidence that the visual performance of most flood warning systems is generally rather poor. The information that they are designed to disseminate fails to reach a large part of the target audience, and the people who make up this audience are not satisfied with the service that they receive. This poor performance is a function of a weak link in the chain that connects the flood forecast with the people who are designated to obtain this information. Without strengthening the visual exploration and dissemination link in the chain, the performance of the whole system is degraded. The advancement of the accuracy of flood forecasting, and better targeting of the dissemination of the ensuing message, is not only for the benefit of the forecasters but of the public and media. A flood warning system can be conceptualised as the combination of a flood forecasting sub-system *"technological part"* and a flood warning dissemination sub-system *"social part"*.

The *technological* subsystem is run by analysts (meteorologists and hydrologists) and consists of precipitation measurement systems, river gauges, and meteorological forecasting systems, which together provide the real-time data input to the FloodViewer for the visual analytics process. Relying on the analysts' results, the operations manager may issue warnings to the civil defence manager. The civil defence manager interacts with the media, the police, other officials and the public, both directly and indirectly. This forms the *social* subsystem of the overall total warning system.

In an emergency situation, we envision the analysts as enablers of complex communications that are appropriate and productive of immediate results. The analyst is expert at assessing data, while the audience for the assessment may not be. The visual presentation of analytic results needs to be clear and concise, and it must take place as soon as possible after the analyst reaches a conclusion. To achieve this, we must equip the analyst with tools to easily create interactive exploration scenarios that reveal what is going on and allow analytical reasoning, presentation and dissemination. It is important to provide the analyst with this capability already during the course of the analysis and be able to share with colleagues, visualizations and associated analytical reasoning that led to the resulting conclusion. These tools will accommodate the sophisticated communication skills of the analyst. Tools will also facilitate communication with a variety of people who have different needs and objectives and who often use different terminology to talk about similar subjects.

FloodViewer provide both dissemination and collaborative tools enabling the meteorologists, hydrologists, operations managers and the civil defence manager to view and discuss the forecasting results in real time across a network *"collaborative network version"* (Section 5.3) before finally interacting with the media, the police, other officials and the public through a *"dynamic document"* (Section 5.2).

5.2   FloodViewer dynamic document "SmartDoc"

Projects in flood forecasting require permanent exchange of information between all involved experts, normally working at different locations and having different specializations, e.g. are meteorologists and hydrologists, operations managers, government officials, media, the police, other officials and the public. Traditional electronic reports are characterized by the paper medium, passive for the reader and normally restricted to static items such as text and imagery.

In visual analytic practice (Thomas & Cook 2005), tools are generally entirely separate from presentation and reporting tools. Analysts explore data and check competing hypotheses against data from a variety of sources using advanced visualization capabilities. For composing a presentation, they must leave their interactive visualization environment and move to, for example, Microsoft PowerPoint to portray their analytical thinking. An integration of interactive visual analysis tools and reporting tools would improve the communication and dissemination process.

NVIS has been exploring the potential for integrating VA components into common productivity applications, such as Microsoft Office or Adobe PDF to support embedded interactive visualization (Jern, Palmberg, Ranlöf, Nilsson 2003). Integrating interactive geovisualisation technology with traditional documentation technology can provide better understanding of emergency situations. Our dynamic documents "SmartDoc" incorporate not only text and static images but also the entire interactive data visualisation and navigation process. "SmartDoc" provides a platform for structured visual argumentation that exists in the background of the analyst's familiar environment and delivers shared real-time situational awareness to other teams of analysts.

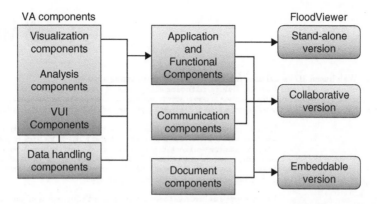

VA components   FloodViewer

Figure 16.   FloodViewer support three VA scenarios: Stand-alone, Collaborative and Dynamic "Embed-dable" version. A set of task-specific VA components are put together to form application or functional components. Special communication components are used by the Collaborative version and document components provide necessary document components for the embeddable version.

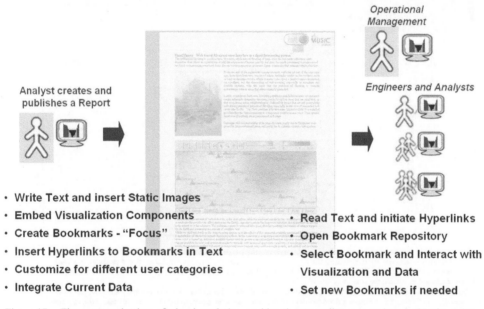

- **Write Text and insert Static Images**
- **Embed Visualization Components**
- **Create Bookmarks - "Focus"**
- **Insert Hyperlinks to Bookmarks in Text**
- **Customize for different user categories**
- **Integrate Current Data**

- **Read Text and initiate Hyperlinks**
- **Open Bookmark Repository**
- **Select Bookmark and Interact with Visualization and Data**
- **Set new Bookmarks if needed**

Figure 17.   The communication of visual analytic consideration to colleagues and operational manage-ment needs to be, visual, clear and interactive, and provide mechanism for quick feedback. To achieve this, we have equipped the analyst with novel communication tools that create interac-tive assessment scenarios and reveal what is going on. "Bookmarks" and "dynamic documents" will accommodate sophisticated communication skills of the analyst and allow visual analytical reasoning, embedded in an electronic report.

Functional and application VA components (Figure 16) can be embedded in, for example, a Word document allowing an analyst "author" to prepare an electronic document that collaborate and share dynamic and interactive flood forecasting results with other hydrologists, operational managers and the civil response managers (Figure 17). Snapshots of visual insights and reasoning are embedded in the Word document structure together with the VA application component FloodViewer. The author creates argument structures based on snapshot scenarios within Word comprising claims supported

FloodViewer - Web based 3D visual user interface to a flood forecasting system

The problem of flooding is as old as time. However, while natural flooding of large areas did not create situations more dangerous than others in a prehistoric world, the expansion of human activity and cities has made preventing damage caused by floods or harnessing over-bank flows for one's own purposes as in ancient Egypt a necessity that remains vital to this day.

From the end of the eighteenth century onwards, with the advent of the industrial age, there have been two courses of action: hydraulic works on the territory, such as land reclamation works, which in many cases upset a land's balance dependent on overflow, and the channelling of watercourses, especially in mountain and foothill sections, with the result that the problem of flooding is brought downstream even to areas that were originally protected.

Lastly, recent years have seen booming population and indiscriminate urbanisation create extremely dangerous situations, with floodplain areas that are inhabited, or that even house entire neighbourhoods sheltered by levees that are not particularly safe during prolonged periods of flooding, especially in the case of suspended bed rivers like the Po. The flood problem is by no means limited to Italy. It is a global problem that has been increasing at a worrisome pace in recent years. This upward trend was if anything more pronounced in Europe.

Damages will rise inexorably in the years to come, partly due to the greater risks posed by larger urbanised areas, and partly due to climatic changes taking place.

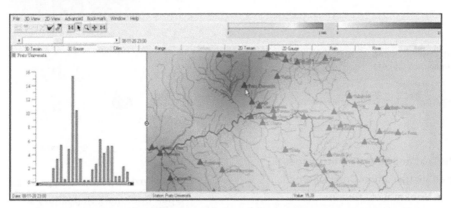

The considerable amount of carbon dioxide in the atmosphere, which is produced mainly by the combustion of hydrocarbons, exacerbates the greenhouse effect increasing the Earth's capacity to retain the long-wave radiation that it produces as a black body subject to solar radiation. This radiation cannot be released into space, thereby upsetting the balance of energy trapped by the Earth and increasing the amount of available heat.

While we shall not dwell on the long-standing dispute as to the effects of this increased available energy and heat, one issue is nonetheless of interest as regards the flood problem. In the same way as evaporation and turbulence increase when the fire under a pot is turned up, increased available energy leads to greater evaporation (and therefore a greater quantity of water vapour available for rain) and greater atmospheric vorticity with increased space-time variability of rain frequency. In other words, the greenhouse effect will result in longer and more frequent rainy and/or drought periods, and rainfall and discharge

Figure 18.   FloodViewer application component embedded in a Word document. "SmartDoc" incorporates the entire interactive data visualisation and navigation process. Text strings can be hyper-linked to pre-defined bookmarks that highlight data discoveries of particular interest to different readers, who use these descriptive bookmarks to quickly locate key information.

or rebutted by evidence. SmartDoc enables analysts to merge analysis and production into a seamless process that dramatically accelerates the formation of coherent arguments for or against particular courses of action.

The "author", usually the analyst, sets bookmarks (Figure 18) that highlight data discoveries of particular interest to different readers, who use these descriptive bookmarks to quickly locate key information. The author has selected suitable data dimensions, display properties, zoomed data

Abstraction Layer that assures media independence

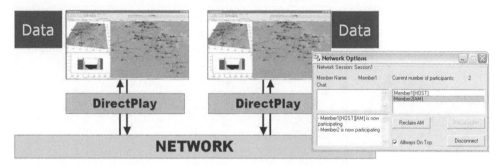

TCP/IP or IPX protocols on LANs or MODEM connections

Figure 19. Our implementation of a collaborative FloodViewer provides a layer that largely isolates the user
from the problems of an underlying network. With a multi-user application sharing session, each
user's FloodViewer is synchronized with that of the other user(s) in the session. A continual
stream of messages flows between the members of the collaborative session. For example, every
time an active member (player) rotates a 3D object or changes an attribute, a message is sent to
update this user's position on the other passive member's FloodViewer application. Application
sharing supports efficient messaging between all members in a session. The GUI panel shows a
collaborative session with two members.

focusing on the risk area and highlighted the "discovery" from a certain view. The bookmarks are
stored as part of the document. When the reader opens the document, it will start the 3D interactive
FloodViewer process based on the author's bookmarks. The visualisation will revert to exactly the
same status as defined by the author. A special "link component" allows the author to link text to
defined bookmarks. The author could also use these tools to create a dynamic reading path.

For example, a flood-forecasting document reports an interactive rainfall scenario for a certain
time period. Target groups for example are the project engineers and the decision-makers of flooding
operations. The engineers normally require detailed information, e.g. single flow states of physical
calculations. The operation management needs condensed information to support their decisions
and may issue warnings to public. So both target groups need different information, which either
cannot be documented in one single traditional document, or which restricts the readability due to
the linearity in documents containing information for more than one target group (Figure 17).

5.3 *Collaborative flood forecasting system*

To support real-time emergency response and decision making, one of the most important challenges
involve distributed visual analytics, enabling geovisualization across software components, devices,
people, and places. The last application scenario of FloodViewer is indeed a collaborative version.

There are multiple ongoing or completed projects in the area of collaborative visualization
systems (Wood, Wright, & Brodlie 1997 and Wood 1998 and Jern 2000). Unfortunately, many
of these frameworks are using an architecture that makes them hard to modify, reuse, or extend
with new functionality. NVIS has not tried to invent yet another collaborative system and network
protocol. The FloodViewer networking architecture is based on Microsoft's DirectPlay API, which
originally was designed for computer games. This API is a part of DirectX library and provides
classes for networking. Since interactive performance is very important in computer gaming, a
shared event model is used since transmitting the entire screen to all game participants would make
game-play practically impossible.

We have developed, demonstrated and validated that an emergency visualization application can
benefit from a conceptual distributed computing infrastructure such as Microsoft's high-performing
networking game technology. DirectPlay layer separates the network layer from the application

235

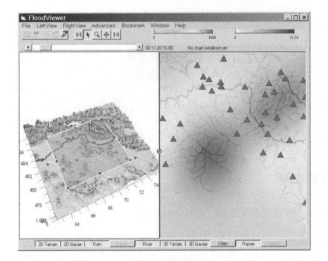

Figure 20. Civil protection agencies need to know what is happening during a flood and what is going to happen even if they are away from the operating centre. The two figures show the similarity between the FloodViewer C#-based DirectX desktop version and the Flash PDA/Mobile version. You can see that the two different implementations of a "context-map". Both displays show hourly rainfall radar data, represented in blue contouring, super-imposed on the river network, DEM data and gauge stations. The user can also select a gauge station and analyse data.

layer, i.e. the application itself does not have to communicate with the network, which makes the implementation easy and provides optional collaborative visualization to the visualization application. FloodViewer supports a collaborative environment where two or more users can meet together in cyber space displaying and interacting with the same dataset remotely. Bookmarks and annotations can also be used in the collaborative session.

The collaborative version of FloodViewer supports most generalized communication capabilities shielding users from the underlying complexities of diverse connectivity implementations, freeing them to concentrate on the real-time navigation scenario. The integration of a network abstraction system, "Application Sharing" and a "Data Navigation Protocol" provide the foundation for real-time collaborative geovisualisation. With a multi-user application session, each user's visualization application is synchronized with that of the other user(s) in the session. A constant stream of "messages" flow to and from each user. For example, every time a user interacts with the visualization, a message is sent to update the corresponding visualization on the other application clients in the collaborative session.

Collaborative flood forecasting is important to improve decision support by better communication between meteorologists and hydrologists, operational manager and the civil defence manager. Experts and politicians normally don't work in the same offices and an integration of video-conferences and collaborative visualisation will allow faster decisions and save valuable time in a critical situation.

5.4 Mobile communications

Hand-held, mobile devices like PDAs and standard mobile phones are becoming increasingly important in emergency response. Trying to bundle all possible visual information into a small window, is a great challenge for the interface designer to deal with. We demonstrate a simple FloodViewer prototype (Figure 20) based on platform-independent Flash, a vector graphics implementation that provides some important advantages when designing a visual user interface for a mobile device, such as zooming, panning and basic pick functions. To zoom in, drag the scrollbar handle up and

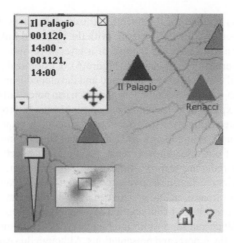

Figure 21. Mobile VUI based on Flash. Pick a gauge station "Il Palagio" (blue triangle) results in a pop-up menu with rainfall data. Focused contour map is shown in context in a separate view 00.

use a pointing device to pan the image. These user interface controls work in the same way as on the desktop where the mouse is used instead. There is an "overview map" view to the right of the zoom scrollbar with a thin frame that shows the position. This "focus & context" feature gives the user an indication of the position of the zoomed map (Figure 21).

6 CONCLUSION

The increased numbers of users with very different backgrounds, who will be using water resource data to make important decisions, elevates the importance of finding reliable and easy-to-use methodologies for flood forecasting. Analysing, visualising, and visual user interfaces are important for spatio-temporal data in general, but it is especially important for water resource data where a small local change may have a dramatic impact.

FloodViewer had adopted the requirements to visual analytics with some of the following visual representation and interaction schemes:

- Space and time awareness;
- Integrated views of large-scale information spaces, supporting coordinated viewing of information in context and provision of both overview and detailed information;
- Schemes for management and exploitation of multiple views and encourage complementary views of the same data;
- Visual representations and interactive schemes from a cognitive perspective;
- Methods for navigation in high-dimensional, multivariate and temporal data;
- Interactive performance that can support analytic reasoning.

Visual representations leverage on the human capability for spatio-temporal reasoning. The increased numbers of users with very different backgrounds, who will be using water resource data to make important decisions, endorse the importance of finding reliable and easy-to-use methodologies for flood forecasting. Analysing, visualizing, and visual user interfaces are important tasks for spatial data in general, but it is especially important for water resource data where a small local change may have a dramatic impact.

Ensuring that dynamically re-sizeable multiple views does not become a mess is far from a trivial programming task. We propose that a dynamically linked view model can efficiently, quickly and easily be developed. The Windows GUI of .NET based applications most likely represent an appropriate interaction method for the novice end user. Further usability testing is needed to discover how users – expert and novice – interact with the proposed linked views metaphor.

Also important is our aim to promote the use of a layered component-based approach to the development and engineering of applications (Figure 6). Customizable and scalable high-level application and functional components are designed and developed from low-level atomic components. We believe that using a layered component approach can potentially provide better scalability and more customizable exploratory visual analytics components. Interoperability with the server-based real-time flood forecasting database is achieved through a request for data in a dynamic HTML form. Selected data is returned as a XML-formatted file and imported into FloodViewer.

Another overall goal of our research is to make people more effective in their information or communication tasks by reducing learning times, speeding performance, lowering error rates, facilitating retention and increasing subjective satisfaction. FloodViewer provides an environment for real-time situational awareness enabling visual analytical collaboration, improve quality of the production and presentations, integrate the creation of products and presentations into the analytical process and provide guidance and support for interactive types of communication. Such a capacity bypasses the need for lengthy reports and presentations. Although this capability has value in many analytical reasoning settings, it is especially valuable in emergency response situations. It gives emergency operations centres a radically better way to understand critical situations, coordinate their expertise and manage their response.

Local water quality agencies in Italy and Poland (EC MUSIC partners) assessed and validated FloodViewer. We could draw some tentative conclusions regarding 2D versus 3D visualisation including that 2D data visualisation methods are more easily accessible to the user. The 3D data visualisation allow the user to combine more information into a single scene, but these methods are not yet accepted as instruments for decision making among the local agencies. Our next step includes a comprehensive user task analysis based on in-depth interviews with potential users, a set of focus group assessments of our proposed visual user interface and achieved interactive performance, and subsequent controlled experiments to test selected aspects of used methods, such as the multiple resizable views, number of attributes and dynamic range sliders that constrains the animated time-series. The visual user interface will be further customized to meet the requirement from the end users.

ACKNOWLEDGEMENTS

I specially like to thank Andreas Nilsson NVIS for his impressive programming contribution to the FloodViewer project. The MUSIC flood project is collaboration between NVIS at Linköping University, AVS, University of Bologna, ET&P and institute CNR Italy, University of Newcastle, Gematronik Germany and 3 end user agency partners (see Web site). The project was partly funded by the EC Commission EVK1-CT-2000-00058. The FloodViewer web site can be found at: http://vita.itn.liu.se/FloodViewer

REFERENCES

Andrienko, N. & Andrienko, G. 2003. Informed Spatial Decisions through Coordinated Views, Information Visualization, 2(4), 2003, pp. 270–285.

Andrienko, N. & Andrienko, G. 2004. Interactive visual tools to explore spatio-temporal variation, In M.F. Coastabile (Ed.) Proceedings of the Working Conference on Advanced Visual Interfaces AVI 2004, Gallipoli, Italy, May 25–28, 2004, ACM Press, 2004, pp. 417–420.

Hindmarsh, J., Fraser, M., Heath, C., Benford, S. & Greenhalgh, C. 2000. Object-focused interaction in collaborative virtual environments, ACM Transactions on Computer-Human Interaction.

Hochheiser, H. & Shneiderman, B. 2004. Dynamic query tools for time series data sets: timebox widgets for interactive exploration, Information Visualization, Vol. 3, Issue 1, Spring 2004, 1–18.

Jern, M. & Bladh, K. 1980, A Color Plotter System and its Applications in Geoscience, Geoscience and Remote Sensing, IEEE Transactions, July 1980. Volume GE-18, Number 3, page 256–263, 1980.

Jern, M. 1985a. Thematic Mapping, Eurographics '85 Conference, Proceedings, 1985.

Jern, M. 1985a. Raster Graphics Approach in Mapping, Computer & Graphics, Pergamon Press, Volume 9, Number 4, 1985. Page 373–381.

Jern, M. 2000. Collaborative Visual Data Navigation on the Web, IEEE International Conference on Information Visualisation, London, IEEE Computer Science Press.

Jern, M. 2001. Visual Data Navigators Collaboratories – True Interactive Visualisation for the Web. Invited Speaker, Mobile and Virtual Media International Conference 2001.

Jern, M. 2005. Web based 3D visual user interface to flood forecasting system, Reviewed and accepted paper, The First International Symposium on Geo-information for Disaster Management (Gi4DM), Delft, The Netherlands, March 21–23, 2005, published by Springer ISBN 3-540-24988-5, available at http://www.gdmc.nl/gi4dm/

Jern, M., Johansson, J., Pettersson, A. & Feldt, N. 2005. Tailor-made Exploratory Visualization for Statistics Sweden, reviewed and accepted paper, CMV 2005, London, July 2005, IEEE Computer Society ISBN 0-7695-2396-8.

Jern, M., Palmberg, S., Ranlöf, M. & Nilsson, A. 2003a. Web based 3D visual user interface to a flood forecasting system, Reviewed and accepted paper, EuroSDR Commission 5 Workshop Visualisation and Rendering, Holland, January 2003.

Jern, M., Palmberg, S., Ranlöf, M. & Nilsson, A. 2003b. Coordinated views in dynamic and interactive documents, Reviewed and accepted paper, InfViz 2003, London, July 2003. IEEE Computer Society ISBN 0-7695-2001-4.

MacEachren, A.M. 2004. How Maps Work, Representation, Visualization, and Design, 3rd edition, The Guilford Press, New York, 2004.

MacEachren, A.M. 2001. An Evolving Cognitive-Semiotic Approach to Geographic Visualization and Knowledge Construction, Information Design Journal, vol. 10, no. 1, 2001, pp. 26–36.

MacEachren, A.M. & Kraak, M.J. 2001. Research Challenges in Geovisualization, Cartography and Geographic Information Science, vol. 28, no. 1, 2001, pp. 3–12.

Muller, W. & Schumann, H. 2003. Visualization methods for time-dependent data, Proceedings of the 2003 Winter Simulation Conference, 737–746, 2003.

MUSIC, available at http://vita.itn.liu.se/FloodViewer

North, C. & Shneiderman, B. 2000. Snap-together visualization: a user interface for coordinating visualizations via relational schemata, Proceedings of Advanced Visual Interfaces 2000, Italy, pp. 128–135.

Roberts, J.C. 2004. Exploratory Visualization with Multiple Linked Views, Exploring Geovisualization, J. Dykes, A.M. MacEachren, M.-J. Kraak (Editors).

SMARTDOC, available at http://vita.itn.liu.se/SMARTDOC

Steiner, E.B., MacEachren, A.M. & Guo, D. 2001. Developing and assessing lightweight data-driven exploratory geovisualization tools for the web, 20th International Cartographic Conference, Beijing, China, Aug. 6–10.

Todini, E. 1996. The ARNO Rainfall-Runoff model, Journal of Hydrology, 175: 339–382.

Todin, E., Catelli, C., & Pani, G. 2004. FLOODSS, Flood operational DSS, Balabanis, P., Bronstert, A., Casale, R., & Samuels, P. (eds): Ribamod: River basin modelling, management and flood mitigation.

Todini, E. & Bottarelli, M. 2002. ODESSEI: Open architecture Decision Support System for Environmental Impact: Assessment, planning and management, Operational Water Management, Refsgaard & Karalis (eds.): 229–235. Rotterdam: Balkema.

Thomas, J. & Cook, K. 2005. Illuminating the Path: The Research and Development Agenda for Visual Analytics, available at http://nvac.pnl.gov/

Wood, J., Wright, H. & Brodlie, K. 1997. Collaborative Visualisation, IEEE information Visualisation '97. IEEE Computer Society; Proc. Phoenix, Oct. 19–24, 1997, pp. 253–259.

Wood, J. 1998. Collaborative Visualisation: PhD Thesis, Leeds University.

Part 5
Integration of heterogeneous data

The semantic mismatch as limiting factor for the use of geospatial information in disaster management and emergency response

H. Pundt
University of Applied Studies and Research Harz, Wernigerode, Germany

ABSTRACT: Geospatial information (GI) is characterized through heterogeneity. Heterogeneity concerns various aspects, such as heterogeneity of data structures, formats, and semantic heterogeneity. Semantic heterogeneity causes problems for data access and information sharing. The rapid exchange of spatial information is very important in emergency situations. Information that is provided by emergency management centers must be reliable and relevant for those who use the data, e.g. emergency rescue teams or local decision makers. Due to the fact that such people have no time to assess whether data are usable, mechanisms are required that support the identification of the "right" data, thus guaranteeing that only such information is provided that is relevant in a specific situation. Formal ontologies are a means that can support the access to and evaluation of spatial information automatically. They are helpful when a fast access to different information sources is required in situations where decisions must be made rapidly. The article discusses the semantic mismatch as typical characteristic of spatial information and promotes ontologies as a vehicle to overcome some of the problems of data access and information sharing that are caused through semantic non-interoperability. The chapter is aimed at introducing the concept of formal ontologies and some approaches to formalize them. Ontologies are proposed to be used when relevant data must be identified and shared in emergency situations quickly.

1 INTRODUCTION

Disaster Management always has a spatial component. Disasters happen somewhere, at a specific location or within a specific region. A burning house, for instance, represents a disaster that is relatively limited concerning its spatial extent. A flood, however, can concern a complete drainage basin, including the floodplain in which various land uses occur, such as agriculturally used land, urban areas, industrial complexes, other objects and uses. An earthquake can be limited to a part of a town, a complete urban region or a larger area, as seen, for instance, in the north of Pakistan in 2005. The Tsunami in Asia, December 2004, concerned a large part of the continent, various countries were affected. In any case, the spatial extent of the disaster and the damages caused by the event can be described more or less exactly by drawing its "borderlines" on a map, usually using coordinates. The objects affected by the disaster lie within these borderlines, they are spatially fixed. Based on such a fixation, decisions can be made. Based on such decisions, measures can be carried out in response to the disaster. The objects that are relevant within the framework of a disastrous event are unique, each one has individual properties, and relationships to other objects in the neighbourhood. The transformation from "general" objects to "unique" and "spatial" ones turns objects into spatial objects. If such spatial objects are relevant in a specific context and therefore help to answer questions or to solve problems, they turn into geospatial information (GI). The spatial objects themselves underlie temporal changes that are – in the case of a disaster or emergency – short-termed: the landscape that is concerned by a disaster shows specific spatial characteristics *before* the event has taken place. These characteristics change *during* the disaster and show a modified face *after* the catastrophe. The spatially exact description of the spatial objects in all three phases enables humans to carry out a proper assessment of the damages which is a presupposition for an equally proper response. Disasters in all their various forms are events where space and time are crucial. Geospatial information (GI) is the ultimate basis for disaster management and emergency response; it is used during all stages of a disaster (before, during, and after). GI services can be used as a key technology to compile ad hoc the necessary information for decision makers (Bernard et al., 2004).

With the advent of more and more sophisticated GI-technologies, GI-services are increasingly used by various communities in disaster management and emergency response. GI services depend on the availability and quality of spatial data. If data for the disaster region are not available, or if data are of bad quality, the software built upon such data will not work or it will not produce any useful information (Guptill and Morisson 1995).

The access to and quality of spatial data – as two important factors limiting the usage of GI – are two issues that will be discussed more deeply in the following sections. The basic thought for this discussion is that pure availability does not necessarily mean that the usability of data is guaranteed. In other terms: if there are data available, it has to be checked carefully, if these data are relevant to support decision making. The assessment of relevance requires criteria, namely quality criteria. But data quality cannot be described in general. *Quality is contextual*, which causes not only technical, but more importantly semantic problems for the usage of GI sources. Such sources are spread over numerous servers, in a world wide network, provided by different geospatial information communities (Harvey 1998).

These thoughts should be seen within the framework of the three temporally separated periods described above. The situation after the event is not the same as before the event. In all three phases, GI is required by different people, e.g. emergency teams, in a timely manner. How can such people identify the adequate data and how can they assess, whether such data are qualitatively "good enough" to support their decisions? Such decisions, however, have to be made, in some cases, within hours, minutes, or less.

The next section will focus on the availability and usability of data. Both aspects play an important role within the framework of a key issue, semantic interoperability of data, as well as services. Therefore the third section will briefly introduce the "Model of Relevance". The fourth section will discuss pathways to overcome the limitations that still exist due to semantic non-interoperability. Finally, a critical conclusion will be drawn and reasons are given that formal domain ontologies can serve as a vehicle to support the identification of adequate and relevant data including the evaluation of quality. They can help in situations, where time critical decisions have to be made.

2 AVAILABILITY AND USABILITY OF SPATIAL DATA

2.1 *Heterogeneity of spatial data*

Users of GI tend to act in contradiction to other scientists. The latter try to develop "common" models of the reality that are unique and accepted in general. The providers of geographic data specify different models for the same objects and with regards to their specific point of view and understanding of the reality (Giger and Najar 2003). The usability of information that is created in one context is often of limited use in others contexts. This special property of GI substantiates the *semantic mismatch* that occurs between data provided by different geospatial information communities (GICs), as described earlier in Bernard et al., 2004, Bishr et al., 1999, Moellering 1996, and others.

For the usage of GI this means that the same object, represented in different ways, is possibly relevant for one application, but not for the other. This depends on a specific view of reality that is consistent within one geospatial information community. Spatial data used in disaster management must be assessed carefully concerning what they represent: the meaning of an object can be manifold (Xu and Lee 2002).

A simple example clarifies the problems of availability and usability of spatial data for disaster management. A flood occurs in a river catchment due to heavy rain over many days. The dykes, some of them more than 60 years old and never reconstructed, threaten to break at some points due to increasing water pressure. A central emergency management group has been set up. The time to make decisions about the next steps decreases, simultaneously the weather service announces increasing rainfall. The management group is supported by a small team of IT specialists that have set up a communication infrastructure between the central office and the local teams that monitor the situation in the concerned region that is threatened by flooding. This group, supported by a few GI-experts, has the task to provide general information as well as spatial data to all emergency teams and the local decision makers. These parties require information quickly to react adequately when conditions can change within a few minutes.

The IT-group provides, among other information, digital topographic maps. Different sub teams are staffed with mobile GIS to enable them to use data outdoors. The IT group has to ensure the provision of data via wireless connections to the sub groups. The latter, however, must send current data directly back to the management group to ensure that such current information is included in analyses and forecasts.

The IT-group gets topographic data through the national mapping agency. An example for such a national topographic information system is the German ATKIS (Official Topographic and Cartographic Information System). The basis of ATKIS is an object catalogue that defines exactly the meaning of various spatial objects. The catalogue is organized in an object-oriented, hierarchical way. The digital landscape model (DLM) includes the vector data of objects as well as a clear description of specific attributes linked to each topographic object. The information collected in ATKIS is ordered in different scales, e.g. 1:25,000 or 1:50,000. Disaster management requires such "views" that differ in scale. In our case, an overview of the threatened area is required to orientate roughly and to plan where emergency vehicles can navigate over streets, farm tracks or other pathways to specific points. A more detailed view is needed when taking action in a spatially limited area, to navigate through a village and finally to an area that is not reachable via fastened roads. For the first case, 1:25000 might be the right scale, for the latter 1:5000 or even larger. Maps in both scales must be available. In our example this is the case. The IT-group receives the data via the server of the national mapping agency, and forwards them to the sub teams at various sites (including the field) considering that every team is provided with data in the "right" scale for its specific purposes. The IT-group in the central office is convinced that they did something useful in a crucial situation; nevertheless some helpers, outside in the field, sent some questions by email, SMS or via other channels that were hard to answer for the IT-group members:

- Do the maps show all roads, tracks, pathways to the dyke?
- Which of the roads, tracks, pathways shown in the map are usable by cars or other vehicles needed for saving the dyke or evacuating people?
- Are farm tracks always paved, or only roads (what's the difference between roads, tracks, paths, ways, routes, …)?
- Is there information about the soil conditions available?
- How old is the dyke at point xy and which part is mostly threatened not to resist water pressure?
- Where is the next fire station?
- What is the shortest pathway to the most threatened part of the dyke?
- From where can I get information from water gauges?
- What's the difference between values delivered by gauges, and water level, mean water level, and stage?
- Are gauge values, delivered by the local water authority the same as water levels delivered by the regional hydrologic service?

Few of these questions could be answered satisfactory. The IT-team-members had to give names and phone numbers of colleagues in other institutions that possibly knew more, but in most cases the askers did not try other information sources due to the limited time. The IT team could only send topographic data hoping that these will help somehow. The evaluation of usability of such data was completely left to the emergency teams. But these weren't able to carry out such evaluations carefully and comprehensively for two reasons:

1. No further information about the data (e.g. metadata),
2. No time to access and assess metadata (if available).

This situation shows that data availability and data provision represent only one side of the coin. The other side is characterized by the question *do the data delivered to the sub teams satisfy their needs*?

A further development of the scenario is as follows. The maps provided by the central IT team are not completely up to date. Some of them are two, some of them three years old, and some even older. "Currently not available" was the answer of the WWW-based geoportal when asking for more current maps. After some time, another phone call comes in from a local emergency team: "we tried to use the track from farm 1 ("Millers") across farm 2 ("McDoggerthys") to the dyke. This track was shown in the map you sent us, but it does not exist anymore. It is arable land now, quite wet, no chance to get through with cars, lorries, any vehicles. The information you sent us is not usable, our decision was completely wrong due to the data you sent us. Now our situation is even worse than before, because we lost time…" And a further call comes in: "We tried the small road from Watertown to the dyke near Stream Village. Now we want to go further on via the track shown in the map you sent us. Is this possible with our lorries carrying sand sacks? What about the trafficability of the tracks and the small roads, shown in the map?"

Data availability and data usability are different aspects. Spatial data are only usable if they are relevant in a given situation. It depends on the context to assess if data are relevant, or not. Data can be relevant in one case; the same data can be irrelevant in another. This is mainly due to quality parameters (such as currentness) and to the semantics of spatial data (are tracks and roads the same? Are they stable, or paved? Do "gauge", "stage", and "water-level" mean the same?). This reminds us to what was said before: The providers of geographic data specify different models for same objects with regards to their specific understanding of reality.

3 THE MODEL OF RELEVANCE

3.1 *Context-sensitivity*

An etymological interpretation of the term *relevance* concerns the description of the correctness of a theory or the ability of a theory to explain specific things or processes. Information that supports the explanation of things or processes makes this information *relevant*. Relevance is not purely objective. It underlies objective, as well as subjective criteria. If something is relevant, or not, is dependent on the person that explains and interprets things and processes. That is why relevance is a notion that only makes sense within a specific context. The same information might be relevant in one context, but useless, or irrelevant, in another. The specific context of a spatial decision that has to be made, is the main viewpoint from which it can be decided whether data are relevant, or not (Frank and Grünbacher, 2002). Two issues are important, when spatial information is considered to be relevant:

- the definition of the context in which the information is used,
- the evaluation of the relevance of the information within this context.

3.2 *Relevance of GI*

At the first international symposium on GI for disaster management, 2005 in Delft, The Netherlands, the *model of relevance*, shown in figure 1, was presented in a slightly different way. The model

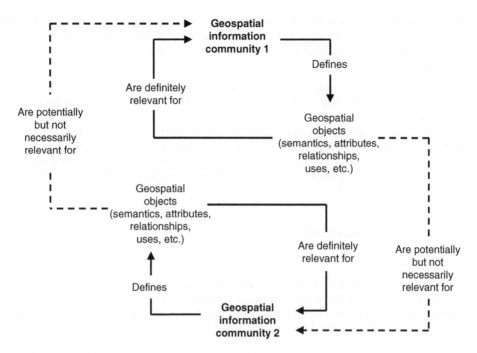

Figure 1.   The model of relevance (Pundt 2005, modified).

supports the definition of the semantic mismatch in the sense defined before: the same objects are modelled differently in different geospatial information communities.

It should not be discussed here, if there is a general model or general ontology that could match somehow the various views on the (same) spatial objects completely. Figure 1 refers to specific spatial domains, each possibly represented by a domain ontology defined by the geospatial information communities. The figure does not consider a top-level ontology.

The model shows that different communities use different terminologies to model their domains. Semantics are specific for information communities which causes semantic heterogeneity. The way to deal with this problem is to include information about the differences into data (model) descriptions. Humans, however, can interpret data taking into account experience as well as various knowledge sources they have in mind or even can look for (such as metadata). For humans, data (which are represented by terms that describe entities, attributes, relationships, potential uses of the information, etc.) have an *implicit* meaning. Referring to a disaster situation, where time is often a limited factor, data interpretation is not possible by human users. Supportive tools have to evaluate themselves, whether data are usable in a given situation. But data basically have no *explicit* meaning, which is necessary to make them understandable to machines (such as the computer) used by GI-services.

Such an evaluation must be based on metadata aimed at describing data more or less comprehensively so as to enable users to assess data usability. But metadata alone are no solution in a disaster situation, especially if they are thought to be interpreted through a human user. The goal must be to enable computers to *understand*, to a certain extent, the information that is provided by different information communities. This would enhance the chance that only relevant data are provided which really help in the disaster or emergency situation. Therefore, metadata and semantically enriched information must be formalized in a manner that makes them readable for computers. This can be achieved by using formal ontologies.

These considerations are related to the discussion of the two terms brought into the spotlight before. The search for data (availability) and the evaluation, if such data are relevant to support decision making for emergency response (usability) leads, so far, to two conclusions:

- if spatial data are used in disaster management, an automatic support to identify *relevant* data sets is needed. This requires the inclusion of metadata that are readable and interpretable automatically.
- Additionally, the consideration of data semantics is required. Ontologies – developed in artificial intelligence to facilitate knowledge sharing and reuse – provide a machine-processable semantics of information sources that can be communicated between different agents, software and humans (Fensel 2004). Both, metadata and ontologies, can be formalised using XML grounded languages to make them machine-processable.

4 FORMAL ONTOLOGIES FOR EMERGENCY RESPONSE

4.1 *Foundations*

The availability of metadata for spatial data sets is an ultimate requirement for the identification of "the right" sets for specific purposes. A formal description of metadata is required to automate the identification of data sets. Formal metadata help to identify those data sets that are *potentially usable* within a specific context. To go the step from "potentially usable" toward "usable", semantics of data have to be considered explicitly. The promoters of the semantic web see ontologies as the vehicles that help to overcome the existing hurdles that stand for semantic non-interoperability. Ontologies in this sense are seen in a pragmatic manner. Such ontologies are aimed at converting machine-readable into machine-understandable information by providing well-defined meaning for the content distributed within the WWW (Vögele and Spittel 2004). They represent formal and explicit specifications of conceptualizations of the real world. Conceptualisations are never universally valid; however, different ontologies are required to represent different conceptualisations of the real world. Ontologies in this sense represent the outcome of the process of specification and formalisation of such conceptualisations.

The main task for ontologies is to act as an interface between different domains (Visser et al., 2002), thus being able to differentiate such domains and exclude those that are *irrelevant* to a problem (Pundt 2005). This requires a communication process and a resulting consensus between different actors. Such actors are members of different geospatial information communities that "speak" different languages which means that – according to the model of relevance (see figure 1) – not all terms used in information community A are necessarily known or equally interpreted in information community B (see figure 2). If such a situation occurs, the actors must develop common ontologies. Only such ontologies will support GI-services effectively when searching for data. Such common ontologies are being developed in various domains, but up to now disaster management and emergency response have been underrepresented. This deficit should be eliminated because time critical situations require such an ontological approach. It could help to save time and improve decisions due to the fact that it supports – automatically – the identification, semantic evaluation and following application of (usable) data.

Ontologies can be based on existing object catalogues (e.g. legal data, cadastres, data sets of national mapping agencies [such as the ATKIS mentioned above], national spatial data infrastructures, etc.). In other cases completely new ontologies must be designed. Such a task requires comprehensive conceptual work through specialists from different domains. Based on such work, ontologies represent a shared and common understanding of some domain that can be communicated between people and applications (Fensel 2004, Fonseca et al., 2000). Here, the application of ontologies to spatial data comes along with their significance for the more general discussion about the semantic web: Search engines today, as powerful as they are, return too often too large or inadequate lists of hits. What is needed is machine-processable information that can

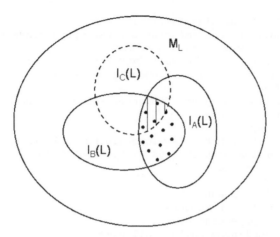

Figure 2.   Three geospatial information communities I_A, I_B, and I_C each one using a specific "language",
            (L). $I_A(L)$ and $I_B(L)$ (as well as $I_C(L)$) are part sets of M(L). A proper communication – infor-
            mation sharing instead of "pure data exchange" – between such communities will be enabled, if
            the development of a "common ontology" is possible. Such a common ontology is based on the
            overlapping area between $I_A(L)$ and $I_B(L)$ (dotted area). If several geospatial information commu-
            nities identify such overlapping areas, then an additional $I_C(L)$ (possibly more) can be integrated
            in the development of such a common ontology (area with dots and lines). The ontology develop-
            ment process may lead to some information loss referring to each single community. But it will
            enable communication of relevant information (referring to figure 1) that is required by all three
            communities (Basic model: Guarino 1998; modified).

point a search engine to the *relevant* pages, or data sets, and can thus improve both, precision and
recall (Berendt et al., 2002, the accentuation has been inserted by the author).

4.2   *XML derivates*

Currently, XML-based languages are used in many different application areas for the design of
ontologies. The World Wide Web Consortium (W3C) proposed these languages to provide formalised specifications of conceptual models (it was Gruber [1993] who defined ontology as a
"specification of a conceptualisation"). XML, propagated as supporting the interoperability of
applications in general (explicitly including semantics) is the basis of GML (Geographic Markup
Language) which has been specified by the Open Geospatial Consortium and is now in the state
of a recommendation for the version GML 4.0. However, GML 3.1 is already in a state that could
be standardized officially but will rest in the state of a recommendation due to the parallel process
within ISO/TC 211, where GML plays a role under the project number 19136. The publication of
the ISO-GML-project is envisaged for 2006 and within the framework of the harmonisation of the
activities of OGC and ISO both drafts will be certified officially by their committees (Fitzke 2005).
GML enables a certain stage of interoperability for GIS applications. The description of geographic
features, based on XML/GML enables various systems to deal with spatial information, including
most of the current database systems as well as GIS, WebGIS, OpenSource GIS, etc. Within the
OGC Initiative CITE (Compliance & Interoperability Testing & Evaluation) an example has been
published that describes a ships emergency on the North Sea, in GML (figure 3). The document is
fully XML, typical GML expressions are usable due to the opening of the GML namespace (line
3: xmlns:gml=http://www.opengis.net/gml). Namespaces are "the means" to support the understanding of languages spoken in specific geospatial information communities. This connects with
what was represented in figure 2.

```xml
<?xml version="1.0" encoding="UTF-8"?>
  <ships_emergency_message xmlns=http://www.opengis.net/examples
  xmlns:gml="http://www.opengis.net/gml"
  xmlns:xsi="http://www.w3org/2001/XMLSchema-instance"
  xsi:schemaLocation="http://www.opengis.net/examples
    havarie.xsd">
    <gml:description>Emergency freighter Pallas in the North Sea
    1998</gml:description>
     <gml:boundedBy>
      <gml:Box srsName=http://www.opengis.net/gml/srs/epsg.xml#4326>
        <gml:coordinates>6.0916,54.51 7.265,55.3583
        </gml:coordinates>
      </gml:Box>
           </gml:boundedBy>
            <gml:featureMember>
             <ships_emergency>
               <gml:name>Pallas</gml:name>
               <state>Bahamas</state>
                 <ships_coarse>
               <gml:linestring
               srsName="http://www.opengis.net/gml/srs/
               epsg.xml#4326>
               <gml:coordinates>
                   6.0916,55.2567 6.925,55.2567 7.595,55.3333
                   7.8917,55.3583 7.8533,55.2083 7.945,55.2025
                   7.9033,55.1253 7.95,55.0916 8.0053,54.9316
                   7.785,54.5983 7.8683,54.51 7.9933,54.5216
                   8.025,54.5366 8.0933,54.5566 8.1533,54.5733
                   8.255,54.5483 8.265,54.5516
                </gml:coordinates>
                </gml:linestring>
             </ships_coarse>
          </emergency_begin>1998-10-25</emergency_begin>
        </emergency_end>1998-10-29</emergency_end>
      </ships_emergency>
    </gml:featureMember>
</ships_emergency_message>
```

Figure 3. A piece of XML/GML-code: a description of a ships' emergency on the North Sea (Fitzke 2005).

In the following example, XML and GML are used to describe (partially) a ship's emergency. The case of the freighter "Pallas" is based on a real emergency on the North Sea that happened in 1998. Apart from the xmlns: tags that open a specific namespace as mentioned above, the gml: tags refer to the Geographic Markup Language that provides specific "spatial" tags. Such a GML-tag is, for instance, "gml:boundedBy" which describes a bounding box to minimize the area where the emergency happened. Another example is "gml:coordinates" which, in this case, describes the course (see the tag ships_course) that the "Pallas" took in disabled conditions. Note that the XML/GML description in this stage is far away from representing a complete domain ontology. A domain ontology must include more elements, which is explained later. The code example here is only meant to get an impression of how the spatial properties are linked to "an object" which, in this case, plays a role in an emergency case. The part of defining the meaning of terms used in this description, as well as of finding relationships to similar ontologies, is not included yet.

The code in figure 3 represents a basic approach toward the XML/GML-representation of a geographic feature. This guarantees a certain degree of interoperability, whereas full semantic interoperability is still lacking here. The enhancement by different elements, such as metadata and – almost important for proper communication – a taxonomy of used terms, as well as the inclusion of information about relationships to other ontologies must be integrated. When we do

```
<owl:Class rdf:ID="Disaster_Flood_Dyke_Break">
    <rdfs:subClassOf rdf:resource="#Disaster_Flood"/>
        <owl:oneOf rdf:parseType="Collection">
            <owl:Thing rdf:about="#Location"/>
            <owl:Thing rdf:about="#Spatial_Extent"/>
            <owl:Thing rdf:about="#Endangered_Area"/>
            <owl:Thing rdf:about="#Evacuation_Area"/>
        </owl:oneOf>
</owl:Class>
```

Figure 4. A piece of RDF/OWL-code, representing some basic aspects within a disaster ontology, describing flood disasters. OWL is based on RDF and XML. An integration of such XML dialects (including GML and possibly SVG, the Scalable Vector Graphics Language which enables the description and sharing of vector based graphics) can lead to an integrated description of graphics, attributes, and semantics of spatial objects aimed at spatial information sharing via the Internet.

this, we approach ontology instead of a pure data model of simple features. To reach such a goal more sophisticated languages are needed to fulfil the requirements of a semantically enriched data description in the sense of ontology. The Resource description Framework (RDF) and Web ontology Language (OWL), also based on XML, are under development at the W3C. RDF and OWL have already been used for ontology generation in the spatial domain (see, for example, Pundt and Bishr 2002, Hart et al., 2004, or Redbrake and Raubal 2004). OWL has been developed in vision of the semantic web, in which information is given explicit meaning, making it easier for machines to automatically process and integrate information available on the web. The semantic web therefore builds on XML's ability to define customized tagging schemes and RDF's flexible approach to representing data (McGuiness and van Harmelen, 2004). Due to the fact that RDF does not provide enough "functionality" to perform reasoning tasks, which is an important presupposition for identifying and understanding "relevant" data among various data sources, OWL has been developed to provide a language that enables humans to formalize ontologies. Nowadays, there is already a large number of ontologies available via the WWW. For example, a DAML ontology library exists which contains about 280 examples written in DAML+OIL (a language that is based on RDF and which is the predecessor of OWL), but also in OWL. Some projects have been carried out on ontology development in the field of GIS (Linková et al., 2005). Up to now, there are few attempts to apply these concepts in the fields of emergency response and disaster management.

4.3 Ontology editors

There are several tools available that support the definition and formalization of ontologies, for example Ontolingua (http://www-ksl-svc.stanford.edu:5915/doc/frameeditor/guided-tour/index.html), Chimaera (http://www.ksl.stanford.edu/software/chimaera/). Another tool is the open software product Protégé (http://protege.stanford.edu/). The Protégé ontology editor supports the Web Ontology Language (OWL), which is the most recent development in standard ontology languages, endorsed by the World Wide Web Consortium (W3C) to promote the Semantic web vision (figure 5). Based on such systems, ontologies can be developed and examined; however, the development of concrete domain ontologies for emergency response is still at the beginning. A Protégé-based definition of domain ontologies could be one for floods, earthquakes, hurricanes, traffic accidents, ships emergencies, and other disasters. All such events would require specific (domain) ontologies. Up to now it is not clear, if such "disaster categories" are specific enough. Probably more concrete and detailed ontologies are needed to represent different flooding or earthquake types, or to distinguish specific fire emergency types, for instance "fires in buildings", "forest fires", and "fires in traffic accidents", etc. An unanswered question is also, whether such ontologies are usable in general, or if regional and local spatial properties must be considered and how such information can be included dynamically into an ontology during an emergency. At this stage, research has been done in only the first, small steps. Protégé supports effectively the

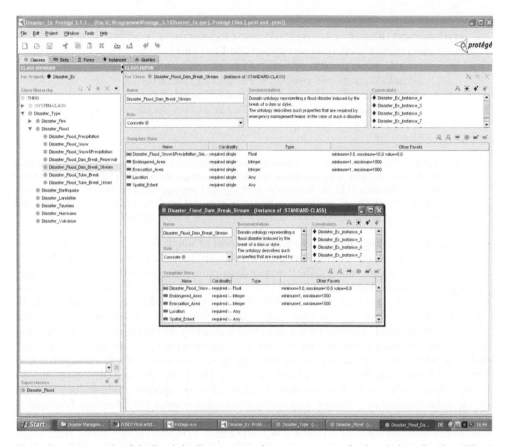

Figure 5. An example of the Protégé editor, representing some aspects of a description of a class "Flood disaster" (which is a sub-class of "disasters" in general).

development of ontologies using classes, objects, slots, and properties. Such an approach can also be implemented using "pure" RDF/OWL, the latter providing more flexibility. The Protégé editor builds a good base for ontology development in any case (Noy and McGuinness 2002).

Reconsidering the example from the beginning that referred to a flooding scenario, we can now develop some thoughts about the usability of ontologies within such a framework. During the flood comprehensive information about the infrastructure within the drainage basin was needed by the emergency teams. They got data from the IT-group, but most of the data was not usable at all. The emergency teams had to identify suitable roads, pathways or tracks on which they can get rapidly to those points where their help is needed. But there was not enough time to make proper decisions about the data they got from the central office. The outcome was that the emergency teams had various data and had to try out which information about roads, tracks, etc. fitted their purposes. This was not only a time consuming procedure, but also a source of potential failure: if decisions are made on the basis of "wrong" data, they can be useless, or even harmful. If tools were available that support the decision about usability of data (especially on the side of the central IT-group, but possibly also on the side of the emergency teams) the chance of using only *relevant* data would have been increased.

Protégé and RDF/OWL are the tools to be used to reach such a goal. They could be used to implement ontologies that define clearly and in a hierarchical order the different data sources. Before sending data to users, an intelligent search machine, or geo-portal, could interpret such formal ontologies and exclude those that were considered as "non-relevant". In the case used this would have meant that only those roads, tracks and further pathways would have been identified that are really usable for various kinds of emergency vehicles.

Protégé is a useful tool for prototyping, but does not include the complete functionality of "OWL full". RDF/OWL should be used if comprehensive ontologies of trafficable ways (in this examples; more generally: of data models of different geospatial information communities) must be implemented. Such ontologies must clearly define what the term "road" means and if "road" in one data set is the same as "street" or "track" in another set. RDF/OWL-based ontologies are machine-readable and therefore, to a certain extent, interpretable. The search machine that identifies several potential data sources is able to decide, which data are usable for the emergency teams. Those data that are identified as "non relevant" can be excluded before they are provided to users. Of course, such an automation of decisions requires more comprehensive requests for data than a simple term entered in a search machine. However, one advantage of using ontologies is that the decision about relevance and non-relevance must not be made by the emergency teams themselves, as described in the use case discussed. This decision is (partly) automated based on ontologies using the technologies described above.

4.4 *Multiple ontologies*

Following the arguments mentioned before, it is not intended to argue for the development of an [*one*] ontology for [*the wide field of*] disaster management. As mentioned within the discussion of figures 1–3, this would not represent adequately the various *different* possibilities to model one and the same geoobject, or class of objects. As said before, in contrast to the philosophical term "Ontology" (which exists uniquely), various application areas in information technology can afford different ontologies. This is also true for spatial applications: within the "spatial domain" various ontologies will be produced to support decision making. This is due to the various existing geospatial information communities working in specific domains on the one hand, and the numerous different application areas on the other. There are many examples that different domain ontologies are needed to support the modelling and analysis of spatial data in one application. Fonseca, Egenhofer et al., to mention only one of several examples, give special emphasis to the case of remote sensing systems and geographic information systems. Here, the levels of ontologies can be used to guide processes for the extraction of more general or more detailed information. The authors conclude that the use of multiple ontologies allows the extraction of information in different stages of classification and that the semantic integration of aerial images and GIS is a crucial step towards better geospatial modeling (Fonseca, Egenhofer et al., 2002).

Such multiple ontologies help to overcome the problems that occur due to semantic heterogeneity. The only way to support information access and sharing is to make data sets understandable not only for humans, but – to a certain extent – for computers. This goal is supported via formal ontologies. Ontologies do not *eliminate* semantic heterogeneity of spatial information. They *support* the adequate use and integration of spatial information by considering explicitly data semantics. In future an increasing number of ontologies will appear, especially *domain ontologies* that capture the knowledge within a particular domain e.g. electronic, medical, mechanic, traffic, urban and landscape planning, and possibly disaster management and emergency response.

5 CONCLUSIONS AND NEXT STEPS

In emergency response, datasets must be available at the right time, at the right location. This requires tools that are able to evaluate the usability of datasets in a given disaster situation. The *relevance* and therefore the *reliability* of spatial data used in emergency response require computer tools able to carry out such evaluations where humans have no time to assess or discuss comprehensively. The latter have to concentrate on avoiding damages and saving lives.

Within the data evaluation process, however, data semantics have to be considered explicitly. Data have to be delivered in short terms and humans must rely on such data to support the decisions that have to be made in hours, minutes, or seconds. Formal ontologies support intelligent data discovery, together with metadata, and enable data providers to model spatial data, their properties, quality parameters, relationships, potential uses, and semantics (Wilde and Pundt 2004).

The starting point is the fact that geoinformation in general is an area where semantic non-interoperability hinders a problem-free information sharing. The model of relevance supports the definition of the semantic mismatch that occurs due to differences in domain descriptions. From this point of view the awareness for the use of formal ontologies for the identification of relevant data increases.

XML grounded and W3C conformant standard languages such as the Resource Description Framework (RDF) and the Web Ontology Language (OWL) enable data providers to formalise ontologies based on their specific views of a domain. This is due to the fact that data become machine-readable and interpretable to a certain extent. The set of "hits" in an emergency situation can be reduced by enabling an appropriate, ontology-based, service that is aimed at discovering the *relevant* data. Due to the fast development of the W3C- and OGC standards the technological basis makes formal ontologies a feasible approach in various areas where GI is used, including emergency response.

The next step to be done must bridge theory and practice. Meanwhile, several examples exist where formal ontologies are used to support decision making in general, but also for spatial problems. Disaster management and emergency response are fields where this approach could also be helpful in supporting *rapid* decision making.

The availability of semantic reference systems will play an important role within this framework (Kuhn and Raubal 2003). This requires much work on ontologies by data providers (supported by GI scientists). The providers, however, should be interested in such methods, because they could enhance the confidence in their data. It would improve qualitatively the decisions that are based on such data, and therefore support the transparency of decision making.

The use of standardised languages contributes to interoperability and would mean that not only data, but also ontologies can be "reused". The current activities to realise the semantic web within the W3C and the computer science community (Fensel 2004, McGuiness and van Harmelen 2004, W3C 2005) motivates developers to deal more concretely with ontology based modelling of spatial data. Apart from the W3C standards those of the Open Geospatial Consortium and the International Standards Organization (ISO) metadata standards must be considered in any case.

Tests must focus on ontology-based data set identification, simultaneously accompanied by the evaluation of relevance of data, related to problems of disaster management and emergency response. A test scenario, based on two or more domain ontologies, will be the next step carried out within the framework of research that is aimed at contributing to overcome the semantic mismatch that still represents an obstacle for the usage of GI, especially when rapid responses in emergency cases are required.

REFERENCES

Bernard, L., Einspanier, U., Haubrock, S., Hübner, S., Kuhn, W., Lessing, R., Lutz, M., Visser, U. 2004. Ontologies for Intelligent Search and Semantic Translation in Spatial Data Infrastructures.
Berendt B., Hotho A., Stumme G. 2002. Towards Semantic Web Mining. In Horrocks, I., Hendler, J. (eds.), The semantic web – ISWC 2002: 264–278. Berlin, New York: Springer.
Bishr, Y., Pundt, H. Kuhn, W., Radwan, M. 1999. Probing the Concept of Information Communities-a First Step Toward Semantic Interoperability. In Goodchild, M.F., M.J. Egenhofer, R. Fegeas, and C.A. Kottman (eds) Interoperating Geographic Information Systems: 55–71. Boston: Kluwer Academic Publishers.
Broekstra, J., Klein, M., Decker, S., Fensel, D., van Harmelen, F., Horrocks, I. 2000. Enabling Knowledge Representation on the Web by extending RDF Schema, AIDministrator, The Netherlands, 20 pages.
Fensel, D. 2004. Ontologies: A Silver Bullet for Knowledge Management and Electronic Commerce. 2nd edition. Berlin, New York: Springer.
Fitzke, J. 2005. Die Welt der Features – eine Welt aus Features (The World of Features – one World of Features). In: Bernard, L., Fitzke, J., Wagner, R. (eds) 2005. Geodateninfrastruktur – Grundlagen und Anwendungen

(Spatial Data Infrastructure – Introduction and Applications). Heidelberg, Germany: Wichmann, pp 73–82.

Fonseca, F., Egenhofer, M., Davis, A., Borges, K. 2000. Ontologies and Knowledge Sharing in Urban GIS. Computer, Environment and Urban Systems 24 (3): 251–272.

Fonseca, F., Egenhofer, M., Agouris, P., Camara, G. 2002. Using Ontologies for Integrated Geographic Information Systems. Transactions in GIS 6 (3): 231–257.

Frank, A., Grünbacher, A. 2002. What is Relevant in a Dataset? In: Ruiz, M., Gould, M. Ramon, J., Proceedings of the 5th AGILE conference on Geographic Information Science: 259–263. Universitat de les Illes Balears: Palma.

Giger, Ch., Najar, Ch. 2003. Ontology-based integration of data and metadata. In Gould M., Laurini R., Coulondre S. (eds) Proceedings of the 6th AGILE conference on Geographic Information Science: 586–594. Lausanne: Presses polytechniques et universitaires romandes.

Gruber, Th. 1993. A translation approach to portable ontologies. Knowledge Acquisition 5 (2): 199–220.

Guarino, N. 1998. Formal Ontology in Information Systems. In Guarino, N. (ed.) Formal Ontology in Information Systems. Proceedings of FOIS'98, Trento, Italy, 6–8 June 1998. 3–15. Amsterdam : IOS Press.

Guptill, S.C., Morrison, J.L. 1995. Elements of spatial data quality. Exeter, United Kingdom Elsevier Science: BPC, Wheatons Ltd.

Hart, G., Temple S., Mizen H. 2004. Tales of the River Bank, First Thoughts in the Development of a Topographic Ontology. In Toppen, F., Prastacos, P. (eds) Proceedings of the 7th AGILE Conference on Geographic Information Science: 169–178. Heraklion, Crete: University Press.

Harvey, F. 1998. Quality is Contextual. In Goodchild, M.F., Jeansoulin, R. (eds.), Data Quality in Geographic Information: From Error to Uncertainty: 37–42. Edition Hermes, Paris.

ISO TC211 2004, available at http://www.isotc211.org (page accessed on 2005-12-01)

Kuhn, W., Raubal, M. 2003. Implementing Semantic Reference Systems. In Gould, M., Laurini R., Coulondre S. (eds) Proceedings of the 6th AGILE Conference on Geographic Information Science: 63–72. Lausanne: Presses Polytechniques et Universtitaires Romandes.

Linková, Z., Nedbal, R., Rimnác, M. 2005. Building Ontologies for GIS. Academy of Sciences of the Czech Republic, Institute for Computer Science. Technical Report No. 932: 1–9.

McGuiness, D.L., van Harmelen F. 2004. OWL Web Ontology Language Overview. W3C Recommendation February 2004, available at http://www.w3.org/TR/2004/REC-owl-features-20040210/

Moellering, H. 1996. Spatial Transfer Standards 2: Characteristics for Assessing Standards and full Description of the National and International Standards in the World. International Cartographic Association, 3–13. Oxford: Pergamon.

Noy, N., McGuiness, D. 2002. Ontology Development 101: A Guide to Creating Your First Ontology.

Pundt H., Bishr Y. 2002. Domain ontologies for data sharing – an example from environmental monitoring using field GIS. Computer & Geosciences, 28(1): 95–102

Pundt H., 2005. Evaluating the Relevance of Spatial Data in Time Critical Situations. In: Van Oosterom, P., Slatanova, S., Fendel, E. (eds): Geoinformation for Disaster Management: 779–788. Berlin, New York: Springer.

Redbrake, D., Raubal, M. 2004. Ontology-Driven Wrappers for Navigation Services. In: Toppen F., Prastacos P. (eds), Proceedings of the 7th AGILE Conference on Geographic Information Science: 195–205. Heraklion, Crete: University Press.

Visser, U., Stuckenschmidt H., Schuster G., Vögele Th. 2002. Ontologies for geographic information processing. Computer & Geosciences, 28(1): 103–117.

Vögele, Th. Spittel R. 2004. Enhancing Spatial Data Infrastructures with Semantic Web Technologies. In: Toppen F., Prastacos P. (eds) Proceedings of the 7th AGILE Conference on Geographic Information Science: 105–111. Heraklion, Crete: University Press.

Xu, Z., Lee, Y.C. 2002. Semantic heterogeneity of geodata. Proceedings of the symposium on geospatial theory, processing and applications, Ottawa, Canada, 2002, available at http://www.isprs.org/commission4/proceedings/paper.html#3 (page accessed 2005-11-28).

Wilde, M., Pundt, H. 2004. Development of an ISO-compliant, internet-based metadata editor for the EU project MEDIS. In: Strobl J., Blaschke T., Griesebner G. (eds) Angewandte Geographische Informationsverarbeitung.: 782–787. Heidelberg, Germany: Wichmann.

W3C 2005. available at http://www.w3c.org/ (page accessed 2005-11-14).

Geospatial Information Technology for Emergency Response – Zlatanova & Li (eds)
© 2008 Taylor & Francis Group, London, ISBN 978-0-415-42247-5

CityGML – 3D city models and their potential for emergency response

T.H. Kolbe

Institute for Geodesy and Geoinformation Science, Technical University Berlin, Germany

G. Gröger & L. Plümer

Institute for Cartography and Geoinformation, University of Bonn, Germany

ABSTRACT: Virtual 3D city models provide important information for different aspects of disaster management. In order to ensure the unambiguous interpretation of the represented objects, ontology in the sense of a common information model for urban and regional structures has to be defined. Furthermore, up-to-date of and flexible access to 3D city models are of utmost importance. Spatial Data Infrastructures (SDI) provide the appropriate framework to cover this aspect, integrating distributed data sources on demand. In this chapter we present CityGML, which is in the first place ontology for the three-dimensional, multi-purpose, and multi-scale representation of cities, sites, and regions. The implementation of CityGML is based on the standard GML3 of the Open Geospatial Consortium and thus defines an exchange format for the storage of and interoperable access to 3D city models in SDIs. The class taxonomy distinguishes between buildings and other man-made artifacts, vegetation objects, water bodies, and transportation facilities like streets and railways. Spatial as well as semantic properties are structured in five consecutive levels of detail. Throughout this chapter, special focus is on the utilization of model concepts with respect to different tasks in the context of emergency response.

1 INTRODUCTION

The quality of available geospatial information is decisive for the planning and realization of rescue operations. This does not only apply to spatial resolution, geometric accuracy and topological consistency, but especially concerns the spatial dimensions of the data. Three-dimensional geospatial data, and in particular virtual 3D city models provide essential information for different aspects of disaster management. First, they memorize the shape and configuration of a city. In case of severe destruction of infrastructure e.g. caused by earthquakes, immediate access to this reference data allows to quickly assess the extent of the damage, to guide helpers and last but not least to rebuild the damaged sites. Second, 3D city models enable 3D visualizations and facilitate localization in indoor and outdoor navigation. Augmented reality systems provide helpers with information that is visually overlaid with their view of the real world. Such systems need 3D city models in order to compute the positions and occlusions of the overlay graphics. Third, 3D escape routes inside and outside of buildings can be determined with an appropriate city model. Fourth, in flooding scenarios 3D city models allow to identify even affected building storeys.

In the context of emergency response, up-to-dateness of and flexible access to 3D city models are of utmost importance (Zlatanova and Holweg 2004). Spatial Data Infrastructures provide the appropriate framework to cover both aspects, integrating distributed data sources on demand (Groot and McLaughlin 2000). However, the prerequisite is syntactic and semantic interoperability of the participating GIS components (Bishr 1998).

Syntactic interoperability can be achieved by using the XML-based Geography Markup Language (GML3, see Cox et al., 2004) of the Open Geospatial Consortium (OGC). GML3 is an XML-based abstract format for the concrete specification of application specific spatial data formats. It is open, vendor-independent, and based on ISO standards; it can be extended and specialized to a specific application domain; and it explicitly supports simple and complex 3D geometry and topology. Furthermore, GML is the native data format of OGC's Web Feature Service (WFS), a standardized web service that implements methods to access and manipulate geospatial data within a spatial data infrastructure (Vretanos 2002).

Semantic interoperability presumes common definitions of objects, attributes, and their interrelationships with respect to a specific domain. However, no common semantic model for 3D city models has been established yet. In the following we present CityGML, a multi-purpose and multi-scale representation for the storage of and access to 3D city models. The class taxonomy of CityGML distinguishes between buildings and other man-made artifacts, vegetation objects, water bodies, and transportation facilities like streets and railways. Spatial as well as semantic properties are structured in five consecutive levels of detail (LoD), where LoD0 defines a coarse regional model and the most detailed LoD4 comprises building interiors resp. indoor features. Included thematic objects, which are especially relevant for disaster management, are different types of digital elevation models, building features like rooms, doors, windows, balco-nies, and subsurface constructions. The data model behind CityGML is based on the ISO standard family 191xx. The implementation is realized as an application schema for GML3.

CityGML has been developed during the last three years by the Special Interest Group 3D of the initiative Geodata Infrastructure North-Rhine Westphalia (GDI NRW) in Germany. On the international level CityGML is investigated within a project of the European Spatial Data Research organization (EuroSDR) since 2006, aiming at a further harmonization in Europe and the practical evaluation with respect to large city models. A comprehensive specification proposal is currently being prepared and is scheduled to be submitted to the OGC in the next months.

2 VIRTUAL 3D CITY MODELS

3D city modeling is an active research topic in distinct application areas. Different modeling paradigms are employed in 3D geographical information systems (3D GIS; Köninger & Bartel 1998), computer graphics (Foley et al., 1995), and architecture, engineering, construction, and facility management (AEC/FM; Eastman 1999). Whereas in 3D GIS the focus lies on the management of multi-scale, large area, and geo-referenced 3D models, the AEC/FM domain addresses more detailed 3D models with respect to construction and management processes (Kolbe & Plümer 2004). Computer graphics rather concentrates on the visual appearance of 3D models.

The possible applications of a 3D city model resp. the tasks it can support mainly depend on the concrete development of the four distinct representation aspects geometry, topology, semantics, and graphical appearance. Whereas geometry and topology describe the spatial configuration of 3D objects, the semantic aspect comprises the thematic structures, attributes and interrelationships. Information about the graphical appearance like façade textures, object colours, and signatures are employed for the visualization of the model.

The representation of geometry and topology of 3D objects has been investigated in detail by Molenaar (1992), Zlatanova (2000), Herring (2001), Oosterom et al. (2002), Pfund (2002), and Kolbe and Gröger (2003; 2004). The management of multi-scale models was discussed (among others) by Coors and Flick (1998), Guthe and Klein (2003), Gröger et al. (2004).

The ISO standard ISO/PAS 16739 'Industry Foundation Classes' (IFC, Adachi et al. 2003) is a semantic model for buildings and terrain which has been developed in the AEC/FM domain. It defines an exchange format and contains object classes for storeys, roofs, walls, stairs, etc. Nevertheless, since IFC is lacking concepts for spatial objects like streets, vegetation objects or water bodies, it is not appropriate for the representation of complex city models. Similar problems arise with respect to 'green building XML' (gbXML 2003), an AEC/FM standard for building energy and environmental performance analysis, and 'Building-construction XML' (van Rees et al., 2002), a standard for the mapping of construction taxonomies.

LandXML/LandGML is a standard for land management, surveying and cadastre, providing a semantic model for parcels, land use, transportation and pipe networks (LandXML 2001). Although LandXML supports 3D coordinates, it does not comprise volumetric geometries. Buildings are only represented by their footprints. Further concepts for 3D man-made objects are missing.

Computer graphics (CG) standards like VRML97 (1997) and its successor X3D mainly model the geometry and the appearance of 3D objects. They do not provide concepts for the representation of thematic aspects, attributes, and interrelationships of the graphical objects.

Since thematic information are crucial for disaster management, CG standards are not sufficient. AEC/FM standards concentrate on man-made constructions and are lacking concepts for the representation of natural objects. Furthermore, none of the discussed AEC/FM and GIS standards supports multi-scale models resp. multiple levels-of-detail (LoD).

3 THE SEMANTICS OF CITYGML

CityGML defines a common information model for cities and regions, including their semantic properties, i.e. the generalization hierarchy between classes, relations between objects, and spatial properties as well the appearance of objects.

To achieve interoperability, the formal specification of CityGML is based on the language GML3.1, a standard issued by the OGC and the ISO. GML is an XML based language, which facilitates data exchange by spatial web services, in particular by Web Feature Services (WFS), see Vretanos (2002). GML3.1 provides classes to define the spatial properties – geometry and topology – of objects. Furthermore, concepts for the definition of features including attributes, relations and generalization hierarchies are provided.

Technically, CityGML is implemented as a GML3 application schema. It defines a profile of GML3, since the extensive geometry model of GML3 is restricted to the classes sufficient to represent cities geometrically. CityGML has been derived from models specified in the *Unified Modeling Language (UML)* (see Booch et al., 1997) by applying the transformation rules given in Cox et al. (2004). Thus, CityGML may be processed by standard GML3 readers and visualized by standard GML3 viewers.

The next section presents the general concepts implemented by CityGML, while the following sections 3.1 to 3.6 describe the thematic models of CityGML which are especially relevant for emergency response applications: the building model, the transportation model, and the digital terrain model.

3.1 *General concepts*

CityGML supports different Levels-of-Detail (LoD), which often arise from independent data collection processes and which facilitate efficient visualization and data analysis. In a CityGML dataset, the same object may be represented in different LoD simultaneously, enabling the analysis and visualization of the same object with regard to different degrees of resolution. Furthermore, two CityGML data sets containing the same object in different LoD may be combined and integrated.

CityGML provides five different LoD, which are illustrated in Figure 1, taking buildings as an example. The coarsest level LoD0 is essentially a two and a half dimensional *Digital Terrain Model* (DTM). LoD1 is the well-known blocks model, without any roof structures. A building in LoD2 has distinctive roof structures. LoD3 denotes architectural models with detailed wall and roof structures, balconies and bays. High-resolution textures can be mapped onto these structures. In addition, detailed vegetation and transportation objects are components of a LoD3 model. LoD4 completes a LoD3 model by adding interior structures like rooms, interior doors, stairs, and furniture.

Beyond buildings, the LoD concept applies to other object classes as well. The focus is on model resolution and perceptibility of object parts, but it addresses also geometrical accuracies and

LoD0   LoD1   LoD2

LoD3   LoD4

Figure 1.   Illustration of the five Levels-of-Detail (LoD) defined by CityGML.

minimal dimensions of objects. The classification may also be used to asses the quality of a 3D city model data set. Furthermore, the LoD category makes data sets comparable and thus supports the integration process of those sets.

Another general concept of CityGML is the *TerrainIntersectionCurve (TIC)*, which facilitates the geometrical integration of buildings or other objects with the terrain. This curve explicitly denotes the exact position where the terrain touches the object. It can be used to 'pull up' resp. 'pull down' the surrounding terrain to fit the TIC. For further details see Kolbe et al. (2005).

3.1.1   *Geometrical modeling*

Spatial properties of thematic objects in CityGML are represented by a subset of the geometry model of GML3, according to the well-known *Boundary Representation* (Foley 1995). For each dimension, there is a geometrical primitive: a zero-dimensional object is a *point*, a one-dimensional a *curve*, a two-dimensional a *surface*, and a three-dimensional a *solid*. A solid is bounded by surfaces, a surface by curves, and a curve by points. In CityGML 1.0, a curve is restricted to be a straight line, and each surface must be planar, i.e. their boundary and all interior points are forced to be located in one single plane. Curves, surfaces and solids may be aggregated to *CompositeCurves*, *CompositeSurfaces* and *CompositeSolids*, respectively.

The primitives curve, surface and solids, and the respective aggregates must satisfy a number of integrity constraints, which guarantee consistency of the model. One important constraint is that the interiors of the primitives must be disjoint. These constraints assure that primitives do not overlap, and touch at most at their boundaries. Since solids must be disjoint, the computation of volumes can be performed much easier than it would be the case if solids overlap.

The graphical appearance of surfaces, i.e. material properties like colors, shininess, transparency and textures, is covered by concepts adopted from the graphics standard X3D.

3.1.2   *Coherent semantic-geometrical modeling*

Another characteristic of CityGML is the coherent modeling of semantics and geometrical/topological properties. On the semantic level, real-world entities are represented by features, for example buildings, walls, windows, or rooms, including attributes, relations and aggregation hierarchies (part-whole-relations) between features. Thus the part-of-relationship between features can be derived on a semantic level only, without considering geometry. However, on the spatial level, geometry objects are assigned to features representing their spatial location and extent. So the model consists of two hierarchies: the semantic and the geometrical, where the corresponding objects are linked by relations. The advantage of this approach is, that it can be navigated in both

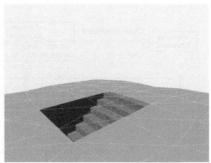

Figure 2.   Passages are subsurface objects (left). The entrance is sealed by a virtual *ClosureSurface*, which is both part of the DTM and the subsurface object (right).

hierarchies and between both hierarchies arbitrarily, for answering thematic and/or geometrical queries or performing analyses.

If both hierarchies exist for a concrete object, they must be coherent, i.e. it must be assured that they match and fit together. For example, if a wall of a building has two windows and a door on the semantic level, then the geometry representing the wall must contain the geometries of both windows and of the door.

3.1.3   *ClosureSurfaces and subsurface objects*
A new concept in CityGML is the *ClosureSurface*, which is employed to seal objects, which are in fact open, but must be closed in order to compute its volume. An airplane hangar is an example for such an object. *ClosureSurfaces* are special surfaces which are taken into account when needed to compute volumes and are neglected, when they are irrelevant or not appropriate, for example in visualizations.

The concept of *ClosureSurfaces* also is employed to model the entrances of *subsurface objects*. Those objects like tunnels or pedestrian underpasses have to be modeled as closed solids in order to compute their volume, for example in flood simulations. The entrances to subsurface objects also have to be sealed to avoid holes in the digital terrain mode (see Figure 2). However, in close-range visualizations the entrance must be treated as open. Thus, *ClosureSurfaces* are an adequate way to model those entrances.

3.1.4   *References to objects in external data sets*
3D objects often are derived from or have relations to objects in other databases or data sets. For example, a 3D building model may have been constructed from a two-dimensional footprint in a cadastre data set, or may be derived from an architectural model. The reference of a 3D object to its corresponding object in an external data set is essential, if an update must be propagated or if additional data, for example the name and address of a building's owner, is required. In order to supply such information, each CityGML thematic object may have *External References* to corresponding objects in external data sets. Such a reference denotes the external information system and the unique identifier of the object in this system. Both are specified as a *Uniform Resource Identifier (URI)*, which is a generic format for references to any kind of resources in the internet.

3.1.5   *Dictionaries and code lists for attributes*
Attributes which classify objects often have values that are restricted to a number of discrete values. An example is the attribute *roof type*, whose attribute values typically are saddle back roof, hip roof, semi-hip roof, flat roof, pent roof, or tent roof. If such an attribute is typed as string, misspellings or different names for the same notion obstruct interoperability. In CityGML,

261

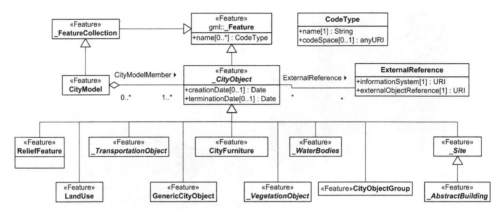

Figure 3. UML diagram of the top level class hierarchy of CityGML. The bracketed numbers following the attribute names denote its multiplicity: the minimal and maximal number of occurrences of the attribute per object. For example, a *name* is optional (0) in the class *_Feature* or may occur multiple times (star symbol), while a *_CityObject* has none or at most one *creationDate*. An *ExternalReference* has exactly one occurrence of the attribute *informationSystem*.

such classifying attributes are specified as *GML3 Code Lists* or *Dictionaries*. Such a structure enumerates all possible values of the attribute, assuring that the same name is used for the same notion. In addition, the translation of attribute values into other languages is facilitated. Dictionaries and code lists may be extended or redefined by users.

3.1.6 *City object groups*

The grouping concept of CityGML allows to aggregate arbitrary city objects according to user-defined criteria, and to represent and transfer these aggregations in a city model. A group may have a name and a type, for example "escape route from room no. 43 in house no. 1212 in a fire scenario" as a name and "escape route" as type. Each member of the group is assigned an optional role name, which specifies the role this particular member plays in the group. This role name may, for example, describe the sequence number of this object in an escape route, or in case of a building complex, denote the main building.

A group may contain other groups as members, allowing nested grouping of arbitrary depth.

3.2 *Top level classes*

Figure 3 depicts the top level of the class hierarchy of CityGML. The base class of all thematic classes is *CityObject*, which provides a creation and a termination date for the management of histories of features, and external references to the same object in other data sets, as described in section 3.1. *CityObject* is a subclass of the GML class *Feature*, thus it inherits metadata (e.g., information about the lineage, quality aspects, accuracy) and names from *Feature* and its super classes. A *CityObject* may have multiple names, which are optionally qualified by a so-called *codespace*. This enables to differentiate for example an official name from a popular name or names in different languages (c.f. the name property of GML objects, Cox et al., 2004).

The subclasses of *CityObject* comprise the different thematic fields of a city model: the terrain, the coverage by land use objects, transportation, vegetation, water bodies and sites, in particular buildings, and city furniture. Generic city objects are explained in section 3.6, and groups have already been discussed in section 3.1.

Thematic classes have further subclasses with relations, attributes and geometry. The ones relevant for the field of emergency response are discussed in detail in the following sections.

Features of the specialized subclasses of *CityObject* may be aggregated to a single *CityModel*, which again is a feature with optional metadata.

262

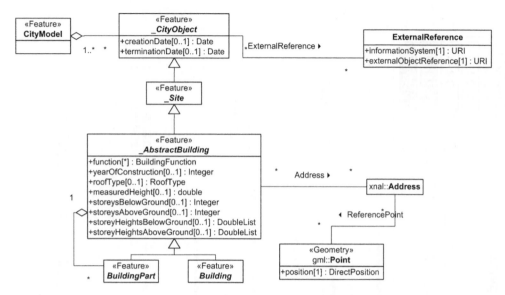

Figure 4. UML diagram (Booch et al., 1997) of CityGML's building model, part one.

3.3 *The building model*

The building model of CityGML allows the representation of thematic and spatial aspects of buildings, building parts and accessories in four levels-of-detail, LoD1 to LoD4. The UML diagrams of the building model are depicted in Figures 4 and 5 (the diagram has been split for clarity reasons). The pivotal class of the building model is AbstractBuilding, which is specialized either to a Building or to a BuildingPart. A simple building is represented by a single building object only, while a complex building is made up of a building object consisting of BuildingParts, which, for example, differ in height, year of construction, or function. Since a BuildingPart is again an AbstractBuilding, an aggregation hierarchy of arbitrary depth may be realized.

A building is described by optional attributes: the function of the building, for example residential, public, or industry; the year of construction, the roof type, the measured height, and the number and the individual heights of the storeys above resp. below ground. The address or the multiple addresses of a building are specified using the *xNAL address and name standard* issued by the OASIS consortium (OASIS 2003), which provides schemas for all kinds of international addresses. An optional reference point denotes the exact location of the entrance of the building, which may be needed for route planning applications.

In the coarsest LoD1, the spatial extent of a Building is given by a *SolidGeometry*, which in this case is a simple block. Since *AbstractBuilding* is a subclass of the root class *CityObject*, the relation to the *ExternalReference* (see section 3.1) is inherited.

Often Buildings are aggregated to larger *Building Complexes*. These can be represented using the grouping concept described in section 3.1. The main building of this group may be denoted by the role name of this object relative to the group.

In LoD2, there are basically two differences compared with LoD1: First, there may be a more detailed geometry replacing the coarser LoD1 geometry, and second the thematic classification of the parts of a building is more detailed. In a LoD2 building, it is possible to distinguish the bounding surfaces as own semantic objects. These surfaces may be classified as *Roof*, *Wall* or *Floor Surfaces*. The geometry of these surfaces, however, is shared with the *SolidGeometry* that defines the shape of the whole building. An opening in a building is modeled by a *ClosureSurface*; this concept was already discussed in section 3.1. A LoD2 building also may have thematic *BuildingInstallations*, for example chimneys, balconies or outer stairs. These *BuildingInstallations* have their own geometry

263

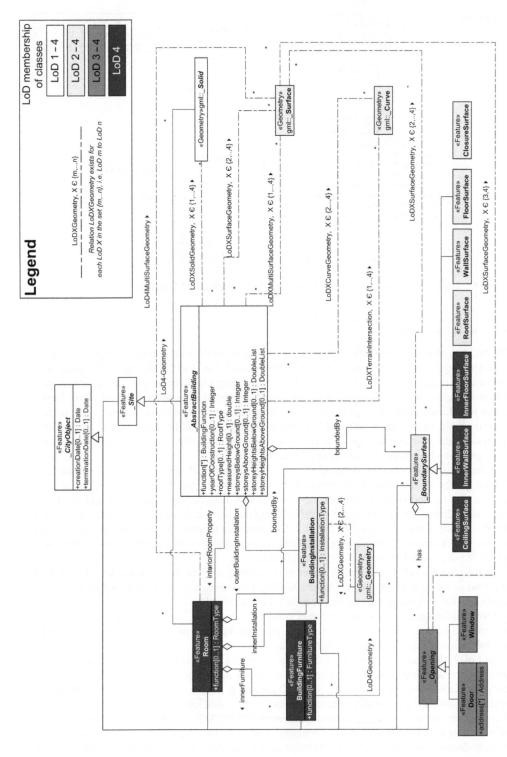

Figure 5.   UML-Diagram of CityGML's building model, part two.

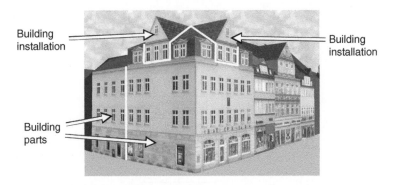

Building installation

Building installation

Building parts

Figure 6.   Illustration of a LoD2 building. It consists of two building parts with different heights. The right part has two dormers represented as building installations.

in LoD2. The type of geometry is not restricted: it is specified by a *ObjectGeometry*, which is the super class of the aggregates *CurveGeometries*, *SurfaceGeometries* and *SolidGeometries*. In contrast to *BuildingParts*, *BuildingInstallations* are smaller and only additional accessories of a building, but not a constituent part of it. Figure 6 illustrates a LoD2 building with two building characteristics and two parts.

The geometry of a LoD2 building is given by *SolidGeometries* and additionally by *Surface-Geometries*, which represent surfaces that are part of the building, but do not bound the solids of the building. The overhanging part of a roof is an example for such a surface.

In LoD3, buildings additionally may have *Opening* features such as *Windows* and *Doors*. The class *Opening* is a subclass of *CityObject*, it is an own thematic object and thus inherits the option to have external references. The geometry in LoD3 is given by separate solids and surfaces, which usually are more detailed than their LoD2 counterparts. In addition, curve geometries may be used to model, for example antennas, if they are not represented as thematic *BuildingInstallation*. As discussed in section 3.1, the accuracy requirements of LoD3 are much higher than in LoD2.

LoD4 complements LoD3 by adding interior structures of buildings such as *Rooms*, which are bounded by *Ceiling-*, *InnerWall-* and *InnerFloorSurface* features. Rooms may have *BuildingFur-nitures* and interior *BuildingInstallations*. A *BuildingFurniture* is a movable part of a room, such as a chair or furniture, while a *BuildingInstallation* is permanently connected to the room. Examples for interior building installations are stairs or pillars. Doors are used in LoD4 to connect rooms topologically: the surface that represents the door geometrically is part of the boundaries of the solids of both rooms. The aggregation of rooms according to arbitrary, user defined criteria is achieved by employing the grouping concept provided by CityGML.

Please note that all these objects inherit the references to objects in external data sets. Important data sources for LoD4 models are IFC data sets (c.f. section 2), which can be converted accordingly (Benner et al., 2005). As discussed in section 3.1, the different accuracy requirements of LoD1 to LoD4 have to be applied to the building model as well.

3.4   *The digital terrain model*

An essential part of a city model is the terrain. In CityGML, the terrain may be specified as a regular raster or grid, as a TIN (Triangulated Irregular Network), by break lines respectively skeleton lines, or by mass points. These four types are implemented by the corresponding GML3 classes. The UML diagram of the digital terrain model is shown in Figure 7. A TIN may either be represented as a collection of explicit triangles (class *TriangulatedSurface*), or implicitly by a set of 3D points (class *TIN*), where the triangulation may be reconstructed by standard methods (Okabe et al., 1992). A break line is a discontinuity of the terrain, while skeleton lines are either ridges or valleys. Both are represented by 3-D curves. Mass points are simply a set of 3-D points.

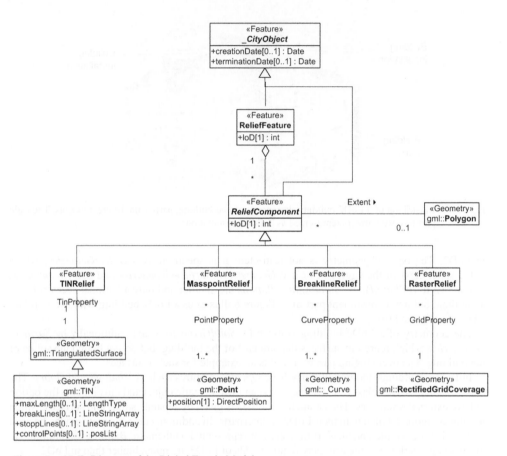

Figure 7.   UML-Diagram of the Digital Terrain Model.

In a CityGML data set, these four terrain types may be combined in different ways, yielding a high flexibility. First, each type may be represented in different levels-of-detail, reflecting different accuracies or resolutions. Second, a part of the terrain can be described by the combination of multiple types, for example by a raster and break lines, or by a TIN and break lines and skeleton lines. Third, neighboring regions may be represented by different types of terrain models. To facilitate this combination, each terrain object is provided with a spatial attribute denoting its *extent of validity*. This extent is represented by a 2-D footprint polygon, which may have holes. This concept enables, for example, the modeling of a terrain by a coarse grid, where some distinguished regions are represented by a detailed, high-accuracy TIN. The boundaries between both types are given by the extend attributes of the corresponding terrain objects. This approach is very similar to the concept of *TerrainIntersectionCurves* introduced in section 3.1.

3.5   *Transportation objects*

The transportation model of CityGML is a multi-functional, multi-scale model, focusing on thematic as well as on geometrical/topological aspects.

Figure 8 depicts the UML diagram of the model. The main class is *TransportationComplex*, which represents, for example, a road, a square or a track. In the coarsest LoD0, the transportation complexes are modeled by line objects establishing a linear network. On this level, path finding algorithms or similar analyses can be executed. Starting from LoD1, a *TransportationComplex* has a

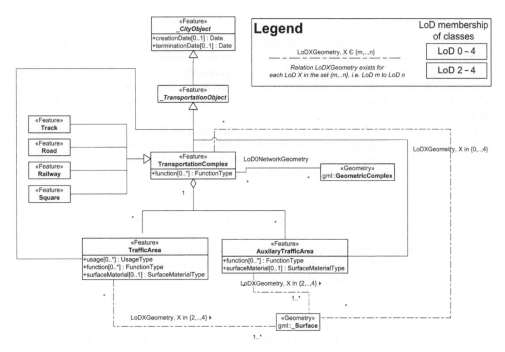

Figure 8.   UML diagram of the transportation model of CityGML.

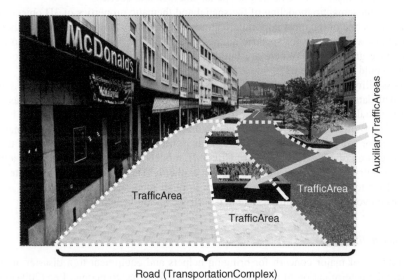

Figure 9.   Example for the representation of a TransportationComplex, a road, which is the aggregation of TrafficAreas and AuxiliaryTrafficAreas.

surface geometry, reflecting the actual shape of the object, not just its centerline. In LoD2 to LoD4, it is further subdivided thematically into *TrafficAreas*, which are used by cars, bicycles or pedestrians, and in *AuxiliaryTrafficAreas*, for example green spaces or flower tubs. The function of both areas may be represented by an attribute, as well as its shape by an areal geometry. An illustration of a Transportation Complex and its parts is given in Figure 9.

3.6 *Further classes*

Besides the transportation model, the building model and the digital terrain model, there are a number of other models covering further aspects of a cities' semantics. The *Water Bodies Model* represents the thematic aspects and three-dimensional geometry of rivers, canals and lakes. *Solitary Vegetation Objects* like trees and complete *Biotopes* like forests are represented by the *Vegetation Model*. *City Furniture* like traffic lights, traffic signs or advertising columns may also be represented by CityGML.

The class *GenericCityObject* allows to model features not provided explicitly by the CityGML schema. This enables the representation of objects which were not anticipated; a situation that is likely to occur in an emergency response phase. In addition, each City Object may be extended by *GenericAttributes* to represent properties of features not covered by the schema.

4 MEETING THE REQUIREMENTS OF EMERGENCY RESPONSE

CityGML was designed as a common information model and exchange format for the multifunctional utilization of 3D city models. In the following, we discuss the potential use of key concepts for tasks in an emergency response phase.

4.1 *Situation analysis, briefing and geovisualization*

Important tasks in the emergency response phase are situation analysis and the briefing of rescue personnel. Immediate access to geometrical and semantically rich geospatial information is essential, as both tasks involve numerous spatial and thematic queries.

The coherent semantic modeling of the spatial and thematic properties of 3D objects and their aggregations is one of the most important features of CityGML. Object classes have thematically rich attributes which allow for specific queries like 'What are the buildings with more than 10 storeys above ground?' or 'Where are buildings with flat roofs which are large enough that a helicopter could land on them?'.

The possibility to provide external references can be used to associate any CityGML object and its parts with data sets of other applications like facility management systems or the cadastre, which is important to determine e.g. the owner of a building. By using external references it is also possible to relate BuildingInstallations (c.f. section 3.3), which are relevant for disaster management like hydrants or fire protection doors with databases that hold the technical data about these.

The generation of 3D visualizations of CityGML models is straight-forward, as the appearance and the geometry of 3D objects are always given explicitly. This means that the concrete 3D shape does not have to be generated or derived from implicit models by using extrusions or boolean operations on volumetric primitives (like in Constructive Solid Geometry).

4.2 *Reachability*

Reachability is of major concern for bringing helpers and equipment to a disaster area. It also influences the determination of possible escape routes for affected people. Lee (2004) and Pu & Zlatanova (2005) propose to derive a network model for building interiors from the geometry and topology of 3D models.

The LOD4 model of CityGML provides explicit information here, as building interiors are modeled by rooms. Their solids are topologically connected by the surfaces representing doors or closure surfaces that seal open doorways. This adjacency implies an accessibility graph, which can be employed to determine the spread of e.g. water, smoke, gas, and air, but which can also be used to compute escape routes using classical shortest path algorithms. The edges of the accessibility graph can be marked by the corresponding distances and types of connection like normal door, fire protection door, open doorway etc. (see Figure 10).

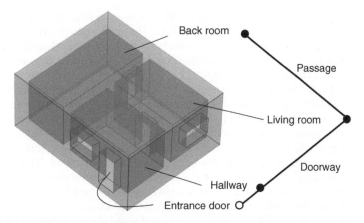

Figure 10. Building interior (left) and accessibility graph (right) derived from topological adjacencies of room surfaces for the determination of escape routes.

Figure 11. 3D city models provide orientation and guidance: determination of the current position using range finders and a 3D city model (left), and background and wayfinding information presented by virtual 3D banners and signposts help to guide rescue teams (right; example taken from pedestrian navigation).

For reachability analyses outside of buildings CityGML's transportation modelling can be employed. It provides a linear network in LOD0 and areal traffic features in higher LODs. As these *TrafficAreas* describe the surfaces of traffic objects explicitly they can be used for trajectory planning resp. geometric route planning. In conjunction with the different types of 3D CityObjects also the extents of vehicles or rescue equipment can be taken into account.

4.3 *Localization*

During rescue operations helpers always need to be informed about their current location. In this context, localization generally comprises two different consecutive steps: First, the carried navigation device has to determine the current position and heading. In the second step the location information has to be communicated to the user.

GPS and a magnetic compass (possibly in combination with a gyroscope) provide sufficient accuracy in many outdoor situations. In indoor operations or in dense urban canyons, lacking visibility of the GPS satellites and multipath effects often cause big position errors or even prevent the achievement of a position solution. For these cases, alternative approaches coming from the field of mobile robotics should be considered.

Autonomous mobile robots often use range finding sensors to determine the distance to their surrounding environment (see Figure 11 on the left). The employed sensors are based on laser, ultrasonic, or radar echo registration, and at least the last two allow position determination in smoke-filled areas. Bayes filters, in particular particle filters, were shown to be very robust concerning the global position estimation with respect to a given model of the environment (*Monte Carlo Localization* MCL, cf. Thrun (2002)). Although MCL mainly needs geometry information about the surrounding surfaces, objects should be thematically divided into immobile or movable objects, because only immobile objects like walls, stairs, pillars etc. provide reliable reference information for position determination. CityGML both provides these geometric and semantic information.

In order to support the perception of one's own position and to provide guidance to the place of action, augmented reality techniques can be applied (Leebmann 2004). 3D labels and virtual signposts which are precisely overlayed in the visual field of rescue workers using head mounted displays (HMD) provide intuitive orientation and navigation cues (see Figure 11 on the right, c.f. Kolbe (2004)). Labels have to be placed at specific, sensible positions and the visual overlay has to consider possible occlusions of these virtual signs by real world objects. While the latter is ensured by occlusion culling using the 3D geometry of CityGML objects, the differentiated building model of CityGML would allow to restrict label placement to specific object types or parts like building walls or traffic areas.

In the case that helpers must move through smoke-filled or dark areas, not only virtual signposts and labels but also the 3D model itself could be overlayed on a Head Mounted Display.

4.4 *Simulation*

In the emergency response phase simulation can be used to assess damaged objects, estimate possible shelters, and to predict the spreading of water, gas, fire, smoke etc. For example, Freund & Roßmann (2003) describe how a 3D simulation environment can be used to investigate different strategies for fire fighting using virtual 3D city models.

In flooding scenarios not only the digital terrain model is important, but also the built-up structures on and below the surface. Especially in urban areas water quickly spreads in subsurface hollow spaces of a city, e.g. in metro tunnels (see Herath & Dutta (2004)). As CityGML also accounts for subsurface structures, it may provide valuable information in this context. Furthermore, the representation of storey heights above and below ground allows to determine to which degree buildings are affected. This information is especially useful for planning evacuations and for damage assessment by aid organizations and insurance companies.

In CityGML the terrain model may consist of neighbored or nested patches having different resolutions. This allows to embed high resolution DTMs for e.g. regions with high flood risk into large area DTMs at low resolution.

Since the geometry of 3D objects should be represented by at least one closed solid, the computability of volumes and masses is facilitated. For example, in flooding resp. fire scenarios it could be estimated how much water resp. smoke or gas will flow into a tunnel, pedestrian passage, or a building. The estimation of masses from volumes is also interesting for planning the removal of debris after an incident.

5 IMPLEMENTATIONS AND DATA ACQUISITION

5.1 *Status of CityGML*

In 2004 and 2005, CityGML has been implemented in the so-called 'Pilot 3D' testbed launched by the GDI NRW. Five project teams realized specific applications of 3D city models, e.g. fire fighting simulation, city planning, tourism, and bicycle route planning. Cross-wise data exchange was demonstrated between different applications and providers. All project teams were organized as public-private-partnerships with participants coming from software manufacturers, academia, and the German cities Berlin, Hamburg, Düsseldorf, Cologne, Stuttgart, and Recklinghausen.

CityGML is now being used for system integration in the official 3D city models of Berlin (see Döllner et al., 2006) and Bonn.

In 2006, CityGML has been brought into the standardization process within the Open Geospatial Consortium. It is currently under evaluation in the OpenGIS Web Services testbed #4 (OWS-4) in an emergency response and homeland security scenario.

Examples giving an impression of the contents and the structure of CityGML data files can be found in Kolbe, Gröger & Plümer (2005). Sample datasets, UML and XML schema files, and free viewer applications are provided on the CityGML website (CityGML 2006).

5.2 Systems and interfaces

A growing number of commercial systems and research tools provide interfaces to read or write CityGML datasets. In the following, some examples shall be highlighted.

One example is the commercial 3D visualization and authoring tool *LandXplorer* distributed by the company 3DGeo. Thematic and spatial properties of the data may be used for the selection of features and to determine their graphical appearance. A viewing-only version of the system can be downloaded for free (3DGeo 2006).

Aristoteles is a free viewer for 3D GML3 datasets which already supports most of the geometry types used by CityGML. It is Open Source software and has been developed at the Institute for Cartography and Geoinformation at Bonn University so far. Besides 3D visualization, the focus is on the exploration of the semantic structure and the hierarchy between features, which can be displayed and queried (Dörschlag & Drerup 2006).

CityServer3D is a client-server GIS application from the Fraunhofer Institute for Computer Graphics Darmstadt (FHG IGD), which supports CityGML as a transfer format between server and clients (Haist & Coors 2005).

Currently, a transactional Web Feature Service (WFS-T) for CityGML is being implemented within the Open Source software framework *Deegree* (Fitzke et al., 2004). The WFS is employed for data access to the 3D city models of Berlin and Bonn. Furthermore, Deegree's implementation of the OGC Web Terrain Service (WTS) can be chained with the WFS in order to produce perspective views of CityGML3D scenes.

5.3 Provision and integration of CityGML data

There are basically two main sources for 3D city models. The first is geodesy and the second the construction domain (architectural design, civil engineering, facility management). As the modeling approach and the data models differ fundamentally in both domains, conversion of models between these domains is not straight-forward (c.f. Kolbe & Plümer (2004)). Whereas the IFC standard covers the aspects of the AEC/FM domain and CAAD systems, CityGML has its roots in the field of geodesy, cadastre, and 3D GIS.

In geodesy, the trend for providing 3D city models is that models are generated from airborne and terrestrial laser scanning resp. photogrammetry (Früh & Zakhor 2004, Kaartinen et al. 2005). Many approaches also take data from 2D cadastre into account and use sensor measurements to obtain building heights. However, the automatic acquisition is still a topic of current research. The semantic properties of buildings and their parts as supplied by CityGML play an important role for e.g. the development of procedures for the automatic reconstruction of semantically rich 3D building models from aerial images (Fischer et al., 1998).

National and regional mapping and cadastre agencies (NMCA's) currently are incorporating 3D geospatial objects into the traditional cadastre and topographic datasets. In the medium term NMCAs will offer 3D city models in the context of Spatial Data Infrastructures. CityGML is intended to support these developments. In the long-term, national mapping agencies as well as 3D cadastres could become the main provider for 3D city models (Stoter and Salzmann 2003).

CityGML models also can be generated from 3D models coming from the AEC/FM domain. Benner et al. (2005) describe how the semantic properties of IFC models can be preserved and transferred during conversion.

In spatial data infrastructures, 3D city models may be assembled from different sources. The integration of these models makes it necessary to resolve geometric and topological inconsistencies. Therefore, homogenization procedures for e.g. the adjustment of the digital terrain model with respect to 3D objects, and the automatic detection and resolution of topological errors like the penetration of solid volumes have to be developed. Possible starting points are the work of Kampshoff (2005) on 3D homogenization and Koch (2005) on the topologically/semantically consistent integration of topographic objects with the DTM.

6   CONCLUSIONS AND FURTHER WORK

Virtual 3D city models provide substantial information for urban disaster management tasks. However, any application beyond 3D visualization like simulation, computation of escape routes etc., requires explicit semantic models. In this chapter we have presented CityGML, a common information model for urban and regional structures. CityGML covers the four main aspects of 3D spatial objects and terrain, i.e. geometry, topology, appearance, and semantic resp. thematic properties. The ability of maintaining different levels of detail makes it suitable for small to large area utilization.

In emergency response, specific tasks like simulation, localization, and reachability analysis are of utmost importance. For each of these tasks it was shown, how required information are provided by CityGML.

The consistent utilization of the ISO 191xx standards facilitate immediate mapping of the data model to an application schema for the Geography Markup Language GML3. Since GML was designed by the OGC to serve as the standard exchange format for spatial data infrastructures, processing of CityGML is immediately supported by corresponding web services like the Web Feature Service (WFS), Web Catalog Service (CS/W), and Web Coordinate Transformation Service (WCTS).

In the future, we will address the dynamic aspects of the represented spatial objects, i.e. movable resp. moving objects like a bascule bridge or the current state of doors and windows (open, closed). Furthermore, the simultaneous inclusion of different discrete water levels, e.g. low and high tide, shall be realized. Also, concepts for the representation of history in the sense of a timeline have to be integrated.

Another topic of future research is the interface between 3D GIS on the one side and CA(A)D models resp. facility management systems on the other. The essential question is how to establish a bidirectional mapping between generative (CSG) and accumulative (B-Rep) 3D geometry models. A first step was made by Brenner (2004) with the introduction of the concept of Weak CSG Primitives.

ACKNOWLEDGEMENTS

We wish to thank the members of the modeling working group of the SIG 3D of the GDI NRW: Joachim Benner, Frank Bildstein, Dave Capstick, Martin Degen, Dirk Dörschlag, Rüdiger Drees, Heinrich Geerling, Ulrich Gruber, Jörg Haist, Peter Henning, Frank Knospe, Andreas Kohlhaas, Kai-Uwe Krause, Ulrich Krause, Klaus Leinemann, Marc-Oliver Löwner, Hardo Müller, Mark Pendlington, Martin Rechner, Wieland Röhricht, and Frank Thiemann. Furthermore, we like to thank our colleagues at the Institute for Cartography and Geoinformation, in particular Viktor Stroh.

REFERENCES

3DGeo 2006. 3D visualization and authoring system landXplorer, available at http://www.3dgeo.de.
Adachi, Y., Forester, J., Hyvarinen, J., Karstila, K., Liebich, T. & Wix, J. 2003. Industry Foundation Classes IFC2x Edition 2, International Alliance for Interoperability, available at http://www.iai-international.org.
Benner, J., Geiger, A. & Leinemann, K. 2005. Flexible generation of semantic 3D building models. In: Gröger, G. & Kolbe, T.H. (eds.) First International ISPRS/EuroSDR/DGPF-Workshop on Next Generation 3D City Models. Bonn, Germany, EuroSDR Publication No. 49, pp. 17–22.

Bishr, Y. 1998. Overcoming the semantic and other barriers to GIS interoperability. International Journal on Geographical Information Science, 12(4).

Booch, G., Rumbaugh, J. & Jacobson, I. 1997. Unified Modeling Language User Guide. Addison-Wesley.

Brenner, C. 2004. Modelling 3D Objects Using Weak CSG Primitives, In: International Archives of Photogrammetry, Remote Sensing and Spatial Information Science, Vol. 35, ISPRS, Istanbul.

CityGML 2006. Homepage of CityGML, available at http://www.citygml.org.

Coors, V. & Flick, S. 1998. Integrating Levels of Detail in a Web-based 3D-GIS. Proc. 6th ACM Symp. on Geographic Information Systems (ACM GIS 98), Washington D.C., USA.

Cox, S., Daisy, P., Lake, R., Portele, C. & Whiteside, A. 2004. OpenGIS Geography Markup Language (GML3.1), Implementation Specification Version 3.1.0, Recommendation Paper, OGC Doc. No. 03-105r1.

Döllner, J., Kolbe, T.H., Liecke, F., Sgouros, T. & Teichmann, K. 2006. The Virtual 3D City Model of Berlin – Managing, Integrating, and Communicating Complex Urban Information. In: Proceedings of the 25th Urban Data Management Symposium UDMS 2006 in Aalborg, Denmark, May 15–17, 2006.

Dörschlag, D. & Drerup, J. 2006. Aristoteles GML3 Viewer, available at http://www.ikg.uni-bonn.de/aristoteles

Eastman, C.M. 1999. Building Product Models: Computer Environments Supporting Design and Construction, CRC Press.

Fischer, A., Kolbe, T.H., Lang, F., Cremers, A.B., Förstner, W., Plümer, L. & Steinhage, V. 1998. Extracting Buildings from Aerial Images using Hierarchical Aggregation in 2D and 3D. Computer Vision & Image Understanding, 72(2), Academic Press.

Fitzke, J., Greve, K., Müller, M. & Poth, A. 2004. Building SDIs with Free Software – the deegree project. In Schrenk, M. (ed.) Proc. of the 9th International Symposion on Planning & IT 'CORP2004', Vienna, Austria, available at http://www.deegree.org.

Foley, J., van Dam, A,. Feiner, S. & Hughes, J. 1995. Computer Graphics: Principles and Practice. Addison Wesley, 2nd Ed.

Freund, E. & Roßmann, J. 2003. Integrating Robotics and Virtual Reality with Geo-Information Technology: Chances and Perspectives. In Bernhard, L., Sliwinski, A. & Senkler, K. (eds.): Geodaten- und Geodienste-Infrastrukturen – von der Forschung zur praktischen Anwendung. Proceedings of the Münster Geo-Information Days (GI-Tage 2003), IfGI Prints, University of Münster.

Früh, C. & Zakhor, A. 2004. An Automated Method for Large-Scale, Ground-Based City Model Acquisition. International Journal of Computer Vision, 60(1).

gbXML. 2003. Green Building XML Schema, available at http://www.gbxml.org.

Gröger, G., Kolbe, T.H. & Plümer, L. 2004. Mehrskalige, multifunktionale 3D-Stadt- und Regionalmodelle. Photogrammetrie, Fernerkundung, Geoinformation (PFG) 2/2004 (in German).

Gröger, G. & Plümer, L. 2005. How to get 3-D for the Price of 2-D – Topology and Consistency of 3-D Urban GIS. Geoinformatica, 9(2).

Groot, R. & McLaughlin, J.D. 2000. Geospatial Data Infrastructure – Concepts, Cases, and Good Practice. Oxford University Press.

Haist, J. & Coors, V. 2005. The W3DS Interface of CityServer 3D. In: Gröger, G. & Kolbe, T.H. (eds) First International ISPRS/EuroSDR/DGPF-Workshop on Next Generation 3D City Models. Bonn, Germany, EuroSDR Publication No. 49.

Herath, S. & Dutta, D. 2004. Modeling of urban flooding including underground space. In Proceedings of the Second International Conference of Asia-Pacific Hydrology and Water Resources Association, Singapore, Volume I, pp. 55–63, July.

Herring, J. 2001. The OpenGIS Abstract Specification, Topic 1: Feature Geometry (ISO 19107 Spatial Schema), Version 5. OGC Document Number 01-101.

LandXML. 2001. LandXML Schema 1.0, available at http://www.landxml.org.

Kaartinen, H., Hyyppä, J., Gülch, E., Hyyppä, H., Matikainen, L., Vosselman, G. et al., 2005. EuroSDR Building Extraction Comparison. In Heipke, C., Jacobsen, K. & Gerke, M. (eds.), Proc. of the ISPRS Hannover Workshop 2005 on "High-Resolution Earth Imaging for Geospatial Information". International Archives of Photogrammetry and Remote Sensing, Vol. XXXVI, Part I/W3.

Kampshoff, S. 2005. Mathematical Models for Geometrical Integration. In: Gröger, G. & Kolbe, T.H. (eds) First International ISPRS/EuroSDR/DGPF-Workshop on Next Generation 3D City Models. Bonn, Germany, EuroSDR Publication No. 49.

Koch, A. 2005. An Integrated Semantically Correct 2.5 dimensional Object Oriented TIN. In: Gröger, G. & Kolbe, T.H. (eds) First International ISPRS/EuroSDR/DGPF-Workshop on Next Generation 3D City Models. Bonn, Germany, EuroSDR Publication No. 49.

Köninger, A. & Bartel, S. 1998. 3D-GIS for Urban Purposes. Geoinformatica, 2(1), March.

Kolbe, T.H. 2004. Augmented Videos and Panoramas for Pedestrian Navigation. In Gartner, G. (ed.), Proc. of the 2nd Symposium on Location Based Services & TeleCartography 2004, January 2004 in Vienna, Geoscientific publication series of the Technical University of Vienna.

Kolbe, T.H. & Gröger, G. 2003. Towards unified 3D city models. In: Schiewe, J., Hahn, M., Madden, M., Sester, M. (eds): Challenges in Geospatial Analysis, Integration and Visualization II. Proc. of Joint ISPRS Workshop, Stuttgart.

Kolbe, T.H. & Gröger, G. 2004. Unified Representation of 3D City Models. Geoinformation Science Journal, (4)1.

Kolbe, T.H., Gröger, G. & Plümer, L. 2005. CityGML – Interoperable Access to 3D City Models. In: Oosterom, v. P., Zlatanova, S. & Fendel, E.M. (eds) Geo-information for Disaster Management. Proc. of the 1st International Symposium on Geo-information for Disaster Management, The Netherlands, March 21–23. Delft 2005, Springer.

Kolbe, T.H. & Plümer, L. 2004. Bringing GIS and CA(A)D Together – Integrating 3d city models emerging from two different disciplines. GIM International, 18(7), July 2004.

Lee, J. 2004. A Spatial Access-Oriented Implementation of a 3-D GIS Topological Data Model for Urban Entities. Geoinformatica 8(3): 237–264.

Leebmann, J. 2004. An Augmented Reality System for Earthquake Disaster Response. In Proc. of the XXth ISPRS Congress in Istanbul, International Archives of Photogrammetry and Remote Sensing, Vol. 35, part B3.

Molenaar, M. 1992. A topology for 3D vector maps. ITC Journal 1992-1.

OASIS. 2003. xNAL Name and Address Standard. Organization for the Advancement of Structured Information Standards, available at http://xml.coverpages.org/xnal.html.

Okabe, A., Boots, B. & Sugihara, K. 1992. Spatial Tessellations: Concepts and Applications of Voronoi Diagrams. John Wiley & Sons.

Oosterom, P., Stoter, J., Quak, W. & Zlatanova, S. 2002. The balance between geometry and topology. In Richardson, D. & Oosterom, P. (eds.): Advances in Spatial Data Handling. Proc. of 10th Int. Symp. SDH 2002, Springer, Berlin.

Pfund, M. 2002. 3D GIS Architecture. GIM International 2/2002.

Pu, S. & Zlatanova, S. 2005. Evacuation Route Calculation of Inner Buildings. In: Oosterom, P. v., Zlatanova, S. & Fendel, E.M. (eds), Geo-Information for Disaster Management, Springer, Berlin.

Stoter, J. & Salzmann, M. 2003. Towards a 3D cadastre: where do cadastral needs and technical possibilities meet? Computers, Environment and Urban Systems, Theme Issue: 3D Cadastres, 27(4).

Thrun, S. 2002. Robotic Mapping: A Survey. In Lakemeyer, G. & Nebel, B. (eds.) Exploring Artificial Intelligence in the New Millennium. Morgan Kaufmann Publishers.

Van Rees, R., Tolman, F. & Beheshti, R. 2002. How BcXML Handles Construction Semantics. In: Proc. of the Int. Council for Research and Innovation in Building and Construction, CIB w78 conference 2002, Aarhus, 12–14 June.

Vretanos, P.A. 2002. Web Feature Service Implementation Specification Version 1.0.0, OGC Doc. No. 02-058.

VRML97. 1997. Information technology – Computer graphics and image processing – The Virtual Reality Modeling Language (VRML) – Part 1: Functional specification and UTF-8 encoding. Part 1 of ISO/IEC Standard 14772-1:1997.

Zlatanova, S. 2000. 3D GIS for Urban Development. PhD Thesis, ITC Dissertation Series No. 69, The International Institute for Aerospace Survey and Earth Sciences, The Netherlands.

Zlatanova, S. & Holweg, D. 2004. 3D Geo-Information in Emergency Response: A Framework. In: Proceedings of the 4th International Symposium on Mobile Mapping Technology (MMT 2004), March 29–31, Kunming, China.

Geospatial Information Technology for Emergency Response – Zlatanova & Li (eds)
© *Crown copyright 2008. Reproduced by permission of Ordnance Survey, ISBN 978-0-415-42247-5*

Integrated emergency management: Experiences and challenges of a national geospatial information provider, Ordnance Survey

C.J. Parker
Research & Innovation, Ordnance Survey, United Kingdom

R. MacFarlane
Training & Doctrine, Emergency Planning College, Cabinet Office Civil Contingencies Secretariat, The Hawkhills, Easingwold, York

C. Phillips
Research & Innovation, Ordnance Survey, United Kingdom

ABSTRACT: The approach to emergency preparation, response and recovery in the United Kingdom is one of Integrated Emergency Management (IEM). The IEM framework is intended to cover the preparation for and to have the capabilities to respond to, and recover from, a range of potential emergencies. The Civil Contingencies Act (2004), the result of a review of previous emergency management experience, delivers a single framework for UK civil protection to better meet new potential emergency situations in a changing global environment. With respect to the UK IEM framework, this chapter has two objectives.

The first objective is to consider the nature of the IEM framework in the UK. It reviews and describes the underpinning processes (anticipate, assess, prevent, prepare, respond, recover), supporting legislation, and organisational command, control and decision-making structures. Importantly, it describes the information requirements and characteristics of information flow and use throughout the framework.

The second objective is to consider the role of a national GI provider, Ordnance Survey, within IEM in working towards providing ease of access and ease of use of GI at the point and time of need. The second part of the chapter summarises Ordnance Survey's national GI provider role and responsibilities within an IEM context. The Mapping for Emergencies scheme and pan-government purchasing, access and data-sharing agreements are described. The Digital National Framework (DNF) – a spatial framework for integrating GI in UK is described. Trends in the development of information and communications technology, GIS and related technologies are considered in terms of recent developments in data capture and representation at Ordnance Survey. Finally, future needs of GI for IEM are considered where research at Ordnance Survey is described into user-centred task analysis, automating change detection, 3- and 4-D data structures, generalization, web services, and semantic reference systems. Three prototypes illustrate how technology and GI may combine in the near future to provide the user within the IEM domain greater ease of access and ease of use at the time and point of need.

1 INTRODUCTION

Preparing for, responding to and recovering from emergencies is both difficult and uncertain for a number of different reasons. These reasons relate both to the causes, circumstances and consequences of the emergency itself (these may be termed 'external' factors) and the working practices of those agencies and individuals responsible for emergency planning and management (which may be termed 'internal factors'). In relation to external factors, it has been observed that 'crises as well as civil turbulences or terrorist actions, can be characterised by 'un-ness' – unexpected, unscheduled, unplanned, unprecedented, and definitely unpleasant' (Crichton 2003), characteristics which are distinctively different from other public-/private-sector decision contexts. These characteristics include:

- uncertainty
- complexity
- time pressure
- a dynamic event that is innately unpredictable
- information and communication problems (overload, paucity or ambiguity)
- heightened levels of stress for participants, coupled with potential personal danger.

All of these characteristics pose challenges, both personal and professional, for the agencies, groups and individuals who have to deal with emergencies, but it is possible to identify some systematic shortcomings, sometimes failures, that constrain their ability to plan for and manage emergencies (defined here as Integrated Emergency Management, IEM, which is elaborated in section 2) both effectively and efficiently. These include:

- IEM is extremely information hungry, and many public sector organisations tend to be relatively weak in their internal management of data and information;
- IEM is necessarily a set of activities that have to be carried out in collaboration with multiple agencies, an area in which many individual agencies do not excel; and
- information must underpin joint working, establishing a Common Operational Picture (COP) of risks, resources, incidents, consequences and their response, so information needs to flow across and between agencies. Non-interoperability of (broadly defined) systems poses a serious impediment to these flows of data and information.

So, those involved in preparing for, responding to, and recovering from emergencies have a need for information. Indeed, that need is more precise, for information that is relevant, appropriate, accurate, timely and delivered in a form that is appreciable under their circumstances. However, this need, or demand, for information is often only partially met as, and when, it is most needed.

Figure 1 illustrates the fact that demand for information, most acutely during an emergency, accelerates at a rate far above that of supply. This leads to what may be termed a demand-provision gap. In most cases this is not because the information, or at least the data from which the information could be generated, does not exist, but because it is not accessible at the point and time of need.

This is of course a generic issue, and one that is far wider than GIS alone, but the need for information is the key driver for the development and implementation of GIS in IEM. The specific value of GIS is that many of the issues that need to be considered in preparing for, responding to and recovering from emergencies are explicitly geographical: roads, rivers, floodplains, industrial hazards, towns and cities are all geographically distributed in a way that is of clear relevance to emergency planning and management. In short, where things are matters a great deal if something may, or does, go wrong there. GIS is a tool that enables us to account for geography, and geography is critical in understanding, planning for and communicating hazards, risks and vulnerabilities.

As introduced above, all of the processes of IEM are 'information hungry' and much of the required information is GI. It is for this reason that GIS represents a significant tool to decision makers at all levels in an IEM context, not only because GIS supports the effective management of existing data but also because analytical and modelling tools support the generation of new information, and permit the integration of data from multiple sources. In an information management

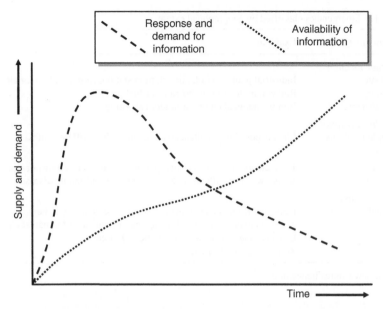

Figure 1. The Information Demand-Provision Gap following an emergency event (based on presentation by Peter Power, Visor Consultants, Royal United Services Institute, July 2004).

context this is termed 'adding value' or 'leveraging' information; in an IEM context it supports evidence-based decision making and the development and maintenance of a Common Operational Picture, the cornerstone of a coordinated approach.

2 EMERGENCIES AND INTEGRATED EMERGENCY MANAGEMENT IN THE UK

Emergencies can, by their very nature, be extremely diverse. Some of the key variables are whether:

- the incident(s) and impact(s) are localised or widespread;
- the cause is simple or complex, which has implications for its management;
- it was a single incident or a repeated incidence;
- the emergency was predicted (and if so over what timescale) or unforeseen;
- it was accidental, deliberate or 'natural';
- it was rapid onset (acute) or slow onset (chronic) in character; or
- they have an identifiable scene or not (see table 1).

As a consequence of this, there are widely varying requirements of planners and responders at different levels of command, and within and between multiple agencies. For instance, a rapid onset emergency such as a serious fire and chemical release demands rapid and decisive action in a timeframe that does not necessarily allow for a highly detailed analysis of potential conse-quences and the implications of different response scenarios. In contrast, a 'creeping crisis' or slow onset emergency, especially where there is prior warning of key characteristics such as magnitude, severity, location and timescale, may permit a detailed analysis of the various options for possible prevention, mitigation and response. Indeed the case for detailed problem analysis and assessment of response options makes very sound business sense. Consideration of the costs and benefits of implementing GIS for IEM is of considerable importance and though that discussion is beyond the scope of this chapter the following example serves to illustrate the benefits to be gained. The School of Veterinary Medicine at Penn University in the US reports that an outbreak of avian influenza in

Table 1. Emergencies classified by geographical extent.

Type of emergency	Example
A. *Single location*	
Fixed site	Industrial plant, school, airport, train station, sports stadium or city centre
Corridor	Railway, motorway, air corridor or fuel pipeline
Unpredictable	Bomb, chemical tanker or random shooting
B. *Multiple locations*	
Multiple locations	Linked, possibly simultaneous or explosions at different sites
C. *Wide area*	
Large area	Toxic cloud, loss of electricity, gas, water, communications or flooding
Whole area	Severe weather, influenza pandemic or foot-and-mouth disease
D. *Outside area*	
External emergency	Residents from one area involved in an emergency elsewhere, e.g. coach or plane crash, passenger ship sinking or incident at football stadium
	Evacuees into one area from another UK area
	Refugees from an emergency overseas

Source: www.ukresilience.info.

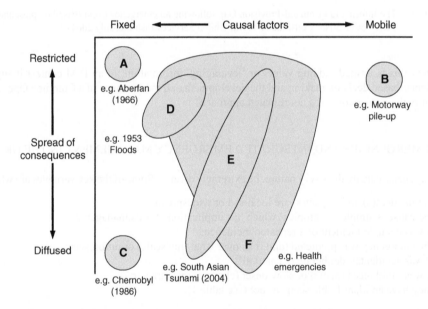

Figure 2. A typology of the geography of emergencies and disasters (see table 2 for explanatory notes).

1997 took several months and cost the State of Pennsylvania $3.5 million to control, and this was before the University had developed a functioning GIS for animal disease control. In 2001, when the GIS was operational, researchers were able to identify the infected poultry flocks, identify surrounding flocks which were at risk by virtue of their location and plan for the transport and disposal of infected carcasses to minimise risk of further infection. The outbreak was controlled in less than a month and at a cost of $400,000 (West, 2005).

From a geographical perspective, different kinds of emergencies have different characteristics, illustrated in figure 2 and table 2.

Table 2. Explanation of the regions identified in Figure 2.

Region	Examples/Characteristics
A	The cause and direct consequences are focused on a small area. The sudden slump of a coal spoil heap onto a school in Aberfan, Wales, in 1966 is an example of this. However, the human consequences of such a disaster can radiate through social networks over a wide area and be very long lasting.
B	An incident such as a multiple-pile up in fog on a motorway is one where some of the causal factors (cars) combines with a temporary fog bank to cause a locally serious emergency with loss of life, with some wider consequences due to road closures.
C	An incident such as the nuclear reactor fire at Chernobyl, Ukraine, in 1986 is one where the causes of the incident are site specific, but where the direct consequences (radiation fallout) were international in scale.
D	The 1953 floods on the East Coast of England were caused by a storm surge combining with high spring tides. Both of these were multiple, widespread causal factors and the consequences were spread from the Humber to the Thames with over 300 deaths.
E	The South Asian Tsunami of 26th December 2004 was initially caused by a sub-sea earthquake off the coast of Sumatra, but the tsunami, itself the cause of the death and destruction of property, was both fast moving and international in scale.
F	Health emergencies can vary from the highly localised (e.g. legionnaire's disease outbreak from a single cooling system, affecting a local community), through to outbreaks of animal (e.g. foot and mouth) or human disease (e.g. SARS), which have the potential to spread between countries and continents.

So, different kinds of emergencies need to be prepared for and responded to in different ways, and recovery from different types and levels of emergencies clearly poses different types and scales of requirements. The UK model of IEM is intended to establish a framework to prepare for and have the capabilities to respond to and recover from such a range of potential emergencies.

In 2001 the UK government undertook a detailed review of previous experience of emergencies, especially during 2000 and 2001, such as severe flooding, disruptions to the fuel supply system, foot- and-mouth disease and industrial action amongst fire fighters. All of these emergencies exposed weaknesses in the frameworks and processes through which hazards and threats were identified, risks assessed and (where possible) mitigated, contingency plans drawn up and emergencies responded to and recovered from. This review ultimately resulted in new legislation – the Civil Contingencies Act (2004) (see www.ukresilience.info) – which places duties on a range of agencies identified as emergency responders, and multi-agency cooperation, collaboration and communication are at the core of the model of pre-emergency, emergency and post-emergency operations. Before discussing this model of IEM, the next section introduces the Civil Contingencies Act.

2.1 The Civil Contingencies Act

The Civil Contingencies Act (2004), referred hereafter as 'the Act', together with accompanying regulations and non-legislative measures, delivers a single framework for UK Civil Protection to meet the new challenges of the 21st century. It is a wide-ranging piece of legislation and only the key elements are summarised here. For the Act itself, the accompanying regulations, issues in relation to the devolved administrations and guidance, consult www.ukresilience.info

Key to the Act is a definition of what constitutes an emergency:

- An event or situation that threatens serious damage to human welfare;
- An event or situation that threatens serious damage to the environment;
- War, or terrorism, that threatens serious damage to security.

It is important to note that the focus is on consequences rather than causes, so the Act applies equally to events or situations that originate outside of the UK as it does for those within UK boundaries.

A detailed coverage of which agencies have which responsibilities is beyond the scope of this chapter, but local government, government agencies, the emergency services and health agencies are defined as having core responsibilities and are termed Category 1 responders. These responders have a range of duties including the following:

- Assess local risks and use this to inform emergency planning;
- Put in place emergency plans;
- Put in place Business Continuity Management (BCM) arrangements;
- Put in place arrangements to make information available to the public about civil protection matters and maintain arrangements to warn, inform and advise the public in the event of an emergency; and
- Co-operation and information sharing.

It should be very clear that information and the effective and efficient flow of information are pivotal to almost all of these duties. Category two organisations primarily include utility, communication and transportation companies, all of which are charged with the duties of co-operating with Category 1 organisations and sharing relevant information.

Under the Act local responders have a **duty** to share information. This information will take many forms, for instance, describing capabilities, resources, processes or contact details for key personnel. Only some of these will be spatial data and information, but these are critical in IEM. In sharing information the Act states that the initial presumption should be that **all** information should be shared, although there are some exceptions to this in relation to national security, personal confidentiality or economic competitiveness. However, these exceptions are limited and, reinforcing this duty to share information, guidance for the Act states that:

> *Information sharing is necessary so that Category one and two responders are able to make the right judgements. If Category one and two responders have access to all the information they need, they can make the right decisions about how to plan and what to plan for. If they do not have access to all information, their planning will be weakened* (Section 3.4), *Emergency Preparedness* (2005) www.ukresilience.info

2.2 *Integrated emergency management*

UK doctrine for IEM identifies a model of six separate but mutually informing and reinforcing processes. It will be clear that this owes much to the widely accepted 'disaster cycle', of which there are variants, but generally comprising mitigation and prevention, preparedness, response, and recovery phases. The disaster cycle and IEM are both premised on foresight, a focus on risks, informed preparation and collaborative response and recovery.

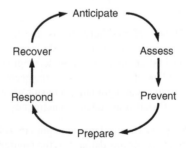

Figure 3. The cycle of IEM.

Note that the elements in figure 3 are processes and activities as distinct from phases. The implication of this is that they are ongoing activities and the model as a whole is dynamic; phases are something that can be regarded as complete. The six processes are:

1. **Anticipate**: knowing what might happen is important in being able to frame and scale an appropriate response. Emergencies arise from either hazards (non-malicious), which may be 'natural' (e.g. severe weather) or human (e.g. industrial accidents), or threats (malicious and deliberate), and their very nature is that they are more or less unpredictable in detail. However, 'horizon scanning' and effective anticipation of hazards and threats is essential.
2. **Assess**: appreciating the spread, severity and consequences of anticipated hazards and threats needs to be set within a risk assessment framework. Risk registers are developed and maintained at the local, regional and national level, and it is important that they reflect the changing nature of hazards and threats and the nature of the population, environment and national security context in their make-up.
3. **Prevent**: it is intrinsically preferable to prevent an emergency than have to deal with its consequences. If an area that has suffered repeated flooding is assessed to be at high risk of flooding on an annual basis and bankside engineering and floodwater storage works have the potential to significantly reduce that risk, this is likely to result over time in both financial savings and reduced potential for loss of life and damage to quality of life.
4. **Prepare**: not all hazards and threats are foreseen, and not all of those that are can be prevented. It is therefore critical to have structures, processes and resources in place to deal with emergencies and mitigate their effects. Central to this is emergency planning, which falls into development (creating, implementing, reviewing and maintaining), and exercising and training processes.
5. **Respond**: emergencies are almost always responded to at the operational level by one or more of the 'blue light' emergency services. In the event of an incident that requires a coordinated multi-agency response, a specialized (e.g. chemical, biological, radiological or nuclear) response or the rapid establishment of a higher level of command, procedures are established to escalate that response in a way that is appropriate. This response will draw heavily on established procedures, frameworks and resources that have been the subject of training and exercising prior to a 'real' incident.
6. **Recover**: although the involvement of the emergency services may be relatively limited in time, the process of recovering from an emergency can take months or years and there are effects – perhaps most notably those of personal loss and trauma – that extend over decades. There are medical, site clearance, decontamination, reconstruction, risk assessment, counseling and many other dimensions to recovery, some of which will overlap with the emergency response phase, others of which succeed it over varying timescales.

It is argued here that IEM should be seen as a 'self-informing process'; that is to say, data, intelligence, situational awareness and other forms of information (see the next section) that are gathered or created at one stage should, under ideal conditions, be available to decision makers at other stages of the process. For example, flood modelling would yield information about buildings, infrastructure, resources and individuals at risk, information that should be available to responders in a flooding emergency. This is not to say that the response should make the same assumptions as the hazard and risk assessment, but that the responders should have access to data describing, for example, demographics, vulnerable facilities, critical infrastructure and resources, all of which are pertinent to the efficient organisation of an effective response. However, a 'silo mentality' that impedes the widespread reuse of data and data flows within and between organisations can all too easily militate against this kind of approach.

2.3 Data, information and decision making

Data and information are of course different things. Data are results, observations, counts, measurements, locations, attributes and other basic descriptive characteristics of things and places in

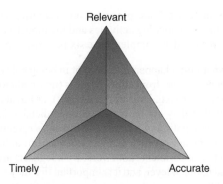

Figure 4. The three dimensions of good information.

a fairly 'raw' format. Information is more processed, ordered, summarised, selective and 'user-friendly', with the intention of assisting correct interpretation. Typically, data are high-volume (a lot of records) whereas information is more tightly geared to the requirements of a specific end-use. One of the key strengths of tools such as spreadsheets, databases and GIS is their ability to transform, if appropriately used, data into information that can be appreciated and acted on more readily. However, it is important to recognise that data are almost universally imperfect, therefore the decisions that are based on them may be misguided; and even when data and information are strong, decisions may still be misguided. Evidence is also widely used as a term and it is defined here as something that is created from information, through further sorting, selection, distillation or triangulation with other sources. In this respect it is similar to the term 'intelligence'; although specifically associated with the work of the security and intelligence services, the term is also widely used in contexts such as local government and regional development in a way that broadly equates with information and evidence.

Data do not just exist, they are created. They are usually created with specific purposes in mind, and for this reason they may be sub-optimal when evaluated for an unforeseen purpose. As emergency management is to a large degree the process of dealing with unforeseen incidents, this is especially pertinent in this context. Data created for the purposes of asset management, public health, community safety or education may be neither structured, appropriately detailed nor attributed for the purposes of emergency managers, but the reality is that they have to work with the available data, while also ensuring that future data are more suited in quality, content, coverage and availability.

Evidence-based practice is central to public-sector businesses, equally underpinning strategy and policy decisions and resource allocation and management decisions. In emergency planning and management, however, the preparation and response processes operate across very different timeframes; if a suitably structured evidence base for response is not available and structured for rapid deployment and application at the time of an incident, the evidence base for operations will be partial and weak.

The emphasis must be on ensuring that (a) data and information sources are valid, suitable and sufficient for all potential applications at different levels in the IEM command and control chain, as well as for agency-specific applications, and (b) that the potential need to use information in an emergency for a legitimate purpose, other than that for which it was originally obtained, is recognised and authorised.

Figure 4 illustrates the three dimensions of good information: relevance, timeliness and accuracy. These dimensions are of course context dependent; information that is necessary and relevant to 'front-line' field responders would be 'noise' to higher-level incident commanders, and strategic issues that may be critical at the latter level could simply confuse those at lower levels of command. GIS and allied technology, including mobile data bearers such as the Airwave™ digital radio system, provide a framework for disseminating information in such a way that it is appropriately targeted to the requirements of different groups, and can support not just incident management and command information but also the 'back-flow' of situational information for tactical management and higher strategic decisions. Too much information can be a bad thing, and the emphasis must be on 'fitness for purpose', a theme that will come up again.

In line with the argument through this chapter that GIS is a tool for managing GI, which should be seen in the wider context of enriched Information (I), all of these handling and communication processes and issues should be seen in the context of Knowledge Management (KM).

Samuel Johnson (1709–84) observed that *'Knowledge is of two kinds. We know a subject ourselves, or we know where we can get information upon it'*. More recently the Improvement and Development Agency (IDeA) has defined KM as:

> *The creation and subsequent management of an environment that encourages knowledge to be created, shared, learnt, enhanced, organised and exploited for the benefit of the organisation and its customers.*

Effective emergency management is clearly something that requires not only the sharing of data and information but also the ability to manage information of different types, so that it can be accessed at the point of need. KM is about both people (ways of working and organisational cultures) and also about technology as an enabler to support people and organisations' requirements for information. In a GI context this embraces data and information that are held within organisations and in other organisations that should be working in partnership in an IEM context.

The key considerations are those of **awareness** (knowing what is available, what the quality and potential applications are, and how and where to access it), **capacity** (the skills base to source, analyse and disseminate data and information), **communication** (the technical and human channels to ensure that awareness is maintained, standards are observed and data and information can move freely as required) and **interoperability** (the ability of technical as well as human systems to work seamlessly together to provide information as, when and where required).

2.4 *Models of decision making*

IEM is to a large degree about decisions, which are in turn about making choices. Effective IEM is about making the right choices. Making the right choices is about (a) approach, (b) information and (c) ability and authority to pursue the determined course of action.

There is of course an almost impossible variety of decisions, but some of the key types are:

- Easy (routine) to Difficult (complex)
- Previously Experienced to Rehearsed to Unforeseen
- Inconsequential to Critical
- Single to Recurrent
- One-stage to Sequential and Contingent
- Single Objective to Multiple Objectives
- Individual to Group
- Structured to Unstructured
- Strategic to Tactical to Operational.

It is clear that decisions relating to emergencies are of the more challenging type. That is, they are complex, contingent, relate to multiple objectives that are defined by a range of groups, they are commonly unstructured at the outset of an incident and a range of levels of Command and Control are involved.

Decision making has been studied extensively and from a variety of different perspectives. Early work established what is known as the rational model of decision making. In this model people are presented with a problem, they gather the relevant data to address it, analyse the data as appropriate to generate supporting information, evaluate the different options and then make what is the optimal decision under the circumstances.

Subsequent work has studied the reality of decision making in different contexts, and while the rational model remains attractive as the basis for making informed and considered decisions under

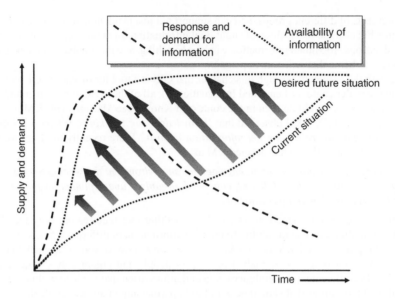

Figure 5.   Accelerating the provision of information to keep closer pace with demand.

ideal conditions, the specific circumstances surrounding an emergency are less than ideal and the assumptions outlined above may be invalid for the following reasons:

1. the problems may be multiple, developing rapidly and contingent on factors that are not yet fully appreciated;
2. the data requirements are not yet fully appreciated for the reasons above, and even core datasets may not be available to incident controllers due to inadequate preparation;
3. the tools to translate data into information may not be available or the available staff may not be sufficiently trained to make correct and effective use of them;
4. identifying options for tackling the incident is far more complex under pressure and where the contingencies are poorly understood; and
5. achieving a common operational picture that is accessible and acceptable to different agencies and levels of operational control is hard to achieve and agreement on actions may be hindered by agency-specific cultures, interests or perspectives.

Decision-making styles under emergency conditions (relating to Crichton's 'un-ness') vary between individuals and organisations, but a general distinction between an essentially intuitive approach and an analytical approach has been identified. The intuitive or 'naturalistic' approach is based on what 'feels right', based on previous experience, training and personal assessment of the observed circumstances and it supports rapid decision making. The analytical approach is inherently slower, more intellectually and resource demanding, but it permits a fuller consideration of the evolving situation, the resources available to address it and the risks associated with different paths of action. Different styles of decision making 'fit' different processes of IEM and different roles and responsibilities. A major emphasis in IEM is on preparing adequately for a range of incidents that vary in their severity, location, contingencies, interdependencies, consequences and time-space overlap with other incidents. Preparation cannot reduce the complexity, time pressure or dynamic nature of emergencies, but it can permit controllers and responders to better appreciate what is happening, what might happen and be able to coordinate and communicate information flows that support effective decision making by all parties, at all levels of control and response. Figure 5 updates Figure 1 to illustrate how information needs to keep closer pace with the demand for information as both the intuitive and the analytical styles require information, although the latter is much more information hungry.

The AAPPRR processes of IEM do not all operate under the stress of emergency conditions and the emphasis in Anticipating, Assessing, Preventing, Preparing, Response and Recovering must be on an analytical approach that is based upon data and information of the required breadth and depth and an appropriate level of quality. Indeed a critical element of preparation focuses on information as a resource that underpins effective response. In responding to an emergency different requirements are observed at different levels of incident control, and these are summarised in Table 3.

The picture is one where there are different kinds of requirements at different levels, but these requirements are more likely to be adequately met if preparation focuses on the core questions of:

- What might happen?
- What do we do if this happens?
- What do we need (to know) to frame and implement an appropriate and flexible response?

Imagination, anticipation and analysis are key to this, not just in horizon scanning or in contingency planning, but also in identifying what it is you need to know. At present information is dangerously underrated by many organisations in relation to emergencies, and this is something that has the potential to severely undermine their ability to conduct effective IEM.

3 REQUIREMENTS FOR GI WITHIN IEM

As stated above, data do not just exist but they are created. However, data are often created for a specific task or purpose, by a defined group within a specific organisation. Access to and use of those data outside of that intended application and organisational context can be remarkably difficult to achieve. Table 4 is a first pass of a definition of core data requirements for IEM. Rather than define specific datasets, the emphasis has been to define thematic areas where data should be sought if local risk assessments indicate they are relevant. It should be clear that while many of these thematic areas are provided by national mapping and information agencies, such as Ordnance Survey or Office for National Statistics, or through data locators and search sites, such as www.gigateway.org.uk or www.magic.gov.uk (multi-agency GI for the countryside), others are created and held, often tightly, by agencies such as local authorities, health agencies, emergency services and environmental organisations. While there are often sound motives for an organisation exercising tight control over their data, there are sometimes less edifying reasons, which may include:

- an organisation not being aware what data it holds itself;
- unwillingness to share data *within* an organisation, let alone with third parties;
- concerns about data protection and a tendency to err on the side of caution;
- financial and/or staff constraints on up-front data processing requirements;
- problems around drawing up and agreeing on data sharing protocols;
- data may be perceived as being of an insufficient quality to share;
- data may be in paper form only;
- data may not be collected at all; and
- disinterest or opposition in data sharing, which may be represented as any of the above.

Although such obstacles remain relatively widespread, as partnership working has increasingly become the norm in many public-sector areas, such as community safety, regeneration, social inclusion, the environment and quality of life enhancements, data and information sharing to develop a common understanding of shared issues and overlapping or common problems is increasingly well established.

Broadly speaking, the use of GI in an IEM context requires that attention is paid to five main sets of tasks:

1. locating (e.g. hearing about it, finding it and assessing it)
2. accessing (e.g. gaining agreement, data transfer and updating)
3. integrating (e.g. fitness for use, formatting, metadata, standards and user awareness)

Table 3. Characteristics of civilian* command and control during an incident response.

Level	Nomenclature	Characteristics
Strategic	**Gold**	Gold command defines 'what to do'. Gold has overall command and responsibility for an incident. In relatively minor incidents gold command may not be formally established, but is just nominally identified. Gold determines policy, overall strategy, resource deployment and the parameters within which lower levels of command will operate. This will include resources from multiple agencies if a single agency is identified as overall commander. This is most likely to be police during a response and local authority during recovery. A multi-agency Strategic Coordinating Group (SCG) may be established at Gold level if required and is intended to complement individual agencies' strategic management structures and procedures, not replace them. The SCG also has a communication and coordination function where an incident crosses boundaries and/or involves government offices, lead government departments or devolved administrations. **Goals** at Gold level may be general, unclear, multiple and implicit. **Decision making** should be analytical, in-depth and broadly referenced, making use of specialist resources, and being able to develop and maintain a Common Operational Picture, identify and assess options and evaluate progress. **Information requirements** are broad and relatively unpredictable, but Gold commanders should avoid the 'long stick' intervention at lower levels of command that too much and too detailed information may encourage. **Information outputs** are varied, including tasking, situational awareness to Silver and meeting public, media and political interests.
Tactical	**Silver**	Silver command defines 'how to do it'. Silver determines and directs the tactics of incident management within the strategy, parameters and with the resources defined at Gold level, which may include resources from multiple agencies. **Goals** at Silver level may be multiple and relatively general, although they should be clear and explicit. **Decision making** needs to identify and evaluate options, which necessitates an analytical approach, although pressure or rapidly changing circumstances may force an intuitive approach. **Information requirements** are more specific than those of Gold, focusing on hazards, vulnerabilities, risks and resources that shape the translation of policy and strategy into practice. **Information outputs** are task specific to Bronze level, concerned with maintaining situational awareness at Silver and Bronze and the upward transfer of changing situational information that is of relevance to Gold.
Operational	**Bronze**	This level of command is concerned with 'doing it'. Bronze commanders work within a functional and/or geographical area of responsibility to implement the tactical plan as defined by silver command. The Bronze commander must have a clear understanding of the tactical plan and have access to information that is critical to its execution on the ground, including the activities of other agencies that may be pertinent to their own goals and actions. **Goals** at Bronze level may be single or fewer in number, but should be specific, clear and explicit. **Decision making** may be characterised by an intuitive approach, based on problem recognition from previous experience, training and exercising. **Information requirements** are task oriented. **Information outputs** are fed upwards to maintain an accurate and relevant Common Operational Picture.

* Military terminology reverses Operational and Tactical in the hierarchy, so that Operational equates with Silver and Tactical with Bronze.

Table 4.   Examples of thematic data sources relevant to emergency planning and management, illustrating the need to develop multi-agency databases.

Hazards	Community and demographics	Built environment and economy	Natural environment	Infrastructure	Resources
River flood plains	Government agency boundaries	Land use density	Terrain	Energy generation and distribution	Emergency services resource sites, service boundaries and control rooms
Coastal flood hazards	Total population and demographic breakdown	Vertical structures	Geology	Telecommunications	
Storm tides	Census data	Centres of employment	Groundwater	Transportation (incl. routes, depots, ports, bridges etc.)	Hospitals
COMAH sites	Deprivation data	Retail units	Nature reserves	Financial services	Suitable buildings (e.g. leisure centres or warehouses)
REPIR sites	Vulnerable individuals (e.g. physical and mental health or mobility)	Service units	National parks	Food supply system	
Pipelines	Schools and colleges	Industrial units	Cultural and historical sites	Water supply system	
Specific building hazards	Care centres	Hotels	Country parks		
Slope instability	Other communal establishments associated with high-density occupancy		Recreation sites		
Fire hazards			Farmland		
Transport-related hazards			Woodland		
			Water and rivers		

Note: The catgories are not of course entirely mutually exclusive, and are used indicatively here.

4. applying (e.g. ontologies, accuracy, spatial framework and metadata)
5. supporting decisions (e.g. map outputs, VR and AR and information outputs).

Relating this to the cyclical model of IEM as a self-informing process (Figure 1) throws up a number of issues for emergency managers. (The term 'emergency managers' is used here to encompass what in reality would be a variety of staff, potentially spread across a range of agencies, that are involved with IEM as a set of processes. Some of these might be information management professionals, others emergency planners; some may be GIS specialists, others analysts with a broader remit. It is their common focus on implementing and supporting IEM that is considered here.) These issues and challenges include the following:

- Emergency managers need to know what they need to know: IEM is promoted as evidence-based practice, so there is a critical need to establish what evidence is required. Given the inherent unpredictability of emergencies, this may seem ambitious to the point of impossibility, but emergency planning is not in the main about the prior definition of a precise response to a range of possible incidents, but in developing an appropriate and flexible capacity to deal with a range of possibilities, at different times, in different places, and with a range of contingencies and consequences.
- Emergency managers need to be able to access appropriate information at the point and time of need: remembering that IEM is a set of processes – Anticipating, Assessing, Preventing, Preparing, Responding and Recovering – different processes have different information requirements. For instance, horizon scanning and risk assessment are less likely to be time critical and pressured activities than decision making in relation to resource allocation to contain a major explosion and chemical fire with atmospheric contamination. However, all these activities require information and the challenge is to make the right information available to the right people at the right time, in the right place, and in a form that is appropriate and appreciable under operational circumstances.
- Access to appropriate information requires significant investment in data, coding, representation, quality, communication and interpretation: experience amply demonstrates that gaining access to data, although often demanding of time and energy, is only the first stage of a lengthy and important process. Attention to classification systems, semantics, accuracy and precision are critical to avoid the all too well known trap of *garbage in, garbage out*. Further to this, the way in which data and information are communicated must be appropriately secure and robust, and preserving their integrity and the appreciability of information, often under difficult conditions is another key consideration.
- Information must be recognised as a critical element of emergency preparation: there is clear evidence that emergency preparation has a tendency to undervalue information, in spite of its demonstrable significance in enhancing both the efficiency and effectiveness of emergency management.

These issues are now examined in relation to emerging technology and the role of a national GI provider.

4 THE NATIONAL GI PROVIDER'S ROLE

Geography and geographic processes are fundamental to IEM, and GI potentially supports or underpins many of the processes of IEM (AAPPRR). What, then, is the role for a national GI provider in an environment where:

- 'un-ness' and uncertainty predominates (but what is certain is that emergencies and IEM processes happen somewhere, at some time);
- the demand for information (much of it geographical) is greater than, and lags behind, the availability of information (Fig. 1);
- emergencies or potential emergencies range from fixed locations to mobile (spreading across or through locations), restricted to diffuse in terms of area and impact, and from local to global geographies (table 1, Fig. 2, table 2);

- emergencies may cause damage to humans and their environment, be natural or man-made, be accidental or malicious;
- processes and activities of IEM include anticipation, assessment, prevention, preparation, response and recovery, all of which require geographic or location-based information to be effectively managed (Fig. 3);
- multiple agencies, multiple domains and the need for co-operation should dominate and, hence, GI needs to be shared and understood;
- skills, experience, and domain expertise and understanding vary and understanding GI is not a core competence of all emergency workers;
- timeliness, relevance and accuracy of information determine its usefulness (Fig. 4) and GI potentially needs to be updated in near real time;
- clarity of message and purpose are critical and GI must be provided in a way that is fit for purpose and clearly understood;
- technological capability varies from minimal to advanced and the manner through which GI is delivered to meet the user's needs to be appropriate to technological and user capability. This may be a paper map or turn directions on an augmented reality screen, for example;
- a variety of decision-making scenarios exist. For e.g. rational, planned to intuitive, gut-feel;
- levels of command are required from strategic, operational to tactical (table 3);
- the goal is to develop and maintain a Common Operational Picture;
- the information, decision-making and task-management environment is dynamic and often rapidly changing;
- the data and information available are not necessarily designed for IEM but for general purpose central and local government, utility and other functions; and
- data and information sharing are governed by legal, policy and commercial considerations.

Effective IEM is determined by making the right decisions at the right time in order to carry out appropriately targeted, scaled, timed and resourced tasks and activities. GI adds value if it enables those tasks and activities to be carried out more effectively. That value of GI to effective task and activity management is determined not only by its timeliness, accuracy and relevance but also by the ease with which GI and related information can be accessed and used at the point and time of need. The role of a national GI provider within IEM, as determined by its governance, must therefore be to work towards providing ease of access and ease of use of GI at the point and at the time of need.

4.1 *Ordnance Survey background, status, aims and objectives*

Ordnance Survey, Great Britain's national mapping agency, was created in 1791 as a military organisation, in order to better prepare Britain's defenses against the posed threat of Napoleonic invasion. From that time until the present day Ordnance Survey maps and GI have been used in support of various emergency management tasks, whether man-made or the result of natural hazards.

The Department of Communities and Local Government (DCLG) has ministerial responsibility for Ordnance Survey, which is now an independent government department, Executive Agency, and Trading Fund since 1 April 1999 (Ordnance Survey, 2004a). Ordnance Survey aims to satisfy the national interest and customer need for accurate and readily available geospatial data and maps of the whole of Great Britain in the most effective and efficient way.

Ordnance Survey's strategic objectives are to:

- collect, portray and distribute the definitive record of the natural, built and planned environment of Great Britain that meets customers' needs and the national interest in the most effective manner;
- improve and maintain the definitive databases in a form that facilitates the association and integration of additional geographic data;

- provide, through the data, the underpinning framework for the government and the private sector to join up its spatial information;
- provide national coverage of medium- and small-scale maps;
- develop a business that focuses clearly on the needs of customers and continuously improves customer satisfaction;
- create, develop and maintain strategic and commercial partnerships that will add further value to Ordnance Survey data and products;
- grow the geographic information market and champion the extended use and sharing of geographic information in the government, business and leisure communities;
- generate profitable revenue that will fund continuous improvement in database content, data structure, data delivery, up-to-datedness, fitness for purpose and accuracy;
- provide a working environment that fosters leadership, personal development, innovation and team working; and
- advise the UK Government on all aspects of surveying, mapping and geographic information.

As a Government Trading Fund, Ordnance Survey is required to generate a return on its assets. Some surveying and mapping activities, required in the national interest but which cannot be justified on commercial grounds, are funded by the Government under the National Interest Mapping Services Agreement (NIMSA)[1]. This would include areas that would not otherwise be mapped if the judgement were based on the revenue generated from sales but for which mapping would be seen as crucial for contingencies and emergencies. Large-scale data, graphics and services are mainly directed to the government and business markets; mid- and small-scale products to government and the consumer markets. Address referencing and height data are also supplied, both sources of GI for IEM.

Ordnance Survey's position in the data-information-knowledge value chain is as a data and information provider. Partners, value-added resellers and customers' hosted applications and services convert that data to information and information to knowledge for use by end-users in their respective domains.

Whilst Ordnance Survey has no statutory role as a first or second responder under the terms of the Civil Contingencies Act, it is interesting to note that many of its strategic objectives echo many of the characteristics required for effective IEM. As a government department under the Act, it has a duty to share information where this does not conflict with commercial interests and indeed is at the forefront of providing mapping and digital data for emergencies (section 4.2.1).

Ordnance Survey's main processes (data capture, database, product and service development, and data delivery) are driven by current and anticipated future needs of customers and users, illustrated schematically in fig. 6.

This activity is underpinned by the principles of the Digital National Framework (DNF) and other industry standards. DNF is described in section 4.2.3 and OS MasterMap, Ordnance Survey's digital database, built and maintained on these principles, is described in section 4.2.4. Ordnance Survey is undertaking a number of current activities to improve these processes. Some of these, with particular relevance to meeting the needs of the IEM domain, are described further in section 5. Recent technological developments enable new ways of capturing, maintaining and delivering GI to the end-user on the ground (see fig. 10), and research of particular relevance to better use and sharing of GI for IEM is described in section 6.

A key requirement for effective and efficient IEM is access to GI, as and when required, and the means by which other information can be related to the geography of the anticipated emergency and shared with other parties involved. Section 4.2 considers the role of Ordnance Survey in

---

[1] The Government terminated its NIMSA arrangements with Ordnance Survey on December 31st 2006. However in June 2007 Ordnance Survey announced that it was confident that a new approach to its rural revision programme would "substantially mitigate the impact" of NIMSA ending."

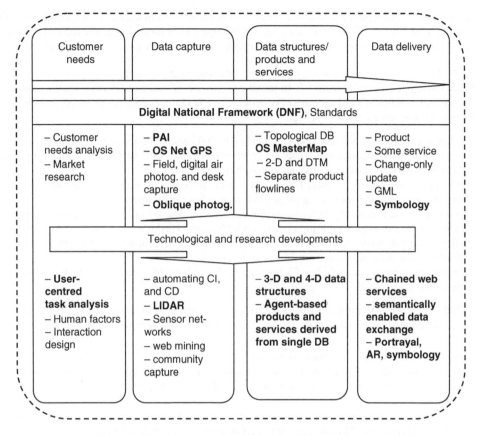

Ordnance Survey scope and role determined by framework agreement

Figure 6. Schematic of Ordnance Survey's main processes, underpinned by DNF principles (see section 4.2.3), and supporting activities being implemented, or researched to improve those processes, in order to meet customers' developing information needs. Items in bold are described further in the text.

contributing to the ease of GI access for IEM through its Mapping For Emergencies (MFE) scheme, Pan-government Agreements, the DNF and OS MasterMap.

4.2 *Contributing to ease of GI access for IEM*

4.2.1 *MFE scheme*

Largely in response to the Lockerbie disaster on 21 December 1988, when an American airliner was blown up over the small Scottish town, east of Dumfries, Ordnance Survey established its MFE scheme as an out-of-hours service to the emergency services and crisis managers. The scheme provides mapping and expertise in response to emergencies such as floods, fires, contamination leaks, disease outbreaks and in searches for missing people. Ordnance Survey co-coordinators and volunteer staff work closely with emergency services and other agencies to ensure the most appropriate data is delivered as rapidly as possible.

More recent examples include mapping, GI and logistical support for the London Bombings on 7 July 2005, Carlisle floods in January 2005, the foot-and-mouth crisis in 2001, and helping the British Transport Police deal with the aftermath of the Selby train crash in 2003. Other agencies receiving help have included the State Veterinary Service, Nottinghamshire County Council, the US

Air Force and various police forces. The scheme is not constrained by current data agreements with customers. During the Carlisle flood event, for example, all information requested of Ordnance Survey for emergency response was supplied irrespective of normal existing agreements.

4.2.2 *Pan-government purchasing, access and data sharing agreements*
A key Ordnance Survey objective is to maximize the use of national GI coverage by ensuring government departments and agencies have better access to Ordnance Survey products and services directly. This is being achieved through:

- a Pan-government Agreement (PGA): launched in 2003 and gives access to all British Government departments, agencies and other government bodies to a wide portfolio of digital map data. Approximately 200 bodies access data through this agreement;
- a Mapping Services Agreement (MSA) gives local government, police and fire services and national parks similar access to a range of Ordnance Survey products;
- as do Collective Purchasing Agreements (CPA), providing access to all ambulance trusts, primary care trusts and strategic health authorities. A pilot National Health Service (NHS™) Agreement – allows over 600 NHS organisations access to support and a wide range of products;
- working with the utilities on their future needs; and
- sharing of information, including GI between public bodies and wider across the boundaries of the public sector CPA.

Under the terms of Ordnance Survey's data licensing agreements to central and local government bodies, data can also be issued to contractors for their use whilst working on these bodies' behalf.

4.2.3 *The DNF – a spatial framework for GI*
According to Andrew Pinder, former Government e-Envoy: 'Geography is one of the key common frameworks that will enable us to link information together and boost efficiency in government.'

Whilst technology now makes it easy to collect information of all kinds, store it and use it in many different ways, Murray and Shiell (2004) point out its very ease of use, the limits and priorities of different organisations, and the lack of effective standards provide barriers to seamless information exchange between agencies and actors and hence effective decision making and response.

Murray and Shiell describe the DNF, launched in response to the need to provide a vision for better GI exchange and integration that: '…provides a permanent, maintained and definitive geographic base to which information with a geospatial content can be referenced' (Ordnance Survey, 2004) and incorporates: '…a set of enabling principles and operational rules that underpin and facilitate the integration of georeferenced information from multiple sources.' These principles are that:

- data should be collected at the highest resolution whenever economically feasible, collected once only and then re-used. (In the context of IEM, it could be argued that this principle should apply to both Ordnance Survey and to customers' geocoding data in an emergency as this could affect the speed to the casualty, proximity to other risks or in post-event analyses);
- information captured should be located to the national geodetic referencing framework for the United Kingdom and Ireland based on the European Terrestrial Reference System 1989 (ETRS89). This enables real-time transformation and positioning information, linking location data to the wider European framework and allowing data exchange with other European countries;
- features are uniquely referenced by a unique identifier, identifying the feature and source using the namespace concept in XML, allocated by a registry;
- composite features are created from existing features wherever possible;
- the existence of data created is registered within a central registry;
- such information may subsequently be used to meet analysis and multi-resolution publishing requirements;

- DNF should incorporate and adopt existing de facto and de jure standards whenever they are proven and robust; and
- GI from any source can be associated and integrated in a 'plug and play' manner.

DNF operates within a service-orientated architecture where technical documents, standards and guidelines and a searchable directory of datasets are made available as a web service, implemented using existing metadata standards. See http://www.dnf.org for further details.

4.2.4 OS MasterMap

Based on DNF principles and the National Grid, OS MasterMap (Ordnance Survey, 2004b) was developed as a single digital map database. Nationally maintained and consistent, it comprises over 440 million objects with unique 16-digit reference numbers (called TOIDs) for all features, providing a common denominator (or relate key) for disparate datasets and database tables held within the public and private sectors in Great Britain (Figure 7) to be shared in other applications and systems. It represents a complete reference system for British geographical data. It is made up of 4 layers: topography, address, integrated transport network and colour, fully orthorectified aerial imagery.

Users can select only the information they require by; geographic areas, layers and/or themes. Selection, quotation and delivery are available online. Currency is maintained through change-only files on a published approximately six-week schedule, managed sets (customers automatically receive updates either quarterly or six monthly), or complete resupply.

This provides the means to accurately georeference uniquely identified geographic features in a coherent and consistent framework in order to integrate disparate datasets, underpinned by location and geography, – a very necessary set of characteristics of GI in the information age.

A critical requirement of IEM is accurate address referencing and related information, free of ambiguity and duplication, fully maintained, easy to access, use and cross-referenced with related information held by different agencies. This is a requirement throughout all the IEM processes of AAPPRR. The OS MasterMap Address Layer improvements will provide just such a national, unique spatial address for all buildings and structures in England, Wales and Scotland, which will be owned by the Government, maintained to agreed and published quality levels and which will improve over time. Within OS MasterMap, Ordnance Survey holds the definitive database of buildings and addresses, with 41 million buildings and 4 million other objects. Of these, 26.7 million postal addresses have already been classified and more than 1 million objects identified without a postal address. The improved OS MasterMap Address Layer will provide:

- national cover of all postal addresses with a high-resolution spatial reference (within the building to which it refers at a resolution of 10 cm). To this will be added:
- geographic alternative addresses in BS 7666 format;
- Welsh alternatives;
- building name aliases;
- objects without a postal address (OWPA), such as churches, community centres, depots and masts (which may have special significance within the IEM context); and
- multi-occupation without a postal address (MOWPA) such as 'Flat A', 21 Any Street, Any Town, which is not currently identified within the Royal Mail® Postcode Address File (PAF®).

Since different buildings are classified by different agencies for different purposes using that agency's unique reference, a cross-referencing link table, licensed free of charge, will link the unique identifier from each organisation. This will allow a user to:

- move directly from the address to the location of the property;
- share information about one address across different organisations; and
- link directly to other OS MasterMap data such as the related property, other features or the road network.

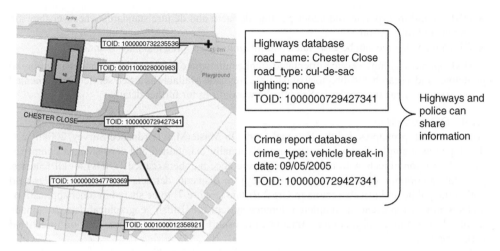

Figure 7.   Excerpt from OS MasterMap Topography Layer. Every geographic object has a unique TOID® for data association.

The benefits to IEM will be: almost zero duplication; consistent, nationally maintained, unique, spatial addressing, accessible and cross referenced, using the address terminology provided to the user. This will enable better preparation and faster response times to incidents.

Why is OS MasterMap significant in the context of IEM? From an emergency manager's perspective OS MasterMap has two key characteristics of real significance: (i) the level of precision and accuracy that allows activities such as damage estimation, building suitability assessment or the identification of helicopter landing sites to be extremely detailed and effective, and (ii) the unique reference number for each object that allows the cross-referencing and linkage of data from multiple departments or agencies that employ this referencing system. It is this combination of precision and accuracy (so spatial searches, for instance, on the basis of topology or proximity will be reliable) with attribute linking that allow the consequences of incidents to be established. Consider an example from the USA that has been slightly simplified here to make the point of contingent effects.

In July 2001 a train carrying chemicals and paper products derailed in a tunnel in central Baltimore, caught fire and, in the ensuing five days, caused a series of infrastructure failures and public-safety problems. The train leaked several thousand gallons of hydrochloric acid into the tunnel, and the fire caused a water main to burst. More than 70 million gallons of water spread over the downtown area, flooding buildings and streets and leaving businesses without water. The fire also burned through fibre-optic cables, causing widespread telecommunication problems. The fire and burst water main damaged power cables and left 1,200 Baltimore buildings without electricity (Peerenboorn et al., 2002). This is a very clear example of interdependencies that are physical, geographical and information related (the loss of fibre-optic cables) and some of the consequences of the initial incident (the train crash), the secondary incidents (fire, chemical contamination, flooding, loss of communications and transportation disruptions) are illustrated in figure 8.

To a large degree, IEM is about establishing contingencies and interdependencies that are critical in the assessment of risks and the management of incidents. As the Baltimore example clearly illustrates, many of these interdependencies are at least partly spatial in nature. Spatial queries can start to identify buildings, infrastructure elements, vulnerable facilities or resources that may be, or have been, affected by an incident such as flooding. Where the spatial database utilises a common spatial framework such as OS MasterMap then the proximate location of infrastructure such as transportation, water, sewerage, gas, electricity and fibre optics can be established. Where agencies with different interests in and responsibility for elements of the spatial framework employ

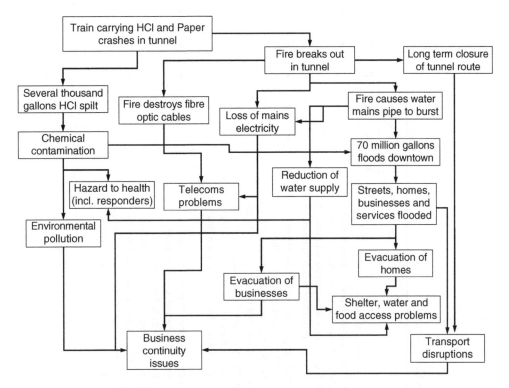

Figure 8.   Schematic representation of spatial interdependencies (based on Peerenboorn et al., 2002).

a common attribute referencing system, then the web of contingencies that determine the severity of an event and the framing of an effective response can be traced with a high level of efficiency. In this respect IEM is little different than other business areas; through the combination of a precise and highly accurate spatial framework with unique 'hooks' for multi-agency data linking emergency managers are better able to significantly slim down the information demand-supply gap (figures 1 and 5). The next question is, of course, whether this opportunity to combine disparate data sources and leverage additional value for decision support is being realized.

Although individual agencies' uptake of OS MasterMap, for instance, within local authorities or police forces, has been widespread, its application in IEM remains limited at the time of writing. Fundamentally, IEM is a multi-agency activity and although OS MasterMap's TOIDs provide the 'hook' for the integration of multi-agency data, realisation of this remains very limited within the IEM domain. Indeed, in a recent UK IEM conference (2004) only 50% of emergency planners currently saw the need for mapping, let alone the value of integrating disparate data and information using the TOIDs provided in OS MasterMap.

To an extent this relates to wider problems of data and information sharing described in section 3, but resourcing issues also loom large; making one agency's data available to others, even where the will to share is strong, is rarely close to the top of any agency's list of priorities. This in turn reinforces an acceptance of approaches that quite adequately serve local requirements but inescapably lead to sub-optimality when cross-border and/or multi-agency Common Operational Pictures are called for. The relevance of vision and leadership (combined with resources and potentially vigorous encouragement) is clear. Some examples of the use of GI, including OS MasterMap are described in table 5.

4.2.5   *Examples of GI across IEM processes*

Table 5. Summary of some emergency incidents and functions using GI supplied by Ordnance Survey.

Incident/ Incident/ Function	Description	IEM Processes				
		A	A	P	R	R
London Bombings July 2005	London underground stations and bus route bombed. MFE team supplied small- and large-scale wall and handout maps of the incident areas to Cabinet Office, combining Ordnance Survey and London Underground data at several scales and formats, that afternoon. Paper copies were printed in London by the Department for Transport and Office of the Deputy Prime Minister; digital copies sent by email.				•	•
Carlisle Floods 2005	Multi-agency flood plan and control centre activated after severe flooding of city centre following heavy storm. Wide range of Ordnance Survey products, including OS MasterMap, accessed through MFE scheme within hours, for use on City Council laptop-based GIS. This was in part a response to the loss of the Council's own GIS capability in the flooding. This included data that was not covered by the local authority agreement but which helped with effective IEM. Street to individual building maps produced for all agencies involved, for search, site risk and recovery operations. Experience resulted in greater assessment, preparation, planning and training with GIS, external data drive back-ups and mobile GIS capability.	•	•	•	•	•
Foot-and-mouth 2001	Department of Environment, Food and Rural Affairs created a GIS application that used Ordnance Survey data. A wide range of maps were produced. Small-scale maps highlighted regional spread patterns to Cabinet Office decision makers and maps of livestock populations and movements for epidemiologists. Mid-scale maps identified land use for burial sites and large-scale maps identified farm buildings and infrastructure for cleansing teams. Digital GI held in a GIS allowed flexibility in updating and maintaining currency in information in a dynamic situation; rapid electronic distribution across the country for display and printing; and easy-to-read maps for public access through a website,			•	•	•

		•	•	•	•	•
	which received over 600,000 hits per day at the height of the outbreak. The GIS is now further developed in the department to support helpline queries, identifying locations and restrictions, and issuing movement licences. The GIS is now used throughout DEFRA to aid emergency planning to man-made and natural disaster management.					
Dorset Fire & Rescue	Dorset Fire and Rescue service working with Dorset County Council, the police, conservationists and other local authorities on the Urban Heaths LIFE Project, which uses a GIS to record heathland damage in an online database that can be accessed by all partners. A 48% reduction in heath-land fire incidents achieved through GI analysis of fires and better use of resources.	•	•	•	•	•
Dumfries & Galloway Police	Using networked OS MasterMap across the force's Intranet allows officers to pinpoint the location of emergency calls rapidly and accurately down to the corner of a building, as well as to identify crime patterns for the effective targeting of resources. Savings of 70% over previous system.	•	•	•	•	•
Greater Manchester Ambulance Service (GMAS)	Link information databases to Ordnance Survey's geographic framework. Use GIS for predictive analysis, plotting where and when emergency calls are likely to be made, thereby reducing response times. Led to national Best Practice in Integrated Cardiac Care Award.	•	•	•	•	•
National Crime & Operations Faculty (NCOF)	Coordinating searches for missing people is a key part of NCOF's role. GIS enables NCOF to decide search areas in minutes rather than days (even before visiting a scene), providing a showcase system that encourages its implementation throughout the police community.	•	•	•	•	•
Royal Berkshire Fire & Rescue Service	The services Integrated Risk Management Project has allowed a more detailed, focused analysis of actual and potential incidents and incident cover through use of OS MasterMap at individual property level. Trends have been identified and resources allocated more efficiently and effectively.	•	•	•	•	•

5 THE CHALLENGES – CONTRIBUTING TO EASE OF USE FOR IEM

5.1 *Trends in ICT, GIS and impacts on provision of GI*

Over the last 10–15 years technological advances offered by GPS, spatial databases, the Internet, wireless, broadband, LBS, sensors and mobile devices have revolutionised the way location-based data is collected, stored, maintained, analysed and delivered to the user.

The map is now just one expression of the spatial database. These advances have brought GI to the information mainstream and are enabling its increasingly pervasive use as a fundamental ingredient to effective decision making and task management. For Ordnance Survey these developments are enabling a move from a product-centric to a database-centric business, from which products and services can be derived.

Technology will enable virtually everything (inanimate and animate objects, including people) to be identified, tagged, sensed and monitored. Miniaturisation will allow processing on the device and geographic intelligence to be held on the network, whilst increased bandwidth and processing power will allow complex distributed analyses across trusted super computing environments. Technological enablers will drive the expectations of GI users within the IEM domain as elsewhere. These are summarized in figure 9. See Parker (2004), Parker & Stileman (2005) and MacFarlane (2005) for further discussion.

These trends in GI reflect the GI requirements of the IEM community as well. In a database-centric environment the overarching challenge for the national GI provider is no longer just the design of mapping products but the design of the whole process of delivery of GI to the user as part of mainstream information delivery. The design of the capture to delivery process from a user-centred perspective is explored in section 6.1.

With increasing demands for real-time data services, greater interoperability of data in services and applications within domain, cross-domain, cross-agency and cross-border situations, greater importance will be accorded to data quality. This will have implications for both the GI provider, the end-user and others in the information service delivery chain. In the IEM domain, where high reliance is placed on the right information at the right time, GI data quality issues can be critical. Data quality aspects, such as lineage, currency, positional accuracy, attribute accuracy, logical consistency and completeness, will all need to be considered. For a full discussion of implications to the data provider and user see Harding (2005).

5.2 *Improving the positional accuracy of geographic features*

Ordnance Survey are undertaking a national positional accuracy improvement (PAI) programme as part of its commitment to improving data quality (Ordnance Survey 2004c). Begun in 2001, the programme will allow Ordnance Survey to capture data at 1:2500 scale to a greater absolute accuracy. The PAI programme will result in an improved and more consistent accuracy standard of mapping data for rural areas (rural towns to an absolute accuracy of ±0.4 metres root mean square error (RMSE) and other rural areas to an overall absolute accuracy of ±1.1 metres RMSE). It will also future proof the data for the addition of new building development and other change, and provide a better relationship between Ordnance Survey 1:2500 scale map data and customers' own GPS-positioned resources.

5.3 *Ordnance Survey GPS network developments*

With the delivery of information through location-based services for emergency service applications now 'into the cab', the registration of GPS-derived positions with large-scale mapping at the user's location, and more precise requirements of engineers and utilities have demanded more accurate GPS networks. OS Net® commercial services, Ordnance Survey's GPS correction network, is the country's most comprehensive framework for correcting signals from orbiting satellites. Over 90 permanent base stations covering the whole of Great Britain contribute data. It is linked in

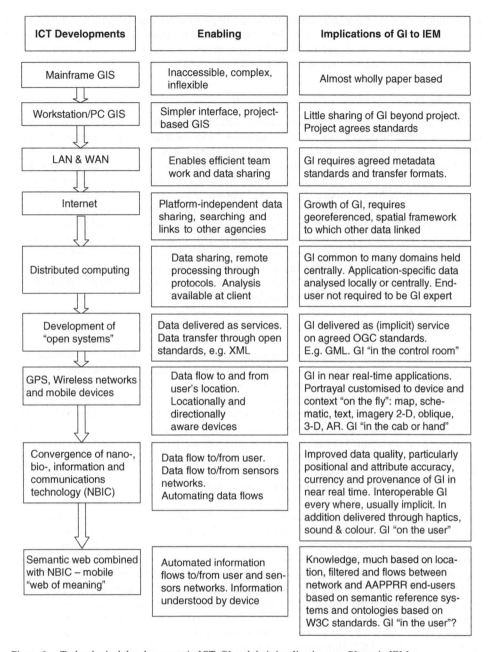

ICT Developments	Enabling	Implications of GI to IEM
Mainframe GIS	Inaccessible, complex, inflexible	Almost wholly paper based
Workstation/PC GIS	Simpler interface, project-based GIS	Little sharing of GI beyond project. Project agrees standards
LAN & WAN	Enables efficient team work and data sharing	GI requires agreed metadata standards and transfer formats.
Internet	Platform-independent data sharing, searching and links to other agencies	Growth of GI, requires georeferenced, spatial framework to which other data linked
Distributed computing	Data sharing, remote processing through protocols. Analysis available at client	GI common to many domains held centrally. Application-specific data analysed locally or centrally. End-user not required to be GI expert
Development of "open systems"	Data delivered as services. Data transfer through open standards, e.g. XML	GI delivered as (implicit) service on agreed OGC standards. E.g. GML. GI "in the control room"
GPS, Wireless networks and mobile devices	Data flow to and from user's location. Locationally and directionally aware devices	GI in near real-time applications. Portrayal customised to device and context "on the fly": map, schematic, text, imagery 2-D, oblique, 3-D, AR. GI "in the cab or hand"
Convergence of nano-, bio-, information and communications technology (NBIC)	Data flow to/from user. Data flow to/from sensors networks. Automating data flows	Improved data quality, particularly positional and attribute accuracy, currency and provenance of GI in near real time. Interoperable GI every where, usually implicit. In addition delivered through haptics, sound & colour. GI "on the user"
Semantic web combined with NBIC – mobile "web of meaning"	Automated information flows to/from user and sensors networks. Information understood by device	Knowledge, much based on location, filtered and flows between network and AAPPRR end-users based on semantic reference systems and ontologies based on W3C standards. GI "in the user"?

Figure 9. Technological developments in ICT, GI and their implications on GI use in IEM.

real time to a server hub at the Southampton head office, where data can be captured and sent from the field to partner applications in milliseconds. The network is designed to help deliver a range of positioning services, both in real-time and post-process applications. The underpinning network now supports a commercial real-time kinematic (RTK) and differential GPS (DGPS) correction service (Leica SmartNet) as well as a free web-based GPS service for non-commercial post-processing and coordinate transformation (www.ordnancesurvey.co.uk 20 December 2005).

Applications, such as emergency vehicle tracking, support for dispersed emergency workers, and accident analysis, can all benefit from improved positional accuracy information. Using GPS to locate emergency personnel and assets within built structures has also developed commercially recently (www.Qinetiq.com – see websites referred to).

5.4 *Using imagery to enhance information quality*

Seamless digital imagery derived from colour aerial photography, or high-resolution satellite imagery is often considered more intuitive to non-GI experts than a map and its symbology. It allows intuitive situational analyses to be conducted rapidly and, if necessary, repeatedly. It can be delivered through the same channels as vector and raster data, or as a web service, and hence play an important role in communicating and maintaining a common operational picture. Recent developments in airborne oblique digital aerial photography systems such as Pictometry® and Oblivison™ assist this. Pictometry, for example, can be used to determine how many floors a building has, it's height, ground area, lines of sight, emergency and regular exits, and all the elevations of a building, including a plan view. This information and visual intelligence can all be obtained ahead of emergency workers responding to incidents on the ground, reducing the information gap and saving time. The OS MasterMap Imagery Layer, combined with OS MasterMap Topography, and/or Integrated Transport Network™ (ITN) and Address Layers, linked to additional information by the TOID offers a powerful visual and analytical tool for IEM.

5.5 *Symbology considerations*

There is a need to ensure clarity and unambiguity of message and purpose in all communications within IEM, including mapping. One critical element in ensuring that the message is not only received but understood is map symbolisation. Many of the examples of thematic data sources relevant to emergency planning and management listed in table 4, are symbolized on Ordnance Survey paper 1:25,000 scale OS Explorer Maps, 1:50,000 scale OS Landranger Maps and 1:10,000, 1:25,000, 1:50,000 scales raster products. However, significant proportions are not. Many of the geographic features required in analyses and decision making in IEM processes have no nationally recognized map symbol. Certainly there is no national standard map symbology accepted across the domain of IEM, and this needs to be addressed.

With the need to address issues of data interoperability, not only between local and national agencies but also across national borders and cultures, through both national and international initiatives to develop spatial data infrastructures, such as INSPIRE (2004), addressing standards for map symbology for IEM is even more apparent. In one example, related to the authors, a symbol consisting of an upright blue triangle with a white dot in the centre indicated, to the Polish Airforce, airborne ice particles (a reference to wing leading edges and 'icing'). To UK forces this is an anti aircraft battery observation point. In the United States the Federal Geographic Data Committee (FGDC) Homeland Security Working Group has been developing a standard set of symbols for point features for use by the Emergency Management and First Responder communities at all levels of need (i.e. national, state, local and incident) FGDC (2005). It represents work in progress and is subject to the standardisation process. Federal, state, and local agencies worked together under the auspices of the FGDC's Homeland Security Working Group to develop the proposed symbology. Symbols and their definitions have been developed for incidents, natural events, operations, and infrastructures at a level to provide immediate and general understanding of a situation. Its suitability for the Hurricane Katrina was questioned as it was felt to be complex and insufficiently intuitive.

Within Ordnance Survey there are some differences in feature symbology between paper and raster product scales. For example, on 1:50,000 scale products public rights of way are depicted in green, secondary roads in yellow and level crossings in yellow. On 1:25,000 scale products these are shown in magenta, orange and magenta, respectively, and work has been initiated to resolve discrepancies. Within digital vector products such as OS MasterMap, of course, the user has complete flexibility as to how the information is symbolised. However, within the OS MasterMap user guide a set of recommended styles and symbols are supplied as XML style sheets, transforming the Ordnance Survey-supplied GML to produce scalable vector graphics (SVG) developed by the World Wide Web Consortium (W3C). See Ordnance Survey (2005) for more details.

Aware of the importance of clarity of message within IEM, Ordnance Survey's cartographic design lab has been working with the Hampshire Ambulance Trust to produce 'versatile maps', where the cartographic representation of OS MasterMap has been manipulated to represent more of the look and feel of mid-scales raster products used by the Trust. Appropriate visualisation of information helps provide what the user needs, when they need it, saving time during emergency response.

Appropriate visualization and representation of GI is important to the GI user and provider. From the provider's viewpoint it is important, firstly, to ensure the information so depicted meets the users' needs, but secondly there are large costs in capturing information at one scale and representing it at others – a process known as generalization. This is discussed in section 6.4.

6   ADDRESSING FUTURE GI NEEDS OF IEM AT ORDNANCE SURVEY

Figures 6 and 9 indicate the future challenges, driven by technological developments, in providing ease of access and ease of use of GI at the time and point of need. The implications of this are described with some examples of innovative research being carried out at Ordnance Survey. Figure 9 indicates that technological developments have enabled GI to move out of the office, into the cab and now into the user's hand or, in future, onto the user's body through the development of web service and wireless mobile devices. Potentially, the user can be bombarded with information. It is even more important therefore that we understand how to provide the user with information they want when they want it, at the time and point of need. This next section describes how we are addressing that challenge through research into user-centred task analysis, LIDAR, 3- and 4-D data structures, generalization, web services, semantic reference systems and ontologies. Finally, three example prototype applications illustrate how GI may be accessed and delivered within IEM in the near future.

6.1   *User-centred task analysis*

Within the IEM domain actors and decision makers use GI to carry out tasks and activities. They are not interested in GI as an end in itself but as a means to an end. GI is one ingredient required to make an informed decision in order to carry out a task. Other ingredients may include resources, costs and time. GI adds value when it contributes to informed decision making, allowing more effective execution of tasks and activities and the saving of time, costs, resources and lives. GI loses value if it adds nothing to the decision making or when there are barriers to accessing the right information at the right time, for example, in the time taken to manipulate the required data, inappropriate scales, poor accuracy or over-complex symbology.

Determining what, when and how to portray the GI that will *make a difference* is a design issue. In determining how to ensure GI adds value within the IEM domain we can learn from the approaches of user-centred design in the consumer world (for example, Marzano, 1998) and from human-factors integration within the defence research environment. Technological advances, have made data collection easier but have contributed to an explosion of potential information availability, making the information management task more complex. Information management frameworks, standards and governing principles are required to manage the complexity. Within the IEM domain users want the answer to their question, the whole answer and nothing but the answer! Anything else complicates the picture and absorbs precious time. A highly complex information environment requires high design to provide simplicity for the user (Aarte and Marzano, 2003).

At the UK Ministry of Defence Human Factors Integration Defence Technology Centre the approach is to design the system around the people rather than requiring the people to adapt or work around poor designs and working environments that make operation difficult or even dangerous. In summarizing their research, they ask: Can this person, with this training, do these tasks, to these standards under these conditions? (Goom, 2003). The question is just as relevant in understanding the GI needs of IEM actors. A further consideration must be the resilience of the system in coping with all the most likely eventualities. Will, for example, wireless GI applications provide sufficient resilience? During the Carlisle flood event in January 2005, when the power went down, the mobile phone network was also out of action. Centrally stored GI information was also under water and was retrieved from a laptop.

Within Ordnance Survey Research & Innovation, a team is investigating the usability of GI and is applying user-centred design principles to the design of product and the design of the information service to the user appropriate to their need (Davies, Wood & Fountain, 2005). This user needs and usability research focuses on *what, when, where* and *how* GI is important in people's tasks and decision-making processes. A user-centered task analysis methodology has been developed that is carried out in a semi-structured interview that takes about an hour with the end-user of the information. The outcomes of the process are:

- identification of where the task could be more effectively addressed with GI;
- the nature of the GI requirement; and
- how that informs Ordnance Survey capture, product and service development and delivery processes.

6.2   *Research into capture technologies*

Ordnance Survey's Research & Innovation unit is engaged in a considerable body of work looking at the effectiveness of capture technologies, including LIDAR and mounted on terrestrial and airborne platforms, and how these approaches may be used to derive height information and 3-D models combined with other techniques. For example, combining georeferenced LIDAR data with 360 degree photography and OS MasterMap to capture scenes, geometry and attribute information more rapidly. The analysis of error incurred in LIDAR capture and analysis processes is reported by Smith, Holland & Longley (2005) for example. More recent work is looking at means to detect and capture changes in the natural and built landscape, using automated processes where possible.

6.3   *Research into 3- and 4-D data structures*

Developments in spatial database and visualization technologies are stimulating user awareness of the potential uses of topologically structured digital data beyond 2-dimensional planimetric views. Integrated emergency management occurs in the real world, in all its dimensions where there are requirements not only for conventional 2-D expressions of the spatial database but also 2.5, 3 and 4 dimensions as well. For example, flood modelling, blast, and pollution plume modelling, mission rehearsals, contingencies and evacuation planning. Better ways of modelling the real world to reflect potential user requirements in the integrated emergency management domain are needed. These include better data models that can be more effectively used to create required products or views of the data.

This challenge poses a number of questions:

- What are the features that need to be considered for modelling the diverse aspects of the real world?
- What are the theoretical data models that allow the identified features to be modelled effectively?
- What are the practical issues involved with the use of the data models to handle live data?
- What generic data framework could encompass the full range of features?

Research sponsored by Ordnance Survey is exploring; a framework for feature-based digital mapping in three-dimensional space (Slingsby, Longley & Parker, 2007) and externally, the relationship of 3-D models to building information models, and means of capturing 3-D data, including the use of LIDAR, data modeling issues associated with terrain and the built environment, integration with internal database structure, database maintenance, semantic descriptions of the built environment, and the use of CityGML (Kolbe et al., 2005) as a means to store and exchange virtual 3-D city models. Large-scale 3-D data integration is explored by Zlatanova and Prosperi (2005) and 3-D data models for emergency response are explored further by Lee, Zlatanova, Kolbe et al. in this publication.

Emergencies and disasters are dynamic events (or a series of dynamic events) that occur at particular locations over time. The management of emergencies also requires actions at particular locations over time and therefore the incorporation of time as the fourth dimension in GI is important for effective IEM. A spatio-temporal data model should facilitate:

- an understanding of the rules that govern real-world change;
- explanation of the state of the world at the current time (now) and previous times; and
- predictions of the state of the world in the future.

Worboys and Hornsby (2004) have been working on a well-founded model that encompasses both geographic objects and events, and have applied this thinking to modelling time aspects of Ordnance Survey's ITN layer within OS MasterMap. Their model, the geospatial event model (GEM) is based on three principal components: geospatial objects, events and their settings.

A further research collaboration involving Oracle®; University of Maine, USA and University of Melbourne®, Australia is exploring the extension of Ordnance Survey's OS MasterMap ITN Layer to include temporal, dynamic and event-based components ITN + T (2005). The overall objective is to establish the data and system requirements to support travelers' and transport coordinators' future needs. The project uses the ROMANSE (ROMANSE: Road MANagement System for Europe) traffic management system as a test bed. The ROMANSE project (www.romanse.org.uk) currently provides some real-time information on travel and traffic, incidents and congestion, and car park status, for selected parts of the south of England. The technology will have as its core the Oracle Spatial DBMS. The project will also consider near-future sensor networks and embedded computing capabilities within transportation systems and, in particular, the impact of these technological developments on Ordnance Survey's data capture to data delivery processes.

6.4 *Automatically deriving data at smaller scales*

Within the IEM domain GI is required at the resolution appropriate to the levels at which decisions and tasks are being carried out, be this at strategic, tactical or operational or at all of these levels. The challenge for a national GI provider is to be able to capture data at the most appropriate resolution demanded by national circumstance but to provide it at the level appropriate to the task at hand. At Ordnance Survey our generalization research is working towards the holy grail of automating and customising the generalization process using agent-based approaches, which will derive medium- and small-scale maps from a single large-scale digital database. In addition the aim is to develop a strategy and the appropriate tools to build on-demand generalization applications, running as automatically as possible. Significant progress has been made in automating the generalization process using agent-based algorithms (Revell, Regnauld and Thom, 2005). With the development of web- and location-based services, this now offers the opportunity to utilise generalization algorithms 'on the fly' as part of delivering the right information in the right way to the mobile worker or users based on how they want information pertinent to their tasks to be displayed.

6.5 *Web services*

Web services is the term given to linking information, information systems and business processes through web-based protocols. The UK Government sees web services as a major enabler in linking and sharing information and coordinating activity amongst public-sector bodies in order to achieve greater interoperability; an essential requirement for effective IEM. GI web services or 'on-demand GI networks' (Vowles, 2004) allow computing devices to connect to GI and other information and

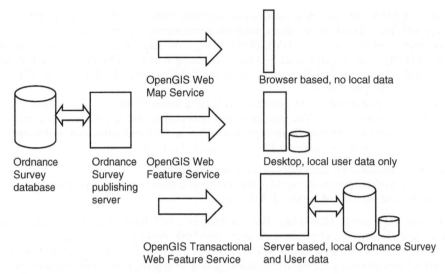

Figure 10. On-demand geo-spatial information services use cases from Vowles (2004) with diagram supplied by Ed Parsons, Ordnance Survey CTO.

data services on the Internet, and to access the data as and when the user needs it, negating the costs and delays of ordering and managing local copies of the data. In the IEM context GI, delivered through web services takes the user a step closer to being able to access the information they require, on demand, when they require it. Successful delivery of GI through web services to end-users in the IEM domain is dependent on the implementation of robust, open standards such as those developed and promoted by the Open Geospatial Consortium (OGC, 2003); robust trading and exchange, provenance, security and access models.

The end-user may require varying levels of access according to process, role and requirement. As a GI provider, Ordnance Survey must anticipate the use cases its customers and partners will require in order to ensure ease of access and ease of use of its information. Figure 10 is reproduced from Vowles (2004) and illustrates three on-demand service use cases where data is maintained in a central database and published from an Ordnance Survey publishing database, through OpenGIS standards-based interfaces.

With the OpenGIS Web Map Service Ordnance Survey data can be viewed as a map rendered as an image or as an SVG, using a standard web browser without a requirement to hold a local copy of the data. The OpenGIS Web Feature Service allows access to the data as a set of individual features, returned to the user's application as GML-formatted data. The user can then work with the local copy of that data. With the OpenGIS Transactional Web Feature Service, the user can retain a copy of the data in local repository. As features periodically change they can be sent to the user's local server and sent to maintain a local, up-to-date repository. Performance and response times can then be improved because the user's application will use the local, periodically updated repository. The services used are dependent on the user's application requirements.

In order to deliver value to the end-user a means to describe the interrelationship and dependencies among these on-demand, cooperating service networks is required. Figure 11, again taken from Vowles (2004), describes these relationships and roles.

In order to provide an efficient and rapid information service to the end-user within the IEM domain, the horizontal flow of data across the diagram needs to be achieved within seconds. Data needs to be propagated automatically down the network to where it is needed, at speeds associated with the frequency with which the data changes. Different organizations and agencies play roles of maintainer, publisher, subscriber and consumer of the information service.

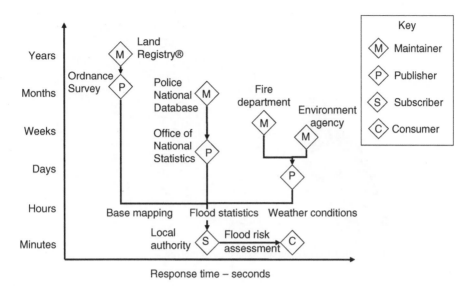

Figure 11. On-demand service proposition diagram.

In addition to the work described by Vowles (2004), Ordnance Survey is also participating in the development of on-demand services as part of the pilot phase of the IST Orchestra project. The overall goal of the ORCHESTRA project (www.eu-orchestra.org) is to design and implement an open-service-oriented software architecture that will improve the interoperability among actors involved in Multi-Risk Management.

Whilst OGC-based standards contribute to the means of achieving syntactic interoperability in GI, semantic interoperability, enabling the meaning of the GI to be transferred between organizations, systems and services, represents a significant area of research at Ordnance Survey, and is summarized below.

6.6 *Semantic reference systems*

Some of the barriers to effective use of shared information can be removed by reducing the time and cost of information processing, which is currently dependent on significant human intervention and manual input. Research at Ordnance Survey is looking at how to encode the meaning of different data sources so that machines can intelligently process data, and combine it with other data to be used across applications in a semantic web environment. Currently, the meaning of our data is implicit only to humans, not computers, which leads to ambiguities of interpretation, inefficiencies in processing, limitations in automated data sharing and interoperability limited to the syntactic and structural levels. The ability to semantically translate between one information source and another may reduce both the time and cost of services better enabling joined-up processes, a significant requirement in the IEM domain. Ordnance Survey research is investigating the development of a topographic ontology to underpin our data and to determine how it may be used to support interoperability with other sources of information and information-based services. Our semantic reference systems research has considered the potential applications of ontologies within Ordnance Survey (Goodwin, 2005), including their use in better data modeling and data consistency, linking to legacy databases to make data more flexible, enabling more effective interoperability, constructing smart queries and in combining semantic and spatial technologies to provide real information power to location-based services. Our research has also investigated methodologies for building topographic ontologies, explicitly developing conceptual

305

Figure 12.   Illustration of the Ordnance Survey Symbolink prototype.

ontologies for human domain expert understanding and a logical ontology derived from conceptual ontologies, expressed in description logic for machine processing (Mizen, Dolbear and Hart,
2005). These ontologies have been tested within the IST ORCHESTRA project. Pundt in this
volume, provides a detailed treatment of the semantic aspects of data with respect to emergency
response.

6.7   Ease of access, ease of use at the time and point of need for IEM

Actors and decision makers in the IEM domain are not necessarily well versed in map-based
information. Providing the right information means considering how GI might be best portrayed
to the user, given their profile and context at the time of need. This need not necessarily be a map,
and developments in wireless and mobile technologies enable more intuitive, ways of providing
the right information at the location of the users' task. Three prototype examples, developed
in Research and Innovation at Ordnance Survey, illustrate the combination of new technology
capabilities with conveying GI at the point of need.

6.7.1   Ordnance Survey Symbolinks
The Ordnance Survey Symbolinks prototype (patent pending) marries the symbology depicted
on a conventional printed map with rich digital information accessed through mobile devices and
location-based services (fig. 12). In the prototype, symbols on Ordnance Survey map sheets appear
with a unique code number. This code can be entered into the Ordnance Survey Symbolinks software
installed on a hand-held device, either manually on the keypad or collected automatically using
the device's camera facility (if so equipped), and sent on to a remote server. The server, depending
on the sending method, reads and processes the code, returning any linked information about the
real-world feature represented by the symbol. This information can take the form as an SMS or
multimedia file. This graphical/digital merger can significantly expand the range of information
imparted by a conventional printed map.

6.7.2   Zapper: pointing application
Clearly GPS positioning continues to be an invaluable tool in emergency planning, response
and recovery but also in delivering situational mapping to field personnel and data collection
tasks. However, by adding a directional component to GPS-derived positional data more advanced
location-based processes are enabled, making it possible to select and query specific features

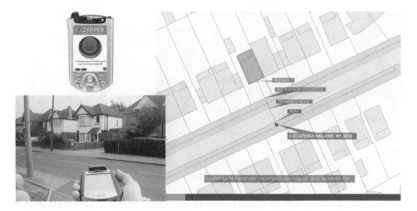

Figure 13.   Zapper prototype pointing application using OS MasterMap data.

in the landscape simply by pointing a hand-held device at them. In essence, the real world is transformed into a virtual desktop and features such as buildings, roads, mountains, and lakes become 'clickable'.

Pointing applications, like the 'Zapper' prototype illustrated in figure 13, utilise a combination of position determined by GPS and a bearing determined by compass. Bearing detection is, of course, possible using GPS alone but requires the user to be in motion. The addition of an electronic compass, however, allows the user to remain in one location, returning a bearing as the user turns in any direction. In the Zapper prototype, these peripheral components are connected to a personal digital assistant (PDA), which the user can aim at the desired feature such as a building. The device can be set up to take account of the user's context, profile and requirements, filtering out unnecessary information not associated with a particular group of tasks.

Once aimed, the user presses a button built into the Zapper software's user interface, which instructs the device to send the user's positional information to a GIS installed on a remote server. The GIS first places the user's location within a 100 metre square of OS MasterMap of the area. Since OS MasterMap is a digital, spatial database of real-world features represented as points, lines and discrete polygons, each of which is uniquely identified by a 16-digit topographic identifier (TOID) the TOID can be used as a primary key in a relational database, enabling additional data to be linked to that feature. After placing the user's location within OSMasterMap, the GIS then 'draws a line' along the reported bearing, querying the classification of objects it intersects along the way. The procedure, for example, searches for a polygon with the classification of 'BUILDING', and discards all others. When it finally encounters a polygon classified as 'BUILDING', it then looks up the information associated with that polygon's TOID in the GIS database or related tables and returns it to the client device.

Pointing applications allow the user to go beyond asking 'Where is …?' and ask 'What is …?', and offer great potential in the areas of asset management, risk and hazard identification.

6.7.3 *Augmented reality applications*
The term augmented reality (AR) refers to the superimposition of a real-world view with computer-generated graphics generating a composite view made up of real and virtual elements, thus 'augmenting' the user's perception of his/her surroundings. This can be of great benefit to users in the field since it obviates the need to transpose a 2-D representation of space or spatial attributes into one's real-world surroundings. AR techniques can attach virtual labels to real-world features, allowing the user to readily identify key buildings, in-bound aid routes, aid stations, landing zones, areas of immediate need, evacuation routes and other need-to-know information. It can also be an invaluable navigational aid to field personnel.

Figure 14. Augmenting reality with an adapted tablet PC displaying points of interest information.

The prototype developed below at Ordnance Survey uses a tablet computer equipped with GPS capability, an electronic compass and an on-board camera to create a 'magic window'. As the device is rotated the camera view can be displayed on the screen. The device's location and bearing is transmitted to the OS MasterMap database residing on a remote server (or stored on board) where attribute information appropriate to the field of view is returned and transmitted to the screen overlaying the camera view. Essentially, using AR technology it is possible to create a virtual 1:1 map.

As with the pointing applications, there is potential for utilising such technology within all field-based emergency-management processes.

7 CONCLUSIONS

IEM is an information-hungry activity. GI is highly significant in respect of that demand for information, primarily because such a high proportion of data and information are spatially referenced, but also because geography is so critical in determining risks, appreciating the extent and significance of an unfolding incident and in framing an appropriate response to that incident. At the present time the uptake of GIS in an IEM context is far from universal: the appreciation of the significance of geography and GI is much lower than it needs to be and leadership within agencies is insufficiently focused on the effectiveness and efficiency gains that appropriate attention to GI can bring. Although the Civil Contingencies Act promotes information sharing, serious obstacles remain in the path of an integrated evidence base for the AAPPRR processes of IEM. In this regard the role of GI in IEM remains far below the level it needs to be promoted to, and the attainment of IEM as an evidence-based practice is a long way from being fully realised.

Against this backdrop, developments in information, communications and related technologies have allowed GI to move out of organizations' centrally located GIS facilities into the hands of the emergency worker end-users through web services delivered over wireless mobile devices. So the ability of spatial information technologies to support this demand for GI is very well developed. The Great Britain DNF establishes the spatial framework for driving up the quality of corporate data and enabling spatial data sharing, which is fundamental to informed multi-agency operations. Web services, coupled with appropriately defined and observed standards, promise greater interoperability, and mobile devices and remote sensing are set to continue the transformation of data capture and dissemination. Future-looking research by Ordnance Survey is establishing tools and techniques to further add value to spatial data, and the potential significance of these developments in an emergency context are considerable. Challenges remain though in the automation of data capture, data analysis, integration and information delivery processes in a way that maximizes ease of access and ease of use at the time and point of need for the end-user.

The efficient use of GI in IEM, as an evidence based practice, is likely to be determined, not by the limitations of technology in the near term though, but by the degree to which the limitations of operationalisation, from a policy, legal, financial and social perspective can be overcome.

ACKNOWLEDGEMENTS

Permission from the Cabinet Office Emergency Planning College for the reproduction of some material from MacFarlane (2005) and the comments from many colleagues at Ordnance Survey is gratefully acknowledged.

This article has been prepared for information purposes only. It is not designed to constitute definitive advice on the topics covered and any reliance placed on the contents of this article is at the sole risk of the reader.

REFERENCES

Aarts, E. & Marzano, S. 2003. The New Everyday View on Ambient Intelligence, 010 Publishers. The Civil Contingencies Act 2004. The Stationery Office Ltd: ISBN 0-10-543604-6, available at http://www.opsi.gov.uk/acts/acts2004/20040036.htm

Crichton, M. 2003. Decision making in emergencies. NATO/Russia Advanced Research Workshop: Forecasting and preventing catastrophes, 2–6 June 2003, Industrial Psychology Research Centre, University of Aberdeen, UK, available at http://www.abdn.ac.uk/iprc/john/Nato_pdf/Margaret%20Crichton.pdf

Davies, C., Wood, L. & Fountain, L. 2005. User-centred GI: hearing the voice of the customer. Paper presented at: AGI '05: People, Places and Partnerships, Annual Conference of the Association for Geographic Information, London, UK, 8–10 November 2005. (London: Association for Geographic Information), available at http://www.ordnancesurvey.co.uk/ partnerships/research/publications/docs/2005/Davies_etal_finalsubmission_geo.pdf

Federal Geographic Data Committee 2005. Homeland Security Working Group Symbology Reference, available at http://www.fgdc.gov/HSWG/index.html

Goodwin, J. 2005. What have ontologies ever done for us – potential applications at a national mapping agency. OWL, Experiences and Directions Workshop 11-12 November 2005, Galway, Ireland, available at http://www.mindswap.org/2005/OWLWorkshop/accepted.shtml

Goom, M. 2003. Overview of human factor integration. Human Factors Integration Defence Technology Centre, available at http://www.hfidtc.com/public/pdf/1%20%20GOOM%20%20OVERVIEW%20OF%20HFI.pdf

Harding, J. 2005. Qualité des données vectorielles: perspective d'un producteur de données. In R. Devillers & R. Jeansoulin (Eds) Qualité de l'information géographique. p.171-192. Paris: Lavoisier

Intelepix, 2004. White paper: An introduction to Oblivision™ Intelepix LLC, available at http://www.intelepix.com/PDF/Intelepix% 20Oblivision%20WP.pdf

INSPIRE, 2004. Proposal for a directive of the European Parliament and of The Council, establishing an infrastructure for spatial information in the Community (INSPIRE) {SEC (2004) 980}, available at http://www.ec-gis.org/inspire/

ITN + T,2005. Integrated Transport Network plus time: project overview, available at http://www.geosensor.net/duitn/index.php?module=pagemaster& PAGE_user_op=view_page&PAGE_id=4&MMN_position=5:5

Kolbe, T.H., Gröger, G. & Plümer, L. 2005. CityGML: Interoperable access to 3D city models. In, P. van Oosterom, S. Zlatanova & E.M. Fendel (eds), Geo-information for disaster management: 883–899.

Kolbe, T.H., Gröger, G. & Plümer, L. 2006. 3D City models for ER. (in this book)

Lee, J. & Zlatanova, S. 2006. A 3D data model and topological analyses for emergency response in urban areas (this book)

MacFarlane, R. 2005. A Guide to GIS Applications in Integrated Emergency Management, Emergency Planning College, Cabinet Office.

Marzano, S. 1998. Creating value by design: thoughts. Royal Philips Electronics, the Netherlands, V & K Publishing, Blaricum.

Mizen, H., Dolbear, C. & Hart, G. 2005. Ontology ontogeny: Understanding how an ontology is created and developed, In, M. A. Rodriguez, et al. (eds) Lecture Notes in Computer Science 3799, proceedings of the First International Conference on Geospatial Semantics: 15–29,November 2005, Mexico City, Mexico, Springer-Verlag.

Murray, K. & Shiell, D. 2004. A framework for geographic information in Great Britain. The Cartographic Journal, (IGC Special Issue) 41,(2), 123–129.

Open GIS Consortium, 2003. OpenGIS Reference Model, available at http://www.opengis.org/info/orm/

Ordnance Survey 2004a. Ordnance Survey Framework Document 2004, available at http://www.ordnancesurvey.co.uk/

Ordnance Survey 2004b. OS MasterMap: definitive digital map of Great Britain designed by Ordnance Survey, Ordnance Survey website, available at http://www.ordnancesurvey.co.uk/oswebsite/products/osmastermap/

Ordnance Survey 2004c. Positional accuracy improvement – background information, available at http://www.ordnancesurvey.co.uk/oswebsite/pai/ backgroundinformation.html

Ordnance Survey 2005. OS MasterMap technical information: style and XML examples. http://www.ordnancesurvey.co.uk/oswebsite/products/osmastermap/ xml/index.html

Parker, Chris. 2005. Research challenges for a geo-information business. The Cartographic Journal IGC Special Issue 41, (2): 131–141.

Parker, C.J. & Stileman, M. 2004. Disaster management: the challenges for a national geographic information provider. In, P. van Oosterom, S. Zlatanova & E.M. Fendel (eds), Geo-information for disaster management: Springer, Berlin, pp: 191–214.

Peerenboorn, J.P., Fisher, R.E., Rinaldi, S.M. & Kelly, T.K. 2002. Studying the chain reaction, Electric Perspectives, January/February 2002.

Pictometry 2006. Pictometry information system for disaster response and preparedness, available at http://www.simmonsaerofilms.com/uploadedFiles/ pictometry/Disaster.pdf

Pundt, H. 2006. The semantic mismatch as limiting factor for the use of geoinformation in disaster management and emergency response (this book)

Revell, P., Regnauld, N. & Thom, S., 2005. Generalising OS MasterMap topographic buildings and ITN road centrelines to 1:50 00 scale using a spatial hierarchy of agents, triangulation and topology. Proceedings of the 22nd International Cartographic Conference, La Coruna, Spain, 9–16 July 2005. CD-ROM.

Slingsby A.D., Longley P.A. & Parker C.J. 2007. A New framework for feature-based digital mapping in three-dimensional space. In Lovett, A.A. and Appleton, Katy (eds) GIS for Environmental Decision Making, Innovations in GIS, Volume 12, CRC Press.

Smith, S., Holland, D. & Longley, P. 2005. Quantifying interpolation errors in urban airborne laser scanning models.

Vowles, G. 2004. On-demand GI networks. In Geo-Solutions, proceedings of the Association of Geographic Information Annual Conference, September 2004, London, UK.

Warboys M.F. & Hornsby K. 2004. From objects to events: GEM, the geo-spatial event model. In M.J. Egenhofer, C. Freska, & H.J. Miller (eds), Geographic Information Science, LNCS 3234, proc. third intern. Conf., Adelphi, MD, USA. Berlin and Heidelberg: Springer-Verlag.

West, N. 2005. Veterinarians Play Key Role in Preventing Influenza Pandemic, Bellwether, 61, 6–7.

Zlatanova, S. & Prosperi, D. 2005. Large-scale 3D Data Integration: Challenges and Opportunities. ISBN 0-8493-9898-3, CRCpress, Taylor & Francis Group, Boca Raton, FL, 245 p.

Websites Cited:

www.dnf.org
www.gigateway.org.uk
www.magic.gov.uk
www.eu-orchestra.org
www.qinetiq.com/home_ep/case_studies/sepura.SupportingPar.0001.File.pdf
www.romanse.org.uk
www.ukresilience.info

Part 6
Applications and solutions

The value of Gi4DM for transport and water management

M.J.M. Grothe, H.C. Landa & J.G.M. Steenbruggen
Ministry of Transport, Public Works and Water management, Department of Geo-Information and ICT, The Netherlands

ABSTRACT: The tasks and responsibilities of the Ministry of Transport, Public Works and Water management in The Netherlands lie in the sectors integrated water management (water quality and water quantity), traffic and transport over waterways, rail and in the air. In the past many calamities have taken place in those sectors. Besides large(r) calamities daily traffic incident management along the main roads and waterways is a major task of the Ministry as well. On such occasions the need for (spatial) information and supporting information systems is great. That need has only grown because attention is especially drawn to a proactive/preventive phase in calamity control and incident management. In this chapter we will outline the use of geospatial information and geoservices in Disaster Management within the field of national transport and watermanagement in The Netherlands.

1 INTRODUCTION

The tasks and responsibilities of the Ministry of Transport, Public Works and Water management (V&W) in The Netherlands are focussed on water management (water quality and water quantity) and traffic and transport over waterways, roads, rail and in the air. In the past several calamities have taken place in those sectors: heavy car collisions due to intense fog nearby the city of Breda (1972, 1990), pollution of Rhinewater due to fire in Sandozplant (1986), airplane crashes near Schiphol and Eindhoven (1994, 1995), derailment of chloride trains in Delfzijl and Kijfhoek (2000, 1986), river floods (1993, 1995), and recently some dike collapses due to heavy drought and rain (2003, 2004). Besides managing these large(r) calamities daily traffic incident management along the main roads and waterways is a major task of the Ministry. On all such occasions the need for (spatial) information and supporting information systems is great. That need has only grown because attention has also shifted to the proactive and preventive phase in calamity control and incident management.

The directorate-general for Public Works & Water Management (Rijkswaterstaat) as part of the Dutch ministry of V&W is since 1798 responsible for maintaining and administering the main roads and waterways in The Netherlands. These tasks include protection of the country against floods from both the rivers and the sea. V&W is an organization having 15,000 employees, an annual budget of approx. €7 billion (US $9 billion), and more than 200 offices throughout the country. Accurate and up-to-date geospatial information has always been a necessity for administering the main water and road networks of The Netherlands. As transport and water management was crucial to survive in the low-parts of The Netherlands the management of the water systems was already conducted in a sophisticated and organized way in the middle ages. There are e.g. map series of thematic maps of the water system on a scale of 1 to 19:000 (15 sheets) dated 1611. V&W thus has a longstanding tradition of mapping and geospatial data processing. Geospatial data management at V&W moves to a new stage in which the demands of today can be met. Key factors in this geospatial data management policy are centralization of geospatial data in centrally managed geodatabases, the use of open standards, the exchange of Windows-clients for small-footprint browser-based web-clients and the use of mobile technology which is seamlessly connected to the main GDI. In this chapter we will outline the Geo Data Infrastructure based on ISO/OGC web services and metadata standards and adopted application architecture and infrastructure.

The department of Geo-information and ICT of Rijkswaterstaat (RWS AGI) is responsible for providing the organization for the IT and the (geospatial) information needed for its tasks. The department of Geo-information and ICT is working with V&W to meet the challenge of reducing IT costs considerably. The strategy to meet this challenge is built on the principles of uniform working models, open standards, server-based computing and central data hosting and maintenance. In this chapter we will outline the Geo Data Infrastructure (GDI) within the ministry of Transport, Public Works and Water management in The Netherlands. Attention is given to the following subjects:

- An actual status description of Disaster management within V&W in terms of strategy, organisation, and working methods;
- Insight in the underlying renewed Geo Data Infrastructure based on ISO/OGC web services and metadata standards and adopted application architecture;
- A descriptive real-world scenario that illustrates the role and importance of geospatial information and Geo Data Infrastructures in calamities of water transport and river flood management in The Netherlands; and
- Some concluding remarks.

2   STATUS DESCRIPTION OF DM WITHIN V&W

2.1   V&W and safety & security policy

In The Netherlands the Ministry of Transport, Public Works and Water management (V&W) has a number of tasks and responsibilities in the field of safety and security. The following domains are distinguished:

- Traffic and transportation by road, rail, water and air: examples are security at Schiphol airport, (inter-)national aviation security, social security in public transport, safety on main roads and rivers and transport of dangerous goods.
- Physical infrastructure: examples are tunnels, bridges, rail shunting emplacements and underground tubes.
- Water management: examples are protection against high water levels of rivers and sea (e.g. dikes and retention areas) (water defence), management of surface water levels in relation to fresh water intake, dumping of cooling water and/or inland shipping (water quantity) and oil and chemical spill detection and recovery (water quality).

The primary policy tasks of V&W are within these domains and the safety and security policy is an integrated part of those tasks. The ministry has formulated a generic view on safety and security which consists of the following four core elements:

- To strive for a permanent improvement of safety and security.
- Within the above-mentioned domains a process of ongoing improvement has been implemented. The process is focusing on permanent reduction of (the chance of) killed and injured people and the (chance of) social disruption. It is evident that in the process of permanent improvement SMART policy objectives and milestones have to be set in order to make progress explicit.
- To weigh measures explicitly and transparently.
- V&W operates in an area of constant tension in which trade-offs between economy, environment and security have to be made. Other policy aims than safety and security must explicitly be assessed, weighed and exchanged against (the chance of) mortalities and injuries and/or material damage. V&W prefers a transparent view of these assessments in order to allow for clear political choices. This is achieved by assembling parcels of measures with varying ambitions by which the positive and negative consequences based on costs, economy, environment and security become obvious.

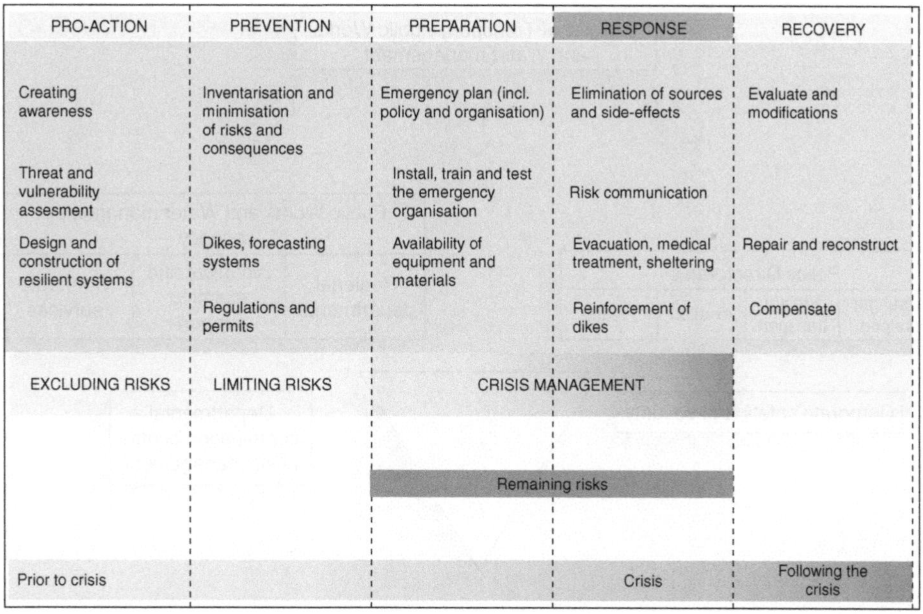

PRO-ACTION	PREVENTION	PREPARATION	RESPONSE	RECOVERY
Creating awareness	Inventarisation and minimisation of risks and consequences	Emergency plan (incl. policy and organisation)	Elimination of sources and side-effects	Evaluate and modifications
Threat and vulnerability assesment		Install, train and test the emergency organisation	Risk communication	
Design and construction of resilient systems	Dikes, forecasting systems	Availability of equipment and materials	Evacuation, medical treatment, sheltering	Repair and reconstruct
	Regulations and permits		Reinforcement of dikes	Compensate
EXCLUDING RISKS	LIMITING RISKS	CRISIS MANAGEMENT		
		Remaining risks		
Prior to crisis			Crisis	Following the crisis

Figure 1. The safety chain.

- To be prepared for inevitable risks.
- Risks are inevitable: even though it is the aim to improve working procedures and results to the maximum and to achieve better safety situations there will always be a risk or chance of misfortune. We don't control nature and it is human nature to create risky situations. Within this context we do not always provide for sufficient financial and technical means to prevent unsafe situations. These so called 'inevitable risks' oblige V&W to take constantly into account the possibility of an accident or calamity, after which it is best to minimize the consequences of such an accident or calamity as much as possible. A useful and perhaps most important method is to practice for such events; this also allows for building and testing the quality of organizational and communicational structures and procedures.
- To obtain and maintain a safety culture within the ministry.
- A necessary condition to achieve long-lasting improvement is to explicitly manage the aspect of safety and security. This can be obtained by the installation of dedicated units within the organization, by explicit reservation of funds and by agenda setting. In this way awareness is created within the organization.

To transform this view into more practical and concrete safety and policies the concept of 'the safety chain' is illustrated (Figure 1). This chain is composed of five phases: pro-action, prevention, preparation, response and recovery.

2.2 V&W and crisis management

Since the end of 1999 V&W has grouped its activities in the field of crisis management in the Departmental Coordination centre Crisis management (DCC). This centre builds on the strategic objective of crisis management: a decisive V&W-organisation that functions administratively, organisationally and operationally in a coherent way during incidents and calamities. From the central objective the following points have been derived:

- Development and implementation of crisis management policy: development of policy on planning, education, training and all other areas related to crisis management.

315

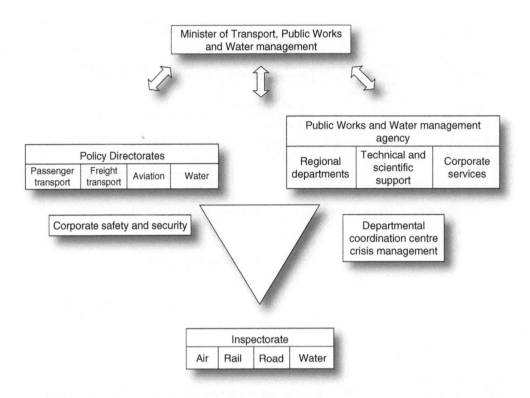

Figure 2.   Organisation of crisis management within V&W.

- Preparing of emergency plans in such a way that a common insight arises in the tasks and responsibilities of the organisational services concerned, insight into the dangers for which the organisational service concerned is responsible for and the insight in relevant networks.
- Training and practising staff so that people concerned with regard to knowledge, skills and experience are well prepared for crises.
- Embedding in the organisational context. The departmental structure and working methods must be coupled to the external network. During crisis circumstances the partners in the network must be able to find each other and know they can count on each other.

2.3 *Organisation*

The generic safety and security policy of V&W is developed by the programme board safety & security; the policy for and the operational crisis management as such is the responsibility of the Departmental Coordination centre Crisis management (Figure 2).

The Departmental Coordination centre Crisis management is the central office for all Directorate-Generals (DG's) and the central services within V&W. In case of a crisis a V&W alarm number is available. A protocol has been developed which describes clearly who is responsible for what and who informs who if there is a problem. This is summarised in network cards for all domains; a comprehensive overview which describes the organisational context of the organisation (Figure 3). Onto these network cards the actors are added for each domain. These network analyses and competence diagrams together with the description of the internal organisation and checklists for decision-making form the handbook of crisis management of V&W.

316

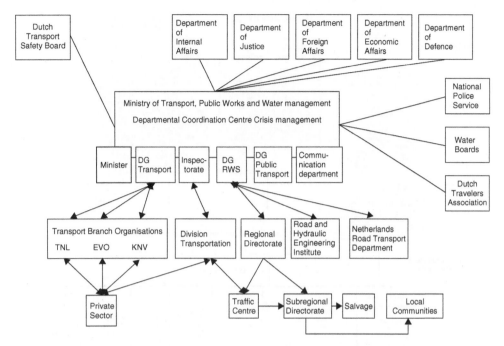

Figure 3. Operational organisation network during crisis events on main roads.

2.4 *Working methods*

At a strategic level the topics of safety, security and crisis management are being developed in a number of ways. Examples are the internalisation of external costs, development of performance indicators for measuring policy effectiveness, developing future scenarios in order to review safety and security plans, developing uniform and accepted models for risk assessment and last but not least working on public confidence. Each of these themes is analysed on several aspects. The economic aspect is considering safety and security as an economic problem; a central question is for example 'how much money does it cost to protect 'X' capital goods or 'Y' lives. The aspect of spatial planning approaches security and crisis management as spatial problem with questions such as 'where can I plan houses or industrial areas and where definitely not'. If security is considered as an administrative problem questions include 'who has which responsibilities and mandates and are they properly addressed?'. The fourth and more generic approach considers security and crisis management as a social problem and aims at developing socially acceptable security level/dangers as parameters for decision-making and aims at a more integrated approach in which the aforesaid aspects are incorporated.

3 V&W GEO DATA INFRASTRUCTURE

3.1 *Towards a V&W corporate Geo Data Infrastructure*

To ensure the adequate use of geospatial information and systems in large organisations a framework is needed. Since the beginning of the nineties within V&W this framework existed and was known as Geo-information Infrastructure (GII). The V&W GII was considered as the complete set of geospatial information, GIS and geospatial applications, standards, policy, IT infrastructure and organisation to ensure an efficient and effective use of geospatial information within the organisation. Nowadays, the term Geo Data Infrastructure (GDI) or Spatial Data Infrastructure (SDI) is internationally acknowledged (Groot and McLaughlin 2000). The main goal of a GDI is ensuring an optimal availability and usability of geospatial information and information systems through

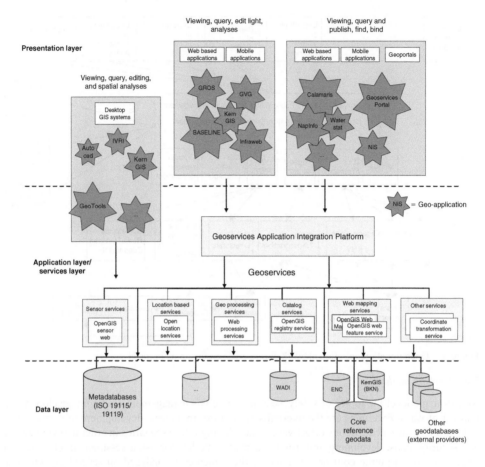

Figure 4.   The V&W GDI-based application architecture.

open interfaces that operate on the web services concept of publish-find-bind. V&W started in 2003 with the adoption of these new interface concepts. This has resulted in a geoservices application architecture (Figure 4).

The GDI application architecture can be considered from different perspectives. Here we consider the GDI from the application perspective consisting of the presentation layer (the applications), the geoservices layer and the database layer. The presentation layer consists of different types of geospatial applications. Within V&W circa 50–60 different geospatial applications exist that are used in the different application fields of the ministry: besides disaster and emergency management there is also public works and asset management, road transport management, water quality management and policy development and evaluation in these fields. Three types of geospatial applications are distinguished; desktop GIS based applications (within V&W software mainly from ESRI and Autodesk), browser based geospatial applications and mobile geospatial applications (also browser based). The geoservices layer consists of different types of services that offer access to spatial data from different sources, perform transformations on spatial data and offer portrayal services for the different types of clients (desktop, browser apps and mobile apps). These geoservices are based on the open standards of Open GeoSpatial Consortium (OGC) and ISO. The geoservices application integration platform offers the instruments to develop applications; from RAD software, geographical user interface components and instruments for services chaining.

At the database level data access is offered to all geospatial referenced information in the organisation. Core reference geospatial information for V&W DM consists of several types of base maps,

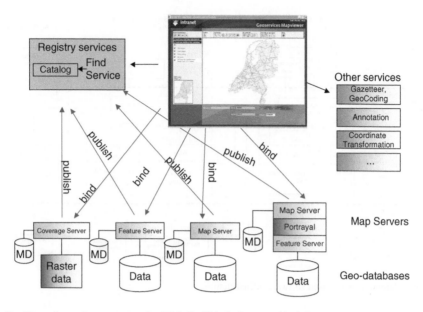

Figure 5.   The web services concept of publish-find-bind of geographical data.

like high scale topographic maps, aerial photo's, satellite images, object information and cadastral data. The core geospatial information is centrally served through web services from different proprietary data formats to these distributed applications. Within the V&W GDI the availability of metadata on data and services is obliged. CEN- and ISO-based metadata standards are used (ISO 19115 for datasets and ISO 19119 for services). Since 1992 V&W has developed and implemented a profile for the CEN98 metadata standard for geospatial information is developed. Recently this profile is translated to the ISO 19115 metadata standard.

The GDI based approach should offer V&W an optimal access to core geospatial information and geo-referenced information needed for applications of disaster management and incident management within V&W. In the next paragraph we outline the adopted parts of the GDI software within V&W.

3.2   The V&W catalogue and web mapping services architecture

Since 2003 V&W is working on the implementation of a central, OGC-based services architecture infrastructure. The most powerful element of the OGC-based web services architecture is the concept of 'publish-find-bind'. This web services concept works as follows (Figure 5). The data manager publishes geospatial information in the form of maps and served by an Internet map server in a registry. The registry is known as the OGC Web Catalogue Services (WCS). The user of an application has a find button, for access to the registry and to answer the questions: which map servers are available and which maps are available? After the maps are found in the catalogue database, the user can bind the maps through the web mapping user interface of the application. The OGC web services concept is based on a distributed systems concept. Data is directly accessible at the source database through an OGC-based services interface. Sharing geospatial information in an open and transparent way is what this is all about. The architecture offers users of geospatial information an easy way to access, publish, find and bind geospatial information through Internet technology.

This geospatial information infrastructure based on the OGC web services architecture has been established using both open source software products and commercial-off-the-shelf components. This infrastructure has already enabled broad geospatial information sharing throughout the organization and has proven to be cost effective. The adopted software architecture is characterised

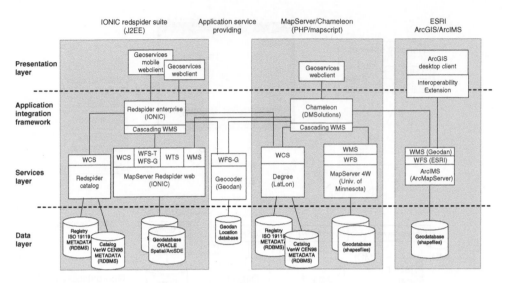

Figure 6. Adopted OGC-based software infrastructure.

by a three layers architecture: presentation layer, services layer and data layer (Figure 6). The architecture is modular and scalable. The kernel of the architecture is the services layer. The modularity of the services layer is illustrated by the fact that several different software components for the Java/J2EE platform as well as the Apache/PHP platform.

For basic web mapping services OGC Web Map Server, Web Feature Server and Web Coverage Server Rijkswaterstaat uses the Redspider suite of IONIC and the Minnesota MapServer (MS4W) of the University of Minnesota (Open Source Software). The Web Catalogue Service is the Open Source Software (OSS) product Degree from the German vendor LatLon. Degree is Java based. A Dutch company Geodan offers a geocoder service based on the Application Services Providing (ASP) business model, to gazetteer. An underlying location database offers access to all addresses, postcodes and municipalities in The Netherlands. The client software platform is based on IONIC's Redspider studio and the Canadian product Chameleon that is based on a PHP server-side scripting language. Furthermore ArcIMS from ESRI is also used for mapping, especially for web mapping in combination with the ESRI's ArcGIS desktop client software.

3.3 The V&W location based services architecture

To carry out the day-to-day activities many Rijkswaterstaat employees are out in the field-inspecting infrastructure, checking permits and regulations etc. To supply these workers with hand-held mobile computers with GPS location and a wireless connection to the office network proved to be very productive. Apart from the time saved also the quality of the decisions made increased because of the information available on the spot. In a pilot project mobile computers running a web browser were connected to several distributed databases and able to do vector editing with WFS-T in these databases all through OGC standard protocols. Expected future implementations include the implementation of a transactional web feature service for mobile clients. Location-based Services (LBS) are information services, accessible through devices such as mobile phones, PDA's, tablet and laptops. Their essence is the ability to locate a user or an object in space, and to use this to make information processes aware of the location dimension. Location-based services are indispensable in emergency services and disaster management.

3.4 Towards a GI4DM architecture

The above-mentioned GDI architecture is not an end in itself. It should support applications with for geospatial information processing capabilities. The illustrated GDI services are geoprocessing

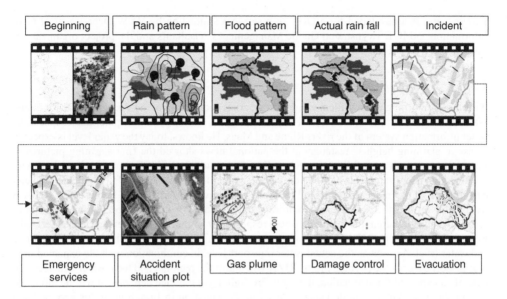

| Beginning | Rain pattern | Flood pattern | Actual rain fall | Incident |

| Emergency services | Accident situation plot | Gas plume | Damage control | Evacuation |

Figure 7.   Geospatial information scenes of the DM flooding scenario script.

components for building risk management applications. However information systems that support DM should have other services available as well. The approach of Orchestra (Annoni et al. 2005) is worth consideration. Orchestra puts forward an architecture that can be considered as a collection of services, tools and methodologies that so-called system users can use to develop disaster management information systems for end users. For instance the above-mentioned mapping and location-based services are providing geospatial information processing capabilities. Additional to these geoservices Orchestra outlines a services framework for risk management that can be applied for different risk and disaster situations with services for formula access, diagramming, documents access, sensor access schema mapping, ontology access and service chaining (Uslander 2005).

4   A DM SCENARIO IN TRANSPORT AND WATER MANAGEMENT

4.1   *Introduction*

Information systems for Disaster Management will benefit from the developments in open Geo Data Infrastructures in an evolutionary way. These benefits can best be shown by outlining a DM scenario. The scenario is an as realistic as possible attempt to demonstrate the usefulness of geographical web services based on the ISO/OGC web services architecture. Information management and information systems for flood defense and disaster management will benefit from these new developments in an evolutionary way. The flood defense crisis is situated near Nijmegen on the river Waal, where flooding conditions due to abundant rain and melting snow coincide with stormy conditions. Four crisis centers communicate using open geographical web services technology provided by a series of governmental organizations resulting in effective and efficient information sharing and quick response to the disaster 'nearly' developing. Here cooperation and information sharing boosts information control.

This scenario consists of a number of scenes in which flood defence is simulated and disaster management in The Netherlands is illustrated (Figure 7). It is a fictive simulation of a ship crash at the Rhine river near Nijmegen in The Netherlands during a critical high water situation. Any resemblance of the events and people in this scenario with reality are due to the merest coincidence.

This scenario is a developed for a OGC web services simulation by V&W Department of Geo-Information and ICT, Twynstra en Gudde Management Consultants and the project organizations Viking, NOAH and HIS.

4.2 DM scenario; the beginning

It is Monday morning when Willem van der Gaag, crisis coordinator of the V&W Institute for Inland Water Management and Waste Water Treatment (RIZA), starts up his computer and inspects the water information system of the rivers Rhine and Maas. He knows, today the water level is expected to rise to alarming heights. Yesterday at the national news showed the first alarming pictures of floods from excessive rainfall in NordRhein Westfalen and the Eiffel (Germany). Combined with extreme high rise water levels of the rivers caused by melting water and rainfall in Switzerland and part of the East of France this will lead to floods in the rivers basins in The Netherlands. RIZA is responsible for the river water level monitoring and discharge (water transport per m³ per second) at Lobith and Borgharen, the locations where respectively Rhine and Maas enter The Netherlands. Besides water quantity, RIZA also monitors water quality. RIZA calculates the expected water rise and water level. As soon as the actual water transport level in combination with the expected time of transport exceeds the defined limits, RIZA will alarm the regional and local water managers. In case of an expected flood situation, the local fire fighting and police department will be informed as a precaution. Willem van de Gaag is checking the situation in Lobith using his geographical information system (GIS); the maps show the situation of the water system Rhine at the location Lobith. He also checks the Rheinatlas (http://www.iksr.org) where he explores the previously predefined flood risk areas.

4.2.1 Scene 1; Mapping the weather forecast
Willem is evaluating the water situation in Lobith. He predicts the expected water levels in the next 48 hours. In order to have realistic water predictions he needs information about the local weather situation in the water system of Rhine and Maas. Therefore he accesses the database of the Royal Netherlands Meteorological Institute (KNMI). While the water rise prediction system calculates the expected water levels, Willem is inspecting the weather maps of KNMI with his GIS. He displays the rain fall maps of the last three days and the rainfall predictions of the next three days. He also takes a look at the maps of the actual and expected wind directions and wind speed. These maps show an increasing rainfall in the river basin area. When he receives the prediction outcomes, he knows that Wednesday morning Lobith is expecting a high water transport level with an expected maximum peak Wednesday evening.

This information leads to the sending of a 'flood warning message' to the regional and local water managers and emergency services (local fire fighting and police departments). This message is an early warning that a flood crisis is to be expected. The message consists of all relevant information including the maps and is offered through geographical web services technology. Especially the map location where the Rhine meets the Waal, Netherrhine and IJssel shows an alarming flood situation. Even taking the uncertainty of the predictions into account, a two-day flood situation is expected with even a chance of extreme floods. Through email, fax, phone and geographical web services all responsible and involved parties are informed.

4.2.2 Scene 2; High water rise
In Arnhem the regional crisis manager Joost Hilhorst receives the incoming flood warning message and is going to measure the 'exact' water rise level per kilometre section. This means that for each kilometre at the waterway the expected water level and discharge is calculated. As long the calculated levels remain under the level of the dikes along the waterway, no floods will occur. However with strong winds and river surges this could lead to floods. Especially at locations with weaker dike stability this can have impacts. The dike management information system of Rijkswaterstaat will inform the regional crisis manager about the status and stability of the dikes. The weak dike spots are mapped together with the detailed, predicted water levels for each river section and located on

a risk map. Joost Hilhorst puts forward the risk map during a meeting with other flood experts. They decide that it is time to express 'Warning phase 1'. The decision is in particular based on the fact that the map shows a maximum of 0.8 meters water rise in the Waalbandijk in the Ooijpolder and 0.4 meters at the lock in this dike. Joost and his colleagues also know that during renewal of the road deck the asphalt at the dike road is removed the last week. This even makes the spot more vulnerable.

4.2.3 *Scene 3; Extreme rainfall*

Tuesday morning Joost Hilhorst is inspecting the latest weather forecasts. These show increased possibility of extreme rainfall in the Rhine water system in Gelderland and Noord Limburg. The weather radar maps show heavy rainfall patterns with strong winds near Wesel and Nijmegen. Messages of water floods in cellars of houses and tunnels in the region give a strong indication that the water level at the waterways will increasingly rise and the emergency services (especially fire fighting units) will be busy in next hours. At the same time the weather forecast shows stormy weather expected on Wednesday evening.

Because of the water level and waves due to the expected stormy weather Joost and his colleagues expect that the transport possibilities at the Waal river will be limited to zero at Wednesday afternoon. A warning to professional shipping companies is sent out; between Wednesday 1200 hours and Friday evening 1,200 hours ship traffic will closed at the Waal corridor between Lobith and Weurt. The margin of the dike stem level for each dike section remains under the critical levels and therefore restricted ship traffic is still possible. The warning message to the ships is accompanied by a map with detailed restrictions.

For the critical situation at the Waalbandijk in the Ooijpolder it is decided to raise the dike stem level with an extra layer of sandbags. Wednesday morning a contractor of Rijkswaterstaat will place a double row of sandbags (40 cm) at a length of 30 meters next to the lock at the Waalbandijk.

4.2.4 *Scene 4; Incident; shipping accident*

The same Wednesday morning 10.00 hours a ship incident message arrives at the emergency room of the regional Fire Fighting Unit. At the location of river section or 'kilometerraaij' 854 two ships have crashed. The captain of one of the ships, loaded with coal, indicates that he manoeuvred his ship into a dike in order to try to avoid the crash. According to the incident protocol Henk Groen at the emergency room collects the necessary information from the captain and tries to figure out if any human victims are involved, whether there is fire or a leak with chemicals or gasoline. The captain of the Rhine vessel loaded with coal is confused and with the second ship there is no contact. Henk Groen decides to contact the river information services system in order to obtain information about the actual traffic at the incident location. At the same time he warns emergency services including a diving team. Boat units of Rijkswaterstaat and police are ordered to the incident spot as well. Henk can track and trace the mobile units on the computer.

The incident spot is marked in the system and because of the expected impact the crisis situation is scaled upwards. First, there is a possibility of a breach in the dike at the spot of the large coal vessel during a flood. Second, it is expected that this ship restricts the water transport with a strong chance of water damming. Henk Groen views the municipality map and informs the crisis coordinator of the municipality. At the same time an inter local warning is dispatched to the surrounding municipalities in case the incident accumulates.

4.2.5 *Scene 5; Crisis coordination*

After a couple of minutes the first police surveillance unit is near the incident spot. Through their C2000 communication system the unit informs the emergency room about the local situation. However, the police surveillance is confronted with the roadblock of sand and other maintenance materials for the renewal of the asphalt layer. The unit is 300 meters away from the incident spot and continues on foot. At first visual inspection they do not observe fire or any other alarming activity. The emergency room decides to order the contractor to clear the road. Also the voluntary fire-fighting unit of the municipality of Ubbergen arrives at the roadblock and cannot access the incident spot. The police unit on foot discovers a small sandy track in the direction of the stranded coal vessel. This small track is not on the map of the emergency room and Henk Groen asks for a more detailed roadmap from the local road network manager. It seems that there is a small track that gives access 300 meters downstream the incident location. The emergency room informs and guides the diving team from Nijmegen to this spot.

4.2.6 Scene 6; The ammoniac tanker and it's gas plume

Then by phone the emergency room Gelderland-Zuid in Nijmegen receives the information about the second ship from the River Information Centre from Rotterdam. There is a strong indication that the second ship might be the ammoniac tanker Diana from Düsseldorf. Up till now it was not possible to reach the ship by phone. According the GPS-based River Information System the last location of the ship is 100 meters downstream of the stranded coal vessel. This indicates that the tanker is damaged. Henk Groen immediately informs the mobile rescue units via C2000 to perform observations at the Waalbandijk of a possible gas leak from the ammoniac tanker. This warning is also communicated with the regional crisis centre of the security region of Gelderland Zuid, which became operational during the last couple of hours. The potential of a gas plume calamity makes this incident an intermunicipal incident being a treat for a large group of the population.

4.2.7 Scene 7; Gas plume evacuation scenario's

Martin Slootsdijk, being responsible information manager at the regional crisis centre Gelderland-Zuid, immediately investigates the wind maps, the actual ship movements at the Waal. Next he calculates a plume using an integrated plume model of TNO-FEL in case of a gas explosion at the ammoniac tanker. The resulting gas plume maps are overlayed with the maps of housing and inhabitants, economic activities, cattle and the road network in order to get an indication of numbers of persons and cattle for evacuation. Because of the fact that the tanker still moves downstream he also copies the plumes to downstream locations and recalculates the number of evacuees. It is obvious that the tanker should be stopped in order to avoid plumes at the larger dense populated city of Nijmegen. He alarms the water police and Rijkswaterstaat to stop the tanker from proceeding downstream.

4.2.8 Scene 8; Prevention

In the mean time at the crisis centre of Rijkswaterstaat East in Arnhem the crisis team is working on a strategy to avoid further damage at the dike by the Rhine vessel. Repairing the damage is one activity of the strategy, the other is trying to avoid further damage because of the fact that the ship's position is on top of the water direction. This might cause in combination with the heavy rainfall and strong winds an additional water rise of 0.15 m as is calculated by one of the information systems. The safety margin becomes very narrow. At the same time there a possibility that the coal ship is floating away, causing more damage at the dike.

In order to the prevent damage the most recent information about industrial activities (dangerous goods that might have environmental impact) and heritage sites (especially monuments) in the area that are threatened by possible floods are collected. At the Risk Map of The Netherlands it is shown that one industrial site is located in the area; it is a stone/concrete factory. Through the permit database it is shown that that the factory has permits for storage of oil spill depot. The crisis centre orders these goods to be removed. The monuments database (or KICH-portal) shows that several monuments are located in the area. However, it is decided that no further actions are necessary to prevent damage of goods in these monuments (all private properties).

4.2.9 Scene 9; High water evacuation

In the mean time at the regional crisis centre of Gelderland-Zuid the crisis team is gathering the necessary information to see whether evacuation of people and cattle will become inevitable considering the potential gas plume and possibility of floods. At that moment a message from the water police arrives at the centre. There has been communication with the ammoniac tanker. The captain of the tanker has confirmed that his gas tank pressure level is OK and he and his crew did not observe any visual damage to the gas tanks. It is decided that the ammoniac tanker will be sent for further visual inspection to the coal harbour of Weurt.

In the mean time several calculations are performed in order to determine the size of an possible evacuation of people, cattle and goods. Through model based calculation with the HIS system (www.hisinfo.nl) it is shown large parts of the Ooijpolder will be flooded. In a period of one hour after a major dike collapse the water will reach the village of Ooij. However the inhabitants will need 3 hours to leave through the southern and eastern route. Ooij has 4 elderly homes with 25 inhabitants, and three public archives of the Municipality of Ubbergen that need special attention.

5 SCENARIO SYNTHESIS

Besides the end of the fictional scenario, because that is a matter of an exercise in collecting and handling information and decision making, this scenario implies that the traditional instruments such as phone and email for handling crisis situations will also change through the use of GDI web services. By plotting information on the map and offering online access to distributed geographic databases, these information services increase speed of handling during crisis situations. Especially, information management can benefit from OGC-compliant web services oriented architectures in crisis situations that have the following characteristics:

- When different organisations with different responsibilities are involved;
- When organisations are located at different locations; and
- When organisations use different distributed information systems.

In table 1 a quick scan of this disaster management scenario in The Netherlands shows the different geodatasets and information systems used and organisations involved. As shown several different types of geodatasets are generated and processed by different information systems owned by different organisations at numerous distributed locations. These are characteristics of situations where a GDI approach offers advantages.

Within V&W the High Water Information System (HIS) is more or less crucial for DM flooding scenarios. HIS has three objectives:

- Perform impact analysis of water dikes during the preparation phase of a high water situation;
- Support communication during a high water situation;
- Perform monitoring of dikes during a high water situation through relevant information.

HIS has a geographical database, different processing modules and geographical interface. HIS is a stand alone information system with specific communication interfaces to other information systems. The distribution of geospatial information is based on data duplication, especially concerning geographic core data (topographic maps, aerial photos, etc). HIS has it's own geodatabase which is regularly updated with new updates of core geodatasets through physical media. It is expected that geographical data exchange in the near future will be based on direct and online access at the core database of the supplier or a broker. Online access will be based on open interface standards of ISO/OGC. The HIS organisation has already in close cooperation with other water partners in The Netherlands and Germany (NOAH and VIKING) initiated the development of an integrated FLood Information and Warning System (FLIWAS) for the large rivers in The Netherlands FLIWAS will have their geospatial information organised and served according to GDI principles. This is part of the newly built Geo data Infrastructure of V&W.

6 FINAL REMARKS

A strategy of uniform working models, open standards, web based computing and central data hosting and maintenance is adopted in geospatial information management of V&W. This is shown

Table 1. Overview of diversity of geospatial information and information systems and involvement of organisations within DM flood scenario in The Netherlands.

Scenes	Geo datasets	Systems	Organisations involved
The beginning	• water system • water levels • flood risk areas	• FEWS • Rheinatlas	• Info centre RWS RIZA
Scene 1 Mapping the weather forecast	• water system • expected water levels • rain fall last 3 days • rain fall prediction • actual wind direction and speed • predicted wind direction and speed • flood warning/high water message	• FEWS • Infocenter.nl • C2000	• Info centre RWS RIZA (KNMI, DWD, ECMWF) • Regional and local water managers • Emergency services
Scene 2 High water rise	• detailed expected water levels per km • detailed expected water discharge per km • Dike protection maps • Weak spots (dike repair)	• Dike Info system • HIS	• RWS ON • Regional and local water managers
Scene 3 Extreme rainfall	• rain fall last 3 days • rain fall prediction • actual wind direction and speed • predicted wind direction and speed • Water transport restrictions • Extra sand layer on dike spot	• IVS'90/RIS • infocenter.nl	• RWS ON • Info centre RIZA • Contractor
Scene 4 Shipping incident	• Information on incident • Ship information • Municipalities map	• GMS • IVS'90/RIS • Sherpa (tracking emergency units) • GIS • C2000 • Infraweb	• Emergency room Gelderland-Zuid • Emergency services and units • RWS ON • Municipality
Scene 5 Crisis coordination	• Detailed roadmap • Incident spot	• C2000 • Navigation systems emergency units	• Emergency room Gelderland-Zuid • Emergency services and units
Scene 6 The ammoniac tanker and it's gas plume	• Ship information	• IVS'90/RIS • C2000	• Emergency room Gelderland-Zuid • Rotterdam Water Traffic Centre • Emergency services and units • Regional crisis centre
Scene 7 Gas plume evacuation scenarios	• Wind maps • Actual ship movements • Integrated plume model • Housing	• IVS'90/RIS • FEWS • GASMAL • HIS	• Rotterdam water traffic centre • Emergency services and units

(Continued)

Table 2. (Continued)

Scenes	Geo datasets	Systems	Organisations involved
	• Infrastructure • Economic activities • Cattle • High risk objects • Road network	• National cattle database • Regional risk atlas	• Regional crisis centre
Scene 8 Prevention	• Dike damage prevention • Industrial activities • Permits industrial activities • Heritage sites	• Risk map Netherlands • KICH-portal	• Regional crisis centre Gelderland-Zuid • RWS
Scene 9 High water evacuation	• Wind maps • Actual ship movements • Integrated plume model • Housing • Infrastructure • Economic activities • Cattle • High risk objects • Road network	• IVS'90 • RIS • Weather database • GASMAL • HIS • National cattle database • Regional Risk atlas • C2000	

to be feasible and has many advantages. This renewed GDI is based on the use of international open standards. These ISO/OGC standards, like WMS, WFS and WCS, are matured so these enable the construction of an enterprise-wide Geo Data Infrastructure. Advantages for the organization are robust spatial data management, the widespread availability of spatial data and low threshold for data sharing. This makes the concept of GDI especially suitable for application in disaster management.

Setting up successful Geo Data Infrastructure for Disaster Management is one of the greatest challenges faced by the geospatial information and disaster management community. This requires considerable effort and commitment by the various users and providers of information as well as clarity on the rights and obligations of all the parties involved. To provide a reliable information infrastructure, the availability of different building blocks is crucial. To adopt successful Geo Data Infrastructures, the availability of geospatial information and geoservices technology and (wireless) communication infrastructure is crucial. There are risks in Geo Data Infrastructure services architectures using for DM as well. One of the main risks is the reliability and availability of the internet and mobile communication infrastructures. The communication network infrastructure is an essential component of such services oriented architectures with online data access at the point of supply. The alarm organizations in The Netherlands (police, fire fighting and ambulances) use their own specific communication network (C2000), but other government organizations that have a role in DM are depending on public or private network communication infrastructures. At the same time the (core data) supplier has the responsibility of offering online data services 7 days a week and 24 hours a day. This means that clear service level agreements are necessary.

The value of geo information for disaster management within V&W is evident: hind and fore-casting, scenario-analysis and not least operational management in case of events are well proven applications. For the near future the focus will be on trend analysis, development and implementation of new location based services, training, simulation and quality management of information and services.

REFERENCES

Annoni. A., L. Bernard, J. Douglas, J. Greenwood, I. Laiz, M. Lloyd, Z. Sabeur, A. Sassen, J. Serrano and T. Uslander 2005. Orchestra: developing a unified open architecture for risk management applications.

In: P. van Oosterom, S. Zlatanova en E. Fendel (eds.), Geo-Information for Disastermanagement, pp. 1–17. Springer-Verlag, Berlin-Heidelberg.

CEN 2005. Geographic information – Standards, specifications, technical reports and guideline, required to implement Spatial data Infrastructure. pr CEN/TR 00287030-1. 2005-09.

ISO 2005. 19119 – Geographic Information – Services.

Groot, R. en J. McLaughlin, eds. 2000. Geospatial data infrastructure. Concepts, cases en good practice. Oxford University Press.

OGC 2003. The OGC Reference Model. OGC 03-040. 2003-09-16. V0.1.3.

Uslander, T. 2005. Reference Model of the Orchestra Architecture (RM-OA).

Geospatial Information Technology for Emergency Response – Zlatanova & Li (eds)
© 2008 Taylor & Francis Group, London, ISBN 978-0-415-42247-5

A decision support system for the preventive evacuation of people in a dike-ring area

K.M. van Zuilekom
Centre for Transport Studies, Faculty of Engineering Technology, University of Twente, The Netherlands

M.H.P. Zuidgeest
International Institute for Geo-information Science and Earth Observation, The Netherlands

ABSTRACT: Optimal evacuation strategies for areas under threat of flooding have to be studied by adopting scientific knowledge from transport engineering. Theoretically speaking full control over departure times, destinations and routes of people will be ideal. In practice the level of control is restricted. The context of preventive evacuation of dike-ring areas is described in this paper. A framework for modelling evacuations is presented. Practical constraints determine the actual outcome of the evacuation. Therefore, several methods are presented, which can assist practitioners in designing evacuation plans. Practical, simple and feasible traffic management solutions are deduced from an optimal solution where there is complete and dynamic control of the destination and routes of evacuees. A static quick-scan is developed as an alternative for time-consuming dynamic model runs. A case study for the dike-ring area of Flevoland, The Netherlands is presented, where several of the methods are demonstrated.

1 INTRODUCTION

Water plays a key role in the safety of The Netherlands. Up to 65% of its area, also an area where most of the economic activities take place, is threatened by either sea or rivers. This is a situation that requires permanent attention. Moreover, the country has to cope with serious consequences of environmental changes. The climate is changing as a result of air pollution and the use of fossil energy. Temperatures are rising, rainfall increases in intensity and frequency, causing extreme fluctuations in river water levels as well as sea level rise. At the same time Dutch soil is sinking because of gas extraction. All these factors together make it increasingly difficult to protect the country against flooding, despite the existing dikes and hydraulic structures.

In view of these problems the Dutch government initiated the research project 'Floris' (Flood Risk and Safety in The Netherlands). This project has four tracks: (1) determining the probability of flooding risks; (2) the reliability of hydraulic structures; (3) the consequences of flooding; and (4) coping with uncertainties. As part of the third track, i.e. the consequences of flooding, a method for describing and analyzing the process of preventive evacuation of people and cattle from a dike-ring area has been developed. The method is implemented as a Decision Support System (DSS), named the Evacuation Calculator. Primarily, the DSS will be used for an ex-ante evaluation of the process of preventive evacuation for some of the 53 dike-ring areas. A dike-ring area is defined as an area that is protected against flooding by a system of dikes and hydraulic structures. Figure 1 shows an overview of dike-ring areas in The Netherlands. The key issue in this respect is to know the total time span needed for a preventive evacuation. An additional benefit of the DSS is that it can be helpful in the design of efficient strategies for organizing the evacuation.

This chapter discusses transport evacuation problems in general and those relate to the preventive evacuation of dike-ring areas in particular. Accordingly methods to calculate optimal evacuation strategies are presented. A practical method to derive a distribution and routing for evacuating trips

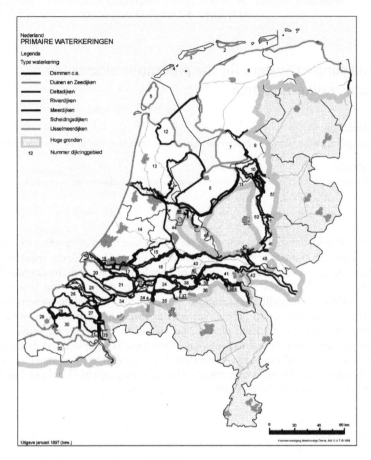

Figure 1.   The 53 dike-ring areas in The Netherlands.

to minimize the total evacuation time is also discussed. By doing so, the potential of a network to accommodate evacuating traffic can be determined. This information can then be used to design the traffic management scheme for the detailed area evacuation plan.

2   EVACUATION OF PEOPLE

2.1   *The process of evacuation*

There is increasing interest for modelling evacuations. Studies exist on risk analysis of nuclear power plants, see Sheffi (1982), and of hurricanes, see Hobeika (1985) or Urbiana (2001). There are many events that require an evacuation. These can be natural phenomena as extreme weather conditions (hurricanes, heavy rainfall, wildfires caused by drought), springtide and geological phenomena (earthquakes, volcanism, tsunami), but also human activities such as industrial accidents, failure of hydraulic structures, accidents with the transportation of hazardous goods and possible politically motivated attacks. The impact of extreme circumstances increases when human activities expand to vulnerable areas. The predictability of time, location, scale and outcome of such dangerous situation differ greatly per type of threat, the area and its characteristics.

In the Dutch situation high water levels of rivers can be predicted several days in advance. Although it is uncertain if and when a dike-ring area will be flooded, there might be enough reason to start preparations. The aim of precautionary action is to reduce the risk and the consequences of a flooding. A way to reduce the consequences is the preventive evacuation of the dike-ring area.

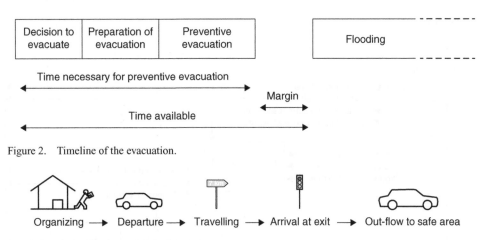

Figure 2. Timeline of the evacuation.

Organizing ⟶ Departure ⟶ Travelling ⟶ Arrival at exit ⟶ Out-flow to safe area

Figure 3. Stages of the evacuation as seen by the evacuee.

It is important that the preventive evacuation is well organized and efficient. A minimum amount of time is needed in order to avoid casualties. An accurate estimate of the evacuation time is helpful in determining the start of the evacuation. It implies that the decision can be taken as late as possible, at a moment when there is a better picture of the threat. A superfluous evacuation should be avoided. The crisis team needs to find a balance between an early decision (when the organization of an evacuation is not yet critical, casualties are unlikely, but an evacuation could be redundant) and a late decision (when the organization of an evacuation is critical and casualties could happen).

The whole process of a preventive evacuation can be outlined in a timeline illustrated in Figure 2.

The evacuee experiences the whole process in the following stages: (1) organization of the departure; (2) departure from home; (3) travelling in the direction of a safe area; (4) leaving the area under threat through one of its exits (in general the Dutch dike-ring areas have several roads to surrounding areas); and (5) continuation of the journey to the destination in the safe area.

In case of a preventive evacuation there is neither actual flooding nor an immediate threat. It is assumed that traffic behaviour is normal. The usual assumptions for modelling traffic behaviour are applicable. In case of an actual flooding, behaviour will change from the 'normal' state to flee or panic behaviour. In the latter situation the usual behavioural assumptions no longer suffice. In that case driver behavioural characteristics in panic conditions akin to those formulated in Hamdar & Mahmassani (2005) should be adopted in the modelling.

2.2 Decision-making in the event of (threat of) a disaster

In the case of a disaster or a pending disaster the authority of the local decision-makers are increased. Depending on the size of the area, a disaster-coordinator is assigned. This could be the mayor, a coordinating mayor (if several municipalities are involved), the province or the Ministry of the Interior. These authorities are entitled to take all the necessary actions within the boundaries of the constitution.

For the organization of a preventive evacuation it means that the enforcement of an evacuation is permitted. Enforcement of time of departure, choice of area exit and route to the exit is possible. The evacuation plan coordinates all the necessary activities.

Such an evacuation plan is usually part of a more general disaster plan. It is therefore only one of the aspects in the complete process of decision-making by the crisis team. Public authorities and functional organizations are all involved and together form the crisis team. Public authorities include municipalities, provinces and central government. Functional organizations include the polder-board, the department of water management and environmental affairs of the province and the directorate general of public works and water management. Prepared plans, as the evacuation plan, are input for the final plan. See for this process Figure 4, based on Martens (2002).

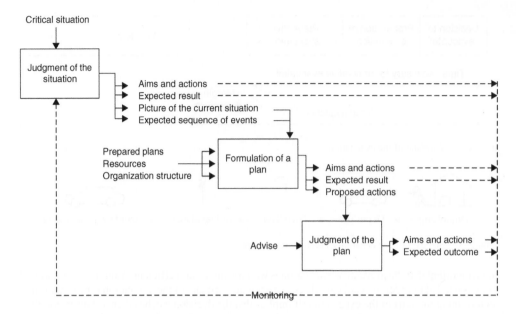

Figure 4.   The process of decision-making under threat of flooding, based on Martens (2002).

Under threat of a disaster like flooding, the crisis team will go through three important process steps: (1) judgment of the situation; (2) formulation of a plan; (3) judgment of the functioning of the chosen approach (Martens, 2002). The quality of the chosen approach depends, partly, on available resources, well prepared evacuation plans, and procedures and commitments about the organizational structure.

Possible actions of the crisis team are:

- influencing the time necessary for preparation (early warning systems, assistance, etc);
- influencing the demand for evacuation in space and time (evacuation in sectors, over time etc);
- providing or organising evacuation assistance to those with specific need (elderly, young, disabled, public transport captives, entrepreneurs with their stocks etc);
- influencing the routing and destination for those who are evacuating.

3   MODELLING OF A PREVENTIVE EVACUATION

3.1   *Literature review*

During a preventive evacuation there is a process of matching supply and demand as in a normal traffic situation, although the setting in case of a preventive evacuation is quite specific. Such process can be modelled by adopting a 'What if' or a 'How to' approach as discussed in Russo (2004), see Figure 5.

In a 'What if' approach, a scenario is modelled and the results are analyzed. The situation is then iteratively adjusted until no further improvements seem possible. The final result is interpreted and translated into an evacuation plan. The final result depends on the interpretation and adjustments of the modeller. The quality of the result is, by lack of a formal objective function, unclear. It is possible to use detailed and complex models in this approach. The modeller will focus on those aspects of the model that are important for the problem.

On the contrary, in a 'How to' approach, the result is determined by the objective function, the constraints and structure of the model. An optimal solution cannot be guaranteed in all cases (due to local optima for example). The objective function, constraints and solution techniques may limit

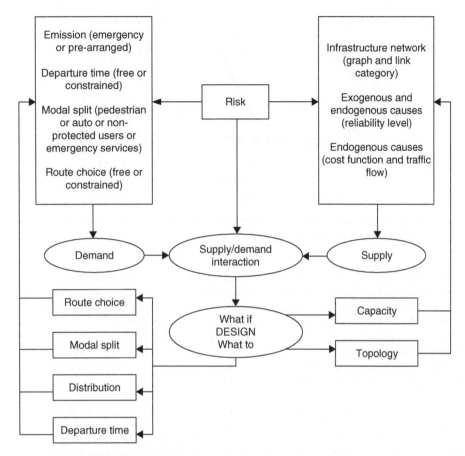

Figure 5. Global procedure for the design of an evacuation plan (Russo, 2004).

the complexity of the model. Moreover, the focus on an objective function can overshadow other difficult quantifiable objectives.

Current risks for natural and man-made disasters have recently resulted in several papers on evacuation modelling. Cova & Johnson (2003) describe a 'How to' procedure to study the trade-off between reduction in the number of crossing and weaving movements on intersections, and the extra travel distance needed in the network. They choose this approach knowing that in many urban networks, the intersections determine the capacity of the network. Elimination of crossing traffic and reduction of weaving is desired. In their approach, the trip distribution and route choice are the determining factors for the objective function.

In addition, Sheffi (1985) discusses the problem of a simultaneous trip distribution (destination choice) and traffic assignment (route choice) in general. Evacuation can be seen as such a type of problem. The objective is to find a distribution and assignment pattern, where a system-optimum is obtained. Every change in the final solution, distribution or assignment, will affect the objective function. It is proven that this problem can be solved using existing techniques by a modest adjustment of the network. This equilibrium assignment is used in an augmented network with one, spanning, destination (see paragraph 3.2.2.). Chiu (2004) uses this solution. Implicitly a perfect control of destination and route choice is assumed. The solution should be considered as a best-case solution that guides (sub) optimal solutions with more realistic constraints.

Sbyati & Mahmassani (2005) simulate evacuations for a selected group of people (those living in certain parts of the disaster area) by applying bi-level programming techniques. In an upper-level program the time-dependent 'shipments' of vehicles are calculated, resulting in recommendations

333

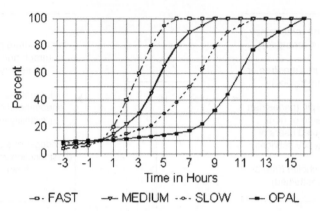

Figure 6. Alabama hurricane evacuation response rates. Estimates (fast, medium and slow) and actual response rate during hurricane Opal (Alabama, web).

on who should evacuate when, to what destination, and following what route. Whilst, in a lower-level program the actual route travel times, incorporating the impact of meeting non-evacuees, are calculated applying a dynamic assignment model. A system-optimal assignment is obtained that minimises the network clearance time, and obstruction for non-evacuees.

Apart from network modelling, several papers study routing in urban transport networks in emergency situations. For example in Zhu et al. (2006), several references are made to routing and dynamic emergency routing algorithms, focussing on the construction of 3D dynamic networks, data integration and network optimization in a GIS environment.

For the 'Floris' project the focus has been on a conservative and realistic estimate of the evacuation time for all people and cattle living in a dike-ring area together with a proposal for a traffic management strategy during the event of an evacuation. Efficiency of the method in terms of data handling and computing time is of importance as several larger dike-ring areas will be investigated. From this perspective it is preferred to use a 'How to' approach. The solution for this problem is found in a method that focuses on the trip distribution and that uses dynamic traffic assignment algorithms to monitor the system performance.

3.2 Formulation of the methods

3.2.1 Introduction
In the situation of a preventive evacuation there are many uncertainties: (1) the number of people and cattle in the dike-ring area during the threat of flooding; (2) the number of cars and trucks involved; (3) the time of departure; (4) the state of the network at the time of evacuation; and (5) the route choice. Discrepancy between expected and actual results is not uncommon. For example, during hurricane Opal in October 1995 in the USA, people left their homes about three hours later than the slowest estimate as depicted in Figure 6.

Therefore it is not functional to focus on a maximum level of model accuracy. It may be appropriate to use a model with limited complexity, but high flexibility. Conducting a sensitivity analysis may then prove to be very helpful in determining the critical processes.

An important requirement for the development of an evacuation strategy is that the strategy itself is feasible in case of a pending disaster, particularly in terms of people and resources.

Several approaches to determine an evacuation strategy can be developed. Theoretically, full control over the departure times, destinations and routes of all people will be most ideal. However, for more feasible and practical approaches decreasing control over destination and route choice processes by the authorities for individuals is required (for example reflected in different evacuation scenarios). Here we present several methods from sophisticated full control down to a quick-scan method with less control of the situation.

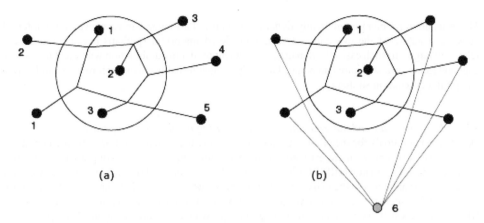

Figure 7.   Evacuation supernode network representation without (a) and with (b) a dummy destination (6) and dummy links connecting this destination node.

3.2.2   *An optimal evacuation strategy using a static distribution-assignment model*

If the total number of evacuating trips originating from each zone in the network in time is known, the question becomes one of how these trips are distributed among the various destination nodes and the routes they accordingly take. The determination of this choice problem is known as the joint distribution-equilibrium-assignment problem. Sheffi (1985) shows how one can solve this problem augmenting the network (the network within the evacuated area) with a dummy destination node and dummy links (with infinite capacity and zero impedance), hence constructing a supernode network as demonstrated in Figure 7. The origin-destination matrix with numbers of trip origins and their destinations will consist of one destination (the dummy destination) only.

This approach will create a solution where travel times are minimal for all users. There are multiple routes form a zone to the supernode.

3.2.3   *An optimal evacuation strategy using a dynamic distribution-assignment model*

In the previous strategy a static equilibrium traffic assignment algorithm is used in a supernode network. This algorithm is dealing with stationary or steady-state flows, the evacuation travel demand and road infrastructure capacity are supposed to be time-independent, meaning that the calculated link flows are the result of a constant demand. Hence, the resulting user-equilibrium flows are average flows for the period of consideration. Some important time-varying congestion effects such as queue spill-back effects are not taken into account because of the assumption of constant link flows and travel times. Congestion is a dynamic phenomenon, whereby its temporal character is not to be neglected.

A dynamic treatment of traffic assignment, again in an augmented network as in Figure 7, might therefore be more adequate. Here, congestion is assumed to have a temporal character, meaning that its build-up and dissolution plays an important role and, the history of the transportation system is taken into account (for example queue spill-back). Dynamic traffic assignment techniques therefore deal differently with:

1. Route choice: each evacuee follows a certain pre-defined route, for example based on a user-equilibrium assumption;
2. Dynamic network loading: resembling the physical propagation of all evacuating traffic in the network. Here, the behaviour of traffic at the nodes connecting the links within this network plays a very important role. Route choice at each intersection is determined using the relative outflow to the different down-stream directions. The percentages are obtained using a static traffic assignment.

The evacuation problem is inherently dynamic in nature. Particularly in cases where the internal network structure of the evacuation area is considered important (perhaps because the area is relatively large). In such cases the dynamic assignment model in combination with the dummy link augmented network structure can be used to analyse the internal network performance, and derive an optimal evacuation strategy.

3.2.4 *Designing sub-optimal evacuation strategies using a quick-scan approach*

One of the main goals for the 'Floris' project has already been mentioned namely, the derivation of a conservative and realistic estimate of the evacuation time of a dike-ring area. To do this, the Evacuation Calculator (HIS-EC) has been developed. For the HIS-EC several scenarios for directing the inhabitants to the exits of the dike-ring area were defined. Estimates of the evacuation time of such a scenario can be derived by the HIS-EC. The traffic management scenario of the HIS-EC reflects a 'How to' model that suggests a most efficient organization of the preventive evacuation within the capabilities and requirements of the crisis team. Important requirements, again, relate to the feasibility of the method in case of a pending disaster, the flexibility, and the usability. Therefore, control of routes is important. Aspects of the static and dynamic simultaneous distribution/assignment models, as discussed in paragraphs 3.2.2. and 3.2.3, should be altered and simplified. Hence, sub-optimal, but realistic estimates of the evacuation time remain. Besides the traffic management scenarios three other scenarios have been developed, mainly to compare the evacuation strategy and time with other business-as-usual, naïve, or local (sub-area) strategies. In the business-as-usual or reference scenario, inhabitants of the dike-ring area are free in their choice of the exit. They follow a 'standard' trip distribution. In the naïve or nearest exit scenario, the evacuees will go to their nearest exit of the dike-ring area, regardless of the capacity and actual use of this exit. In the sub-area scenarios inhabitants are directed to specific exits regardless of the optimal destination calculated, for reasons of resource efficiency (police, army etc.). In general, the scenarios 'reference' and 'nearest exit' represent worst-case situations. The 'traffic management' method will represent a best-case scenario. With the out-flow area method, it will be possible to manually adjust the result of the 'traffic management' method to practical and local constraints.

The capabilities for influencing behaviour are important constraints for an evacuation plan as well as for the development of an evacuation method.

In general the possibilities for manipulation of the evacuation process (control variables) are:

1. Time of departure: This can be influenced by means of information and direct orders;
2. Trip distribution: It is possible to instruct evacuees to go to a specific exit;
3. Mode of travel: In general, people with access to a car will use it. For people without own means of transport, the authorities will be responsible for supplying public transport;
4. Route choice: It will be possible to guide the traffic by means of information and instructions.

In general, it will not be possible to influence the number of evacuees.

The purpose of the evacuation plan, given the departure rate of people, is to distribute evacuees and to determine their routing in such a way that the evacuation time is short and the possibilities of the network are utilized efficiently, while the necessary traffic management can be realized.

Crossing streams of traffic, as demonstrated in Cova & Johnson (2003), are a source of waiting times and disturbances, and should be avoided. Diverging traffic, namely traffic coming from one area is going to different exits, introduces a choice problem for drivers and thus problems for the local traffic management authority. Therefore, diverging traffic is ideally not allowed. Instead, converging flows of traffic are assumed. In the actual implementation of an evacuation plan, exceptions can be introduced if the local situation dictates it. Introduction of one-way traffic through reverse laning or contra flow (Urbina & Wolshon, 2001) will increase capacity.

Using purely converging traffic flows, delays at crossings are negligible. The traffic volume at the exits will be the highest (as a result of the converging flows), and as a result of this it is likely

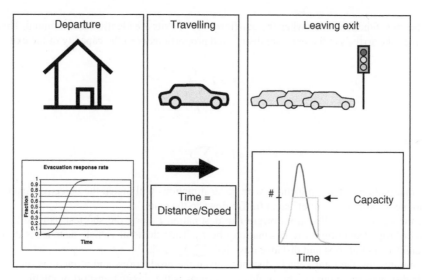

Figure 8.   Conceptual model of the evacuation.

that the exit will be the bottleneck. It is assumed that it is possible to assign a capacity to an exit, which is appropriate for the main route towards the exit.

These assumptions lead to the conceptual model in Figure 8.

The task is now to create a trip distribution, where the traffic flows are converging and the capacities of the exits are used optimally. When the initial trip distribution Table with total numbers of people travelling from origin to destination is determined, it is possible to create the distribution in time by using the evacuation response (or trip generation) rate and the arrival time at the exit, which is dependent on the state of the local network.

Let P_i in Person Car Units, PCU, be the trip production of evacuees from origin zone i, based on social-economic data and trip production rates. The evacuation time is determined by the last car leaving the dike-ring area. The objective function is defined by:

$$\min (\max (\textit{out-flow time}))$$ (1)

where it is assumed that:

1. All evacuees will leave the dike-ring area;
2. The traffic flow to the exits is efficient.

This objective function suits a preventive evacuation only. In situations where the urgency is high and casualties cannot be avoided this objective function is not valid anymore. Other objective functions, as minimising casualties within a timeframe, would then be more appropriate.

The out-flow time U_j [hours] of an exit zone j is determined by the arrivals at the exit A_j [PCU/hour] and the capacity C_j [PCU/hour] of the exit:

$$U_j = \frac{A_j}{C_j}$$ (2)

where:

$$\sum_i P_i = \sum_j A_j$$ (3)

337

The objective function is met when the out-flow time of exits are identical and minimal. This is the case when the arrivals at the exits are distributed proportionally to the capacity of the exits:

$$A_j = T \frac{C_j}{\sum_j C_j} \tag{4}$$

where:

$$T = \sum_i P_i \tag{5}$$

Every distribution of the productions with these attractions will match the objective function, but will not necessarily result in efficient traffic flows to the exits.

Let the distance travelled from origin i to destination j along the shortest path be z_{ij}, let the number of trips from origin i to destination j be T_{ij}. Then the total vehicle distance is defined as the weighted sum of trips and distance travelled: $\sum_i \sum_j z_{ij} T_{ij}$.

By minimizing the total vehicle distance, given the productions and attractions, unnecessary vehicle distances are avoided. This problem is known as the classic transport problem, or:

$$\min \left(\sum_i \sum_j z_{ij} T_{ij} \right) \tag{6}$$

subject to the following set of requirements:

$$\sum_j T_{ij} = P_i$$
$$\sum_i T_{ij} = A_j \tag{7}$$
$$T_{ij} \geq 0$$

This transport problem needs an initial origin-destination (OD) matrix. In the implementation of the HIS-EC the trips form i to j are distributed proportionally to the production and attraction levels, as follows:

$$T_{ij} = \frac{P_i \cdot A_j}{T} \tag{8}$$

When the resulting OD-matrix is loaded onto the network using an All-Or-Nothing traffic assignment, the traffic flows at intersections are in general convergent and only divergent if not avoidable. There are no crossing traffic flows at intersections.

Let F_{tk} be the fraction of evacuee category k, who will leave in time interval t. In the implementation, the user is free to define the fraction for each F_{tk} or by using a logistic function with parameters a_k and b_k:

$$F_{tk} = \frac{1}{1 + \exp(a_k (t - b_k))} - \frac{1}{1 + \exp(a_k ((t - 1) - b_k))} \tag{9}$$

The OD-matrix per time interval is now determined by:

$$T_{ijt} = \sum_k T_{ijk} F_{tk} = \sum_k T_{ij} \frac{P_{ik}}{P_i} F_{tk} \tag{10}$$

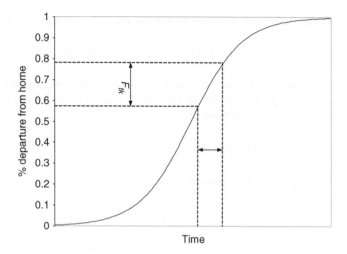

Figure 9.   Use of the logistic function for F_{tk}.

After departure from home i the vehicles will arrive after r_{ij} time intervals at the exit j. The travel time is dependent on the distance from origin to exit, z_{ij} [km], and the average speed, $\bar{v}$ [km/h], in the dike-ring area and the number of time intervals in an hour, I:

$$r_{ij} = \text{int}\left(\frac{z_{ij}I}{\bar{v}}\right) + 1 \tag{11}$$

Now it is possible to calculate the number of arrivals, ARR_{jt} [PCU] at exit j for a time interval t:

$$ARR_{jtk} = \sum_{i} T_{ijk}F_{t-r_{ij},k} \tag{12}$$

thus:

$$ARR_{jt} = \sum_{k} ARR_{jtk} \tag{13}$$

The vehicles that arrive at the exit will leave the dike-ring area for as far the capacity of the exit allows. In a time interval the number of vehicles that leave the exit are limited to:

$$C_j^{\max,t} = \frac{C_j}{I} \tag{14}$$

The out-flow DEP_{jt} at exit j for time interval t depends on the available traffic (delayed and just arrived) and the capacity of the exit:

$$DEP_{jt} = \min\left(ARR_{jt} + DEL_{j,t-1}, \frac{C_j}{I}\right) \tag{15}$$

$DEP_{j,t-1}$ are those vehicles that couldn't pass in earlier time interval(s) using exit j. For the first time interval there are no delayed vehicles, that is:

$$DEL_{j,0} = 0 \tag{16}$$

Vehicles that can't pass in time interval t will be delayed and will use a later time interval:

$$DEL_{jt} = \max\left(ARR_{jt} + DEL_{j,t-1} - DEP_{jt}, 0\right) \tag{17}$$

The number of departing vehicles for category k is calculated by assuming that at arrival the categories are spread homogeneously over all vehicles. A fraction of all delayed and just arrived vehicles will eventually leave the dike-ring area:

$$DEPRATIO_{jt} = \frac{DEP_{jt}}{ARR_{jt} + DEL_{j,t-1}} \tag{18}$$

The departures for category k in time interval t are now:

$$DEP_{jkt} = DEPRATIO_{jt}\left(ARR_{jkt} + DEL_{jk,t-1}\right) \tag{19}$$

The number of delayed vehicles for category k in time interval t is:

$$DEL_{jkt} = ARR_{jkt} + DEL_{jk,t-1} - DEP_{jkt} \tag{20}$$

Central unit of measurement in the model is in person-car-units PCU. The number of evacuees in PCU Q_k can be transformed into the number of persons (or cattle) N_k by using the occupancy μ_k and the PCU-value of the used type of vehicle PCU_k:

$$N_k = Q_k \frac{\mu_k}{PCU_k} \tag{21}$$

For each exit and all exits together the resulting output of the model is:

- Arrivals and out-flow [PCU/time interval].
- Arrivals [PCU/time interval] for all categories together and for each category.
- Out-flow [PCU/time interval] for all categories together and for each category.
- Out-flow [number/time interval] for all categories together and for each category.

The resulting OD-matrix (total evacuation or per time interval) is available for (dynamic) traffic assignment, as will be demonstrated in paragraph 4.

The method has some relationship to the first stages of the classic unimodal traffic model:

- The trip end calculation. Where trip production is determined by the social-economic data of the different zones. While in the calculation for the attractions, the capacities of the exits are leading.
- The trip distribution. Instead of using a gravity model, for this method the trip distribution is reflecting the minimization of total vehicle distance travelled.

The results (both graphs and OD-matrix) are available in seconds. An assignment of the OD-matrix gives a better insight in the resulting traffic flows, and can be used for calibration purposes.

The 'traffic management' scenario uses the 'How to' approach. Other scenarios that have been implemented in the Evacuation Calculator, i.e. 'reference/business-as-usual', 'naïve/nearest exit' and 'out-flow areas', use a 'What if' approach, and will be summarized below.

In the business-as-usual scenario the main differences with 'traffic management' scenario are:

- Trip end calculation, where the user is free in setting the (relative) attraction of the exits. In general these will be determined by the traffic volumes on the exits on 'normal' working days;
- Trip distribution, where the distribution is identical to the initial distribution of the 'traffic management' scenario. There is no minimization of total vehicle distance.

Compared to the 'traffic management' scenario, the naïve scenario uses a:

- Trip end calculation, where the attractions are not explicitly chosen, but are a result of the distribution;
- Trip distribution, where all productions of an origin are allocated to the nearest exit,
- which is:

$$t_{ij} = P_i \qquad (22)$$

for all j where:

$$z_{ij} = \min(z_{ij}) \quad \forall j$$
$$t_{ij} = 0 \qquad (23)$$

for all other j.

Finally, as the 'traffic management' scenario will result in sets of more or less independent out-flow areas, which use one or more exits, it is not likely that this out-flow areas can be transferred to an evacuation plan without change. Geographic, jurisdictional and other local constraints will make adjustments necessary, which are accommodated in the out-flow scenarios. With the out-flow area scenario, the user is free of defining sub-areas with one or more exits. Within every sub area the trip end calculation and trip distribution is solved using the 'traffic management' scenario method, implying that (1) attractions of the area's is proportional to the exit capacities; and (2) minimizing the total vehicle distance travelled given the productions and attractions.

The minimization of the total vehicle distance for all out flow areas is solved by manipulating the distance matrix z_{ij} before the actual minimization takes place. Those relations that are not part of an out-flow area are flagged with an extreme large value. The further procedures are identical to the 'traffic management' scenario. The main difference with the 'traffic management' scenario is the manipulation of the distance matrix z_{ij} in the trip distribution phase:

$$z_{ij} = \infty \forall i \in D_m \wedge j \notin D_m$$
$$z_{ij} = \infty \forall j \in D_m \wedge i \notin D_m \qquad (24)$$

where D_m is the set of i and j defining sub-area m.

3.2.5 Designing sub-optimal evacuation strategies using a quick-scan approach in combination with a dynamic assignment algorithm

In the previous discussion, the out-flow structure produced by the HIS-EC, is assumed to depend on:

1. The time-varying trip generation profile, indicating the number of departing evacuees from their origin zones per time period;
2. The travel time. This is the time necessary to reach the exit of the dike-ring area, assuming an average travel speed and where the location of the exit is dependent on the trip distribution scenario chosen;
3. The delay near the exits.

This implies that delays can only occur near the area exits. This only happens when the number of arriving vehicles at the exits is higher than the number of vehicles leaving the dike-ring area. This may be a valid assumption as long as dike-ring areas are relatively small. In cases where the internal network structure is causing more delays than the external network structure the HIS-EC's assumptions are not valid anymore.

Traffic performance on network structures can be studied using traffic assignment. Such traffic assignment algorithms predicts route choice, network flows, link travel times and route travel costs for a given transport network with a given travel demand. Dynamic assignment models, as discussed in paragraphs 3.2.2. and 3.2.3., also take time-dependencies explicitly into account. Given the inherent dynamic nature of the evacuation problem a dynamic assignment model (in our model using OmniTrans - MaDAM) can be used in combination with quick-scan sub-optimal HIS-EC approach, to study and fine-tune parameters related to the network and network performance.

3.3 *Classification of the modelling approaches*

The presented modelling approaches each have their specific qualities. These specific qualities together with some practical considerations are summarized. A distinction can be made between the type of distribution, the modelling environment, assignment and routing as the key elements in classification of the approaches.

Distribution: An important issue for the authorities is to what extend people can or should be directed to the several exits. Conflicts can arise between the preferred exits of the evacuees and the feasibility of the destination to accommodate these evacuees. At the same time there are limitations for the authorities to direct the evacuees.

Modelling environment: The scale of the dike-ring area, the required level of detailing and accuracy will determine the modelling environment (microscopic/macroscopic, static/dynamic). In the Dutch situation the large dike-ring areas have a scale where microscopic modelling is not feasible. Macroscopic dynamic models combine the advantages of microscopic models (using the dynamic interference of cars) with the ease of macroscopic static models (limited data need and relative fast computation).

Assignment: The nature of a preventive evacuation makes a dynamic handling of time the preferred one. Dynamic assignment gives the best insight into the process and the best estimate of the evacuation time. However, in the process of designing an evacuation plan there is a need to investigate a series of scenarios. Here static assignment will give enough insight to select the most relevant scenarios for further, in-depth (dynamic) analysis.

Routing: Routing will be an important issue in an evacuation plan. The main concern is to avoid unnecessary disturbances and at the same time to best use the capabilities of the network. Here the challenge is to use the limited resources of the authorities and the autonomy of individual drivers' best.

With macroscopic dynamic assignment methods it is possible to determine the outflow in time by analysing the assignment. With a static assignment this is not possible.

By nature of the optimal simultaneous distribution and assignment approach only multi-routing is valid.

For the research project 'Floris' the classifications 4, 6 and 7 were used. For illustration of the approach classifications 1 and 3 have been used in the next section as well.

The HIS-EC has been used in the Dutch-German project FLIWAS (FLood WArning System). A dike-ring is analysed in a workshop for rescue workers e.g. police and fire brigade. Here it is possible to generate evacuation scenarios in full interaction with the rescue worker scenarios, calculate first results (classifications 7 and 6) and discuss the outcomes of the scenarios within an afternoon. Participants gained insight in the scale and complexity of the operation. Most promising scenarios were selected for more detailed analysis afterwards (classifications 5, 4 and 3).

4 CASE STUDY OF FLEVOLAND

Flevoland (Figure 1) is one of the larger dike-rings in The Netherlands. Flevoland is about 50 km long and 20 km wide and has a population of about 258 thousand inhabitants (102 thousand house-holds). The area is not very densely populated, at least by Dutch standards. About 67% of the population live in the two larger cities Almere (48%) and Lelystad (19%).

4.1 *Description of the dike-ring area*

For this case-study just 3 of 8 exits are considered, mainly for demonstration purposes. Limiting the number of exits will make congestion occur (the area normally has a relatively large exit capacity).

The capacities of the three exits are set to 6600 [PCU/hour] for the south-west exit, 4300 [PCU/hour] for the north-east exit and 1500 [PCU/hour] for the south east exit. This leads to a total capacity of 12400 [PCU/hour].

4.1.1 *Trip end calculation*

We define two categories: (1) person cars (2) people who need assistance or public transport to leave the dike-ring. It is assumed that:

1. All passenger cars will leave the dike-ring. Passenger cars have a PCU of 1.0.
2. 10% of the inhabitants in the age of 35 to 64 and 50% of those in the age of 65 and more will use a bus. Busses will transport on average 20 passengers. A bus has a PCU-value of 2.0.

This will result in 93630 PCU for the category passenger cars and 2079 PCU for the category of bus users, 95708 PCU in total. The evacuation will therefore need at least seven and a half hours (95708/12400).

4.1.2 *Departure rates*

For each category the departure rate is defined with a logistic curve as in Figure 9. The characteristics of these curves are: 50% will have been departed in 7 hours, 90% in 10 hours.

During the peak hours about 25000 PCU will enter the network. This exceeds the capacity of the exits about two times (25000/12400).

4.2 *Methods used*

Some of the methods that have been discussed in paragraph 3 will be demonstrated here. We will start with the easier and sub-optimal methods and end with the optimal approaches:

- First: a HIS-EC Quick Scan using the traffic management scenario to get some feeling for the scale of the evacuation time (classification 7 in Table 1).
- Second: a macroscopic single route static assignment. The OD-matrix is created by the HIS-EC using the traffic management scenario (classification 6 in Table 1). This will give an impression of the use of the network.
- Third: a macroscopic dynamic single-routing assignment of the HIS-EC OD-matrix using traffic management (classification 4 in Table 1). With this assignment result it is possible to predict of the evacuation in time more precise.
- Fourth: macroscopic dynamic assignment using multi-routing assignment of the HIS-EC OD-matrix (classification 3 in Table 1). Comparison of this run with previous run will give insight where it is useful to bypass bottlenecks.
- Fifth: optimal simultaneous macroscopic dynamic assignment (classification 1 in Table 1). As previous run, but bottle-necks will be bypassed. Dynamic full use of the capabilities of the network.

For this case OmniTrans is used for the dynamic and static assignment. With the MAcroscopic Dynamic Assignment Method (MADAM) of OmniTrans routing is defined in terms of turn ratios created in a pre-processing stage using an incremental assignment. This means that there is no explicit routing. It also means that during the actual dynamic assignment the direction of the flows are determined by the turn ratios of the static incremental pre-processing stage. Ideally the turn rates (or routes) are determined during the dynamic assignment process. As a result of this the fifth method will be an approximation of the optimal solution.

Table 1. Classification of the modelling approaches for preventive evacuation.

Distribution	Modelling environment	Assignment	Routing	Classification
Optimal simultaneous distribution and assignment	Microscopic or macroscopic	Dynamic	Multi-Routing	1. Full use of the capabilities of the network. Assumes full control of routing during the whole evacuation for all vehicles
	Macroscopic	Static	Multi-Routing	2. As in 1 for the busiest hour as a fast approximation of 1. Illustrates the potential improvement of a single route assignment
Sub-optimal distribution strategies generated by the HIS-EC	Microscopic or macroscopic	Dynamic	Multi-Routing	3. Full use of the capabilities of the network in a situation where the authorities assign areas to particular exits
			Single-Routing	4. As in 3 but limits to the preferred routes (appointed by the authorities) without use of bypasses for particular bottlenecks
	Macroscopic	Static	Multi-Routing	5. As in 3 for the busiest hour as a fast approximation. Illustrates the potential improvement of a single route assignment
			Single-Routing	6. Fast approximation of 4. Limits to the preferred routs (appointed by the authorities) without use of bypasses for particular bottlenecks
	Quick-Scan	Static	Single-Routing	7. Very fast approximation of evacuation clearance times. Particularly useful for sensitivity analysis

4.2.1 *HIS-EC Quick Scan using the traffic management scenario*

For the HIS-EC Quick Scan the average speed is set to 25 km/h. The formal capacities of the exits are not adjusted[1] . With these parameters the evacuation time is estimated at 26 hours by the HIS-EC.

4.2.2 *Macroscopic single route static assignment with HIS-EC traffic management*

Purpose of the single route static assignment is tot get an impression of the use of the network. The result shows that almost all zones are assigned to a specific exit. Some zones are assigned to two or more exits. Crossing traffic at intersections is avoided and in general all traffic flows are converging. These are a result of the traffic management method of the HIS-EC. The static assignment gives information to determine the evacuation routes and out-flow area's. Interpretation of this result will give guidelines for traffic management at crossings, evacuation routes and the assignment of area's to exits.

---

[1] In case heavy congestion and upstream bottlenecks are expected the average speed and capacities can be adjusted.

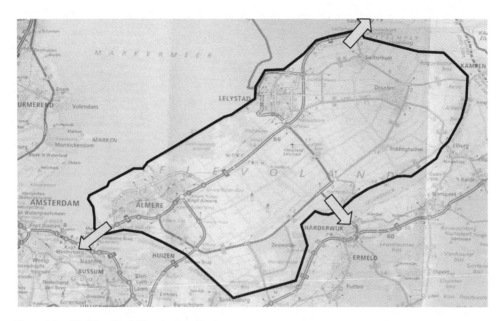

Figure 10.   The Flevoland dike-ring area.

Figure 11.   Static assignment of the HIS-EC Traffic management scenario.

4.2.3 *Macroscopic dynamic single-route assignment with HIS-EC traffic management*

With the previous methods it is assumed that the exits are the bottlenecks in the network (with a known capacity). This is not necessarily the case. Bottlenecks could occur some where upstream in the network and as a result of that the capacities of the exits could be misjudged. Macroscopic dynamic assignment overcomes these problems with static assignment.

With the macroscopic dynamic single-route assignment the dike-ring is empty in 24 hours. The results show that the use of the exits is unbalanced. After 14 hours the south-east exit has processed all assigned evacuees, while the other exits still are heavily loaded. Not all available capacity is used during the whole evacuation process.

4.2.4 *Macroscopic dynamic multi-routing assignment with HIS-EC traffic management*

The previous run restricted to single-routing. This run with multi-routing shows the potential improvement by avoiding bottlenecks. In the case of Flevoland the dike-ring will be fully evacuated

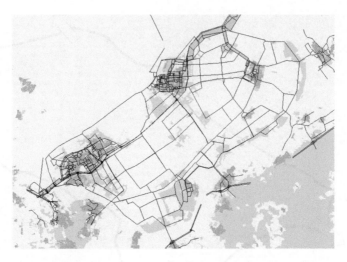

Figure 12. Macroscopic dynamic single-route assignment with HIS-EC traffic management. After 14 hours the south-east exit is hardly used. The south-west exit is used below capacity.

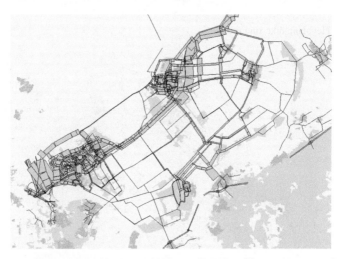

Figure 13. Macroscopic dynamic multi-route assignment with HIS-traffic management at the 9th and busiest hour of the evacuation.

in 20 hours when using muli-routing, while with single-routing at 20 hours still 6% have to leave the dike-ring.

4.2.5 *Optimal simultaneous macroscopic dynamic distribution and assignment*
With the used macroscopic assignment procedure, MADAM, it is only possible to approximate the optimal solution (see paragraph 4.2). This approximation is slower than the actual optimal solution.

With this approximation the dike-ring is evacuated in 17 hours. Bottlenecks in the network are avoided by using alternative routes (if possible). Crossing traffic flows are not avoided systematically (as illustrated in Figure 15 for the traffic flows in the 9th hour).

The situation assumes full knowledge by the evacuees on the conditions of the network during the event of the evacuation.

Optimal multi-routing approach is significant better in the last hours of the evacuation when the balance over the several exits is critical. There is quite a difference in vehicle distance travelled

346

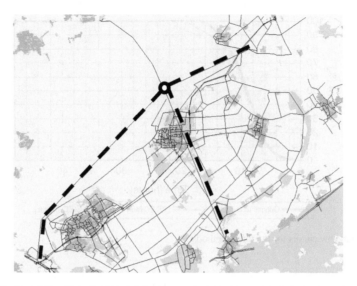

Figure 14.   Super-node network for Flevoland.

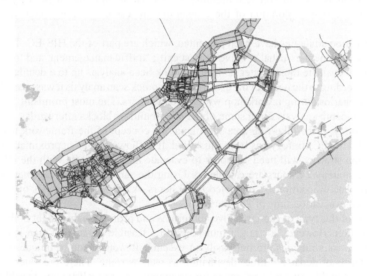

Figure 15.   Traffic load at the 9th and busiest hour of the evacuation using the approximation of the optimal simultaneous dynamic distribution and assignment.

between these two runs. In the HIS-EC multi-routing run 2.4E6 vehicle kilometres are travelled, while in the optimal multi-routing run 3.36E6 vehicle kilometres are travelled.

5   CONCLUSIONS

Efficient evacuation planning is important for protecting lives in case a dike-ring area has to be evacuated under threat of flooding. In the event of a preventive evacuation it is possible to perform detailed analysis of the situation on forehand. This analysis acts as a basis for the development of the evacuation planning.

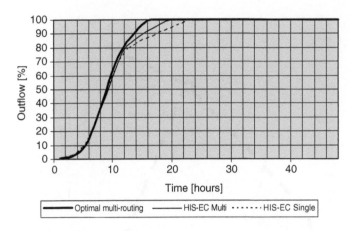

Figure 16. Out-flow curves for the three methods.

In this chapter a framework is presented for modelling such preventive evacuation. This framework is depicted as a set of coherent building blocks. These building blocks vary from a quick scan analysis for first exploration of the evacuation process up to more detailed and accurate approaches.

Four evacuation strategies have been presented, which are part of the HIS-EC. These strategies are: 'reference/business-as-usual'; 'naïve/nearest exit'; 'traffic management' and 'out-flow areas'. It is possible to evaluate these strategies from a Quick Scan analysis up to a detailed and accurate modelling according to the needs of the user. Using the quick scan analysis it was possible to develop evacuation scenarios during a workshop with rescue workers. The most promising scenarios were selected for in-depth analysis using some of the other building blocks afterwards.

In a case study of the dike-ring area Flevoland, the concept of the framework and the 'traffic management' HIS-EC strategy has been illustrated. It is shown that an approximation of the optimal evacuation strategy will need 17 hours to evacuate all inhabitants. With the HIS-EC traffic management strategy 24 hours are calculated. Using the same traffic management strategy, but multi-routing, the evacuation is completed in 20 hours. A quick scan analysis of the evacuation predicts a total evacuation time of 26 hours.

The results of this type of exercises give authorities and professionals insight in the dimensions of the evacuation. However, a straight translation to a feasible evacuation plan is impossible as the needed control strategies are not easily to execute. It would require unique orders to each evacuee in terms of departure time, destination as well as route. Therefore, such ideal evacuation strategies should serve as a reference for designing a more realistic and feasible evacuation approach. From the point of view of the crisis-manager the central question will be: 'Which management will result in the fastest evacuation within the capabilities of the authorities, the available resources and the network?'. The presented models don't give straightforward answers to these questions. Instead, these models are a stereotype reflection of reality. The models will help the crisis manager to search the crucial strengths and weakness of the complex combination of demand (the evacuees) and supply (the network and its management) in the dike-ring area in case of an evacuation. Next logical step is to include the available resources and its management in these models.

An evacuation is a process with many uncertainties due to unforeseeable events (like accidents) and difficulty in controlling the behaviour of people. Therefore, every evacuation will be an unique, complex, and difficult to predict event. Additional, ad-hoc decisions always have to be made. Further development of the HIS-EC should be aimed at supporting the crisis manager in maximising the effects of limited resources, such as personnel, equipment and communication facilities.

REFERENCES

Boetes, E., Brouwers, N., Martens, S., Miedema, B. & Vemde, R. van 2002. Evacuatie bij hoogwater: informatie voor een verantwoord besluit tot evacuatie, Thesis MCDM (Master of Crisis and Disaster Management), Netherlands Institute for Fire and Disaster Management (NIBRA) & The Netherlands School of Government (NSOB).

Chiu, Y.C. 2004. Traffic Scheduling Simulation and Assignment for Area-Wide Evacuation. Proceedings 7th Annual IEEE Conference on Intelligent Transportation Systems (ITSC 2004). Washington D.C., USA.

Cova, J.T. & Johnson, J.P. 2003. A network flow model for lane-based evacuation routing. Transportation Research Part A, 37(A): 579–604.

Hamdar, S.H. & Mahmassani, H.S. 2005. Modelling driver behavior under extreme conditions. Proceedings 12th World Congress on ITS. San Francisco, 6–10 November 2005.

Hobeika, A.G. & Jamei, B. 1985. MASSVAC: A model for calculating evacuation times under natural disaster. Emergency Planning, Simulation Series 15/23.

Martens, S. 2002. Wat maakt een operationeel leider competent; Orientatie op de competenties van operationeel leiders, Thesis MCDM (Master of Crisis and Disaster Management), Netherlands Institute for Fire and Disaster Management (NIBRA) & The Netherlands School of Government (NSOB).

Russo, F. & Vitetta, A. 2004. Models for evacuation analysis of an urban transportation system in emergency conditions. Proceedings 10th World Conference on Transport Research (WCTR 2004), Istanbul, Turkey.

Sbayti, H. & Mahmassani, H.S. 2005. Faster to wait? Development and evaluation of staged network evacuation strategies. Proceedings 12th World Congress on ITS. San Francisco, 6–10 November 2005.

Sheffi, Y., Mahmassani, H. & Powell, W.B. 1982. A Transportation Network Evacuation Model. Transportation Research. 16A(3): 209–218.

Sheffi, Y. 1985. Urban Transportation Networks: Equilibrium Analysis with Mathematical Programming Methods. Englewood Cliffs, New Jersey: Prentice-Hall Inc.

Urbina, E & Wolshon, B. 2001. National Review of Hurricane Evacuation plans and policies: a comparison and contrast of state practices. Transportation Research, 37(A): 257–275.

Zhu, Q., Li, Y., Tor, Y.K. & Li, J. 2006. Multi-dimensional and dynamic vehicle emergency routing algorithm based on 3D GIS. In: Zlatanova, & Li (eds.) Geo-Information technology for emergency response, ISPRS Book Series, Taylor & Francis.

Geospatial Information Technology for Emergency Response – Zlatanova & Li (eds)
© 2008 Taylor & Francis Group, London, ISBN 978-0-415-42247-5

GIS technology and applications for the fire services

R. Johnson
ESRI, Redlands, California, USA

ABSTRACT: The mission of the fire service is to protect life, property, and natural resources from fire and other emergencies. With increasing demands, the fire service must utilize the best tools, techniques, and training methods to meet public expectations. Risk management, preparedness, and mitigation have taken on new importance with challenges facing the fire service today. Effective response cannot be continually achieved without adequate planning and preparedness. One of the emerging tools that is helping the fire service optimize its emergency services delivery is geographic information system (GIS) technology. GIS supports planning, preparedness, mitigation, response, and incident management. GIS extends the capability of maps – intelligent, interactive maps with access to all types of information, analysis, and data. More important, GIS provides the required information when, where, and how it is needed. This chapter will examine how GIS technology is helping the fire service meet the needs of the community more efficiently than ever before.

1 INTRODUCTION

When a fire occurs, any delay of responding fire companies can make the difference between the rescue of occupants versus serious injury or death. The critical time between fire containment and flashover can be measured in seconds. Fast access to critical information is essential. Tools that help firefighters pinpoint the emergency call location, assess the potential consequences, and determine the most efficient strategy will minimize property damage and better protect the safety of occupants and fire service personnel.

Historically, first responders have relied on experience, good equipment, communication, and teamwork to achieve successful emergency response. However, with all of the challenges confronting emergency crews today, effective response requires good planning, risk management, comprehensive training, and intelligent deployment through preparation. Geographic information system (GIS) technology has become a powerful tool for improving all aspects of fire service delivery systems.

As populations and building development increase, the role of the fire service becomes more demanding and complex. As never before, fire departments are being called upon to deliver services with greater efficiency and economy. Citizen tax-reduction initiatives, burgeoning needs for different kinds of local government services, and a host of other factors have brought new demands to the desks of fire chiefs – most notably, the demands to "do more with less" and to do it "better, faster, and cheaper." GIS technology brings additional power to the process whereby hazards are evaluated, service demands are analyzed, and resources deployed. In addition, GIS contributes to the speed with which emergency responders are able to locate, respond, size-up, and deploy to an emergency.

Fire officers are continually collecting data from a wide variety of sources to better perform their jobs. This data may come in many incompatible forms. By utilizing GIS, data can be quickly analyzed and displayed in different arrangements to allow patterns and trends to emerge. Fast access to needed data can save time, money, and lives.

2 WHAT IS GIS?

A GIS is a computer-based technology that links geographic information (where things are) with descriptive information (what things are like). GIS is used to capture, display, and analyze data spatially. GIS combines layers of information about a place to give users a better understanding of that place. Unlike a flat paper map, a GIS-generated map can present many layers of different information that provide a unique way of thinking about a geographic space. By linking maps to databases, GIS enables users to visualize, manipulate, analyze, and display spatial data. GIS technology can create cost-effective and accurate solutions in an expanding range of applications. GIS displays geographic data as "map layers." Some GIS map layers fire departments use include

- Streets
- Parcels
- Fire hydrants
- Utility networks
- Topography
- Lakes and rivers
- Commercial and government buildings
- Fire station locations
- Police station locations
- Hospital locations
- School locations
- Satellite or aerial imagery
- Historical incident or emergency call locations
- Fire demand zones
- Public occupancies

These map layers can be selected and displayed (overlaid) by a GIS user. These map layers are linked to data tables that contain detailed information about the geographic features being displayed. For example, a parcel layer may contain various attribute information such as

- Owner
- Value
- Zoning class
- ZIP Code
- Address

A map layer of historical incidents (represented by points or icons) consists of attribute information for each incident, which may include

- Incident type
- Incident cause
- Date of incident
- Time of incident report
- Units that responded
- Unit arrival times

This attribute data allows GIS users to perform complex analyses. GIS makes map displays interactive and intelligent. For example, a GIS user could begin to analyze and display incident trends. A spatial query could request incident locations by cause, time of day, specific geographic locations, and so forth. GIS searches the data tables, gathers the data that matches the spatial request, and displays it on the map. Incident trend analysis can be done quickly, displayed logically, and understood easily. These types of analyses provide decision support information for issues related to fire prevention, staffing requirements, and apparatus placement/deployment.

Incidents during March

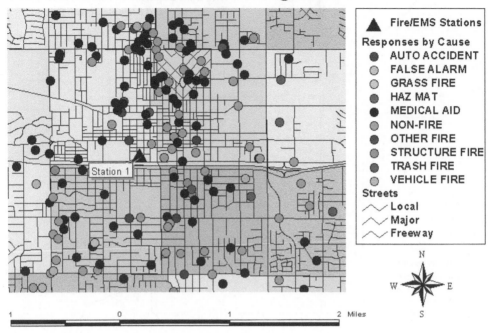

Figure 1. Historical incidents accessed from the records management system displayed in GIS.

2.1 Using GIS for complex analysis

2.1.1 Response time modeling

Utilizing a fire station layer and a street layer, response time analysis can be performed. A street layer is often represented in GIS as a series of lines that intersect on the map, creating a GIS network. Each street line segment between intersections can contain the road type, distance, and travel speeds (miles or kilometers per hour) permitted in the underlying data table within GIS. This allows users to identify a station location, specify a travel time, and run a "network analysis". The result will be illustrated by an irregular polygon around the station that closely approximates where a fire apparatus could travel in any direction for the specified time. This type of analysis could be performed simultaneously on all the department's stations to analyze gaps in coverage, run orders, and so forth.

2.1.2 Incident trend modeling

Incident trend analysis is a common practice by fire departments. With GIS, incident trend analysis can be performed quickly with all of the relevant information. GIS can access and "code" (place a point on the map) historical incidents. This capability can be refined by conducting a spatial query to the records management database that specifies the type of incident, time range, or specific geographic area. For example, a GIS user could request to see arson fires that occur between the hours of 1:00 a.m. and 5:00 a.m. on Saturdays in fire districts 1 and 2. GIS will interrogate the records database and place points on the map that meet this request. The GIS user can access all the information concerning each incident by simply clicking on the incident point. GIS can add additional information by displaying the demographics for each of the two fire districts identified in the spatial request.

2.1.3 Event modeling

Event modeling allows the user to identify a location (factory, hazardous material location, rail track intersection, etc.), place a point on the map, and run a selected model. Models could be

353

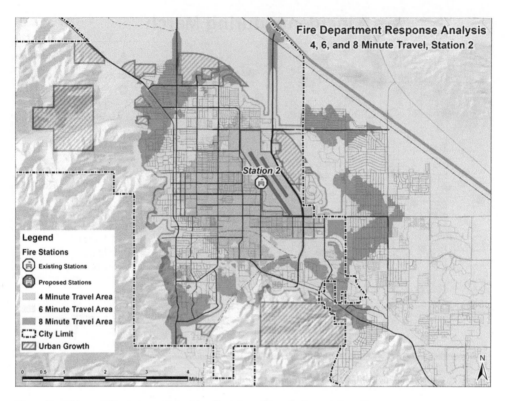

Figure 2.   This map illustrates various drive times by color code from station sites.

anything from plume dispersion to an explosion. GIS can display the model on the map; delineate various levels of danger; identify exclusion zones, infrastructure damage, and population effects. In addition, road closure requirements, safe routes into and out of the hazardous area, and appropriate hospitals that could quickly service the emergency can be displayed along with other informa- tion for emergency decision support. Modeling can be used for analyzing vulnerabilities, preplan development, training, or communicating with the public and policy makers.

2.1.4   *GIS for centralizing data*

GIS can become a central repository for a variety of nonspatial data. Nonspatial data, such as floor plans (computer-aided design drawings), photographs, preplans, and other documents, can be linked to features on the map (documents or photos that pertain to a particular building location or other actual feature location). This information, when configured with mobile computers, can provide first responders with information essential for size-up for deployment.

Historically, GIS required having software and GIS data on a computer with a trained GIS technician. Today, newer GIS application software has evolved and can operate effectively in a networked or Web-based environment. GIS software can reside on a Web server, the GIS data can be in several different locations or other Web servers, and users can access the GIS application through a Web browser. Web-based GIS services make it possible to deploy regional GIS applications and dramatically reduce costs and maintenance.

GIS is rapidly becoming a standard technology for many industries. The remainder of this chapter will examine how GIS can be and is being used in all aspects of fire and emergency services.

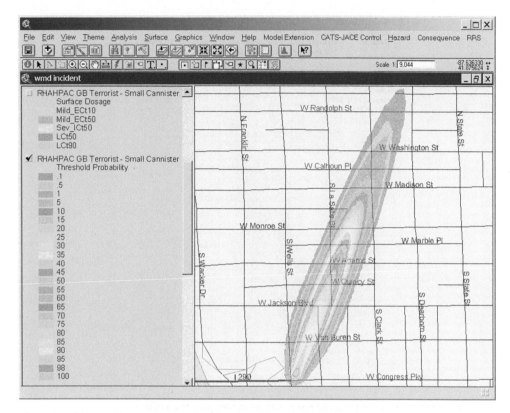

Figure 3.   This map illustrates a Sarin Gas plume model and color-coded areas of concentration.

3 COMPUTER-AIDED DISPATCH

Dispatchers have an important responsibility to process emergency calls and send the appropriate public safety resources to the emergency location based on the type and urgency of the incident. GIS is an important component of the dispatch system. Dispatch systems or computer-aided dispatch (CAD) systems typically contain a file called the Master Street Address Guide, or MSAG. This file contains street address information and service areas for the jurisdiction that the dispatch center services. As emergency calls are received, they may be accompanied with address information from the telephone company's emergency phone record database. This address is entered or electronically transferred to the CAD system, which compares it to the MSAG. When the address is matched, the specific service area is also identified with the specific units that should be dispatched to the emergency. If the telephone company does not provide a digital address with the call, dispatchers must obtain it from the caller and type it into the system. Many computer-aided dispatch systems have begun to integrate GIS technology. GIS takes the address and automatically geocodes the incident and displays it on a map. There are several benefits of having the incident displayed on a GIS map. New calls reporting the emergency may have different addresses but are reporting the same incident that was previously recorded. The GIS map display will illustrate that even though it is a different address, it is in the same proximity as the original call. Other benefits include

– Global positioning system (GPS) – Many public safety agencies are equipping response units with GPS devices. This provides the dispatcher (and perhaps other appropriate public safety managers) the ability to see locations of units through a GIS display and track them to the incident when dispatched. This is important during heavy call volume or for mobile vehicles such as police units and emergency medical units. This provides dispatchers a virtual or near

355

Consequence

Report Type: Population Effects
Hazard: RHAHPAC GB Terrorist - Small Canister EXPLOSION 2.0KG
   7:00 PM Threshold Probability Hazard Bands

Asset: Grid 2K By US County
   Grouped By: County

Summary: 77
Residential POP 2000   1
Number of Hazard Bands:

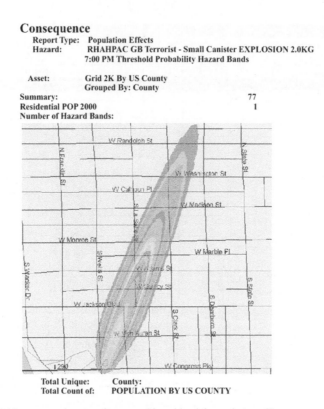

Total Unique: County:
Total Count of: POPULATION BY US COUNTY

Figure 4.   A quickly generated map and report with residential population effects.

real-time view of incident locations and emergency response units to activate an appropriate dispatch based on emergency unit availability.

– Routing – GIS can quickly analyze and display a route from a station or GPS location to the emergency call. This route (depending on the sophistication of the street file) may be the shortest path (distance) or the quickest path (depending on time of day and traffic patterns). This information can be displayed to the dispatcher and on a mobile computer screen in the response vehicle. Vehicles equipped with mobile computers and GIS can also benefit by providing first responders access to preplans, hazardous material locations, photographs, and other location-based documents linked to actual specific locations through the GIS map display.

– Move up and cover – During periods of high volume and simultaneous calls or a complex emergency, GIS can display areas of high risk that are left substantially uncovered. GIS can provide recommendations for re-allocating available resources for better response coverage.

– Emergency wireless calls – Wireless technology and cellular telephone technology has added to the necessity of GIS. Wireless phone-reported emergencies are not associated with an address, and the caller may not know his or her address or street location. GPS-enabled cell phones can provide latitude and longitude coordinates during an emergency call. Other technologies are available that triangulate the call location between cell towers or measure strength of signal to provide approximate location information. These coordinates are relatively meaningless to a dispatcher, but GIS can quickly consume and display a latitude/longitude or other coordinate location. This enables the dispatcher to see the incident location or general area and the closest or quickest response units on the GIS display.

– Records management system – Dispatch systems often have a database to capture, store, and archive incidents. This database may be called a records management system or a CAD records system. The records stored in this database concerning emergency calls usually contain location

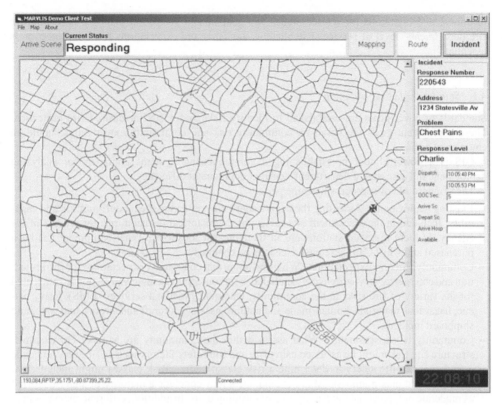

Figure 5.   Address and quickest route to the incident can be displayed on the mobile computer terminal within the emergency medical service (EMS) unit.

information. GIS can link to and geocode (place a point on the map) the location that represents this event. These events can be analyzed and assessed by any field in the records database (time, incident type, etc.). GIS provides a powerful capability to see, understand, and assess a department's volume of business, developing trends and response performance. The investment and information within a records management system can be leveraged for additional purposes when geospatially enabled.

4   PLANNING

The need for comprehensive planning and analysis has been understood for more than a century. Today, GIS technology provides the capability to analyze, dissect, and plan for fire protection problems quicker and with greater detail than previously possible.

Sir Eyre Massey Shaw, London Fire Brigade, authored "A Complete Manual of the Organization, Machinery, Discipline, and General Working of the Fire Brigade of London" (Layton 1876), which states the following:

"If you wish to control a problem, you must know more about the problem than anyone else and if you need to know more about the problem, you must coin a terminology, a lexicon, that allows you to understand it and not use imperial rhetoric."

In 1876, the London Fire Brigade was already developing fire preplans for buildings.

GIS can help define station locations; realign response districts; and identify and better under-stand hazardous materials locations, industrial facilities, commercial occupancies, water supply

357

locations, and high calls-for-service areas. The goal of fire protection planning is to improve fire departments' level of service. Establishing standards and expectations for fire protection is essential.

4.1 *Standards of cover planning*

Standards of cover goals are established to identify risks and needs for desired fire protection. These deployment goals are based on community risk and community expectations should an emergency occur.

GIS is a powerful tool in establishing a comprehensive systems approach for analyzing deployment and assessing a department's current deployment efficiency.

The standard of cover process has nine parts.

- Existing deployment – Assessing the department's current deployment configuration and capability. Included in the assessment is a review of the historical decision making process of the agency. Is there a reason stations are in their present locations? What equipment has been purchased and why? Can these past decisions be changed?
- Community outcome expectations – What are the current community expectations for fire protection and emergency service delivery? Included in this outcome is a review of response anticipated for the variety of fire risks in the community, emergency medical services (EMS), heavy rescue, hazardous materials, human-made and natural disasters, aircraft and airport, and water and shipboard incidents.
- Community risk assessment – What assets within the community are at risk? For example, structure fire risk might be assessed using fire and life safety factors such as fire flow and code compliance for life safety to determine a risk classification. Risk classes might include low, moderate, and maximum risk. Many communities may conclude a majority of areas identified as moderate or typical risks are composed of dwelling units. In addition, different responses may be provided to urban, suburban, rural, or remote areas as defined using census terminology.
- Distribution study – Where are locations of first-due resources?
- Concentration study – Where are the concentrations of incidents and the adequacy of the first alarm assignment or effective response force?
- Historical reliability – Is there a multiple-call frequency issue such as call stacking simultaneous calls within a specific area or areas?
- Historical response effectiveness studies – What percentage of compliance does the existing system deliver based on current performance goals?
- Prevention and mitigation – Are there tactics that can be initiated to strategically impact the outcome of events that do occur? The United Kingdom released "The National Plan" in 2004, which replaced earlier standards of coverage documents. The new report found that without prevention and mitigation, impacting the level of safety for responders and the public would reach a plateau. Using analysis of risk and looking at what strategic actions can be taken may not only prevent the incident from occurring but may also minimize the severity when and if the incident ever occurs.
- Overall evaluation – Propose standard of cover statements by risk type. For example, "in 90 percent of all incidents, the first-due unit shall arrive within four minutes' travel or six minutes' total reflex time. The first-due unit shall be capable of advancing the first line for fire control or starting rescue or providing basic life support for medical incidents.
 "In a moderate risk area, an initial effective response force shall arrive within eight minutes' travel or 10 minutes' total reflex time, 90 percent of the time, and be able to provide 1500 g.p.m. for firefighting, or be able to handle a five-patient emergency medical incident."

GIS can be used to determine if proposed coverage statements may be impacted by events occurring outside the control of the agency. In other words, what will be affected if roadways are closed due to trees or other blockages? Can the response still occur? Alternate plans may have to be developed for risks based on the data generated through GIS analysis.

GIS can also be used to evaluate if the standards of coverage statements are being met and, if not, reveal in what areas, during what times, and may reveal the conditions preventing meeting the level of coverage. Based on the analysis, different station locations may be necessary or alternative means for delivery of service may have to be deployed. A tactic being used in some areas is positioning "roving" units for EMS based on historic call volumes during specific days and times.

Simply creating a coverage statement is not enough; analyzing whether the agency is in fact meeting that coverage is critical to maintaining the credibility of the organization. If the agency states if can provide a level of service and then does not, the safety of the responder and/or public may be at risk.

GIS provides fire personnel a tool that integrates isolated data systems that can be effectively analyzed and displayed for a variety of planning and preparedness functions. GIS enables users to visualize and analyze all aspects of standards of cover.

An effective use of GIS for planning requires

– Data
  - Street data with address points or address ranges
  - Water lines and hydrants
  - Utilities information for electric systems and gas main locations
  - Development services information on buildings and zoning information
  - Hazardous material and target hazard locations
  - Fire department information including fire station locations, historical incident reports, demand zones, and inspection records.
– Software
  - Comprehensive GIS software with analysis capability including spatial analysis and network analysis
  - Geocoding tools and data editing capability.
– Hardware
  - Computer hardware including workstation with sufficient memory, storage, graphics, and processing capability to support comprehensive GIS software, printer, or plotter for map production.
– People
  - Person or persons proficient in GIS technology to manage the fire department GIS projects and conduct appropriate analysis.
– Methods
  - Often GIS technicians who assist the fire department will not be intimately familiar with the requirements and overall responsibilities of the fire department mission. It is important to identify exactly what is required and which GIS information analysis and products are required.

Examples of other functions GIS can perform for fire service planning and analysis functions include

– Display of jurisdictional boundaries including fire demand zones – GIS can display a theme (editable) with an underlying data table describing all the appropriate information for each demand zone and jurisdictional area.
– Layout of streets and local/state/federal highway network – GIS can display streets by type (streets are usually represented by a line). These streets are accompanied by a data table with address ranges. Events or incidents, hydrants, or other features can be added to the map display by entering an address. A point will be added along the street, which represents the address location. Response time analysis from fire station locations can be modeled along the street network to determine coverage for various response time requirements.
– Defining mutual and automatic aid zones – GIS can display a theme (editable) with an underlying table describing all the appropriate information for each mutual aid zone and/or automatic aid area.

- Definitions of geographic planning zones – GIS can display the various land use areas, planning zones, or other regulated use areas. These areas will contain tables with all the appropriate information.
- Locations of buildings and parcels – GIS can display all the parcels within a jurisdiction along with all the pertinent ownership records in an associated table for each property – property values, ownership, property tax, and so forth. Building footprints can be displayed for each parcel. The appropriate information for each building can be contained in an associated table. Information and images, such as blueprint drawings, building values, and owners, can be associated with the building footprint on the GIS display.
- Topographic features – Topographic features can be displayed in GIS. This can include slope, vegetation aspect, soils, rivers, earthquake faults, erosion zones, floodplain, and so forth.
- Demographics – The demographics can be displayed by geographic area (block groups, ZIP Code areas, etc.). This would include income levels, ethnicity, age groups, and so forth.
- Incident trend analysis – GIS can display emergency responses by placing a point on the address or geographic area where it occurred. The underlying table contains associated information about each incident. By clicking on the point, information about the incident type, date and time, response units, damage, victims, and so forth, can be accessed.
- GIS can display travel times along a road network – The user can identify a point (station location) and determine the shortest route to another location. GIS can also identify where a unit could travel within a specific time period from a station in any direction.
- Display of pipeline systems – GIS can identify where a water system or network of pipes for petroleum or chemicals reside within a geographic area. The pipeline can display valves, mains, shut-offs, supply points, and the like. The underlying table can contain information about pipe size, materials, directions of flow, and so forth.
- Location of built-in fire protection devices – GIS can display all buildings and facilities that contain fire protection systems and devices. The underlying table can contain the information concerning the protection system – contact person, number of devices, types of devices by location within the facility, and so forth.
- Locations of fully sprinklered buildings – GIS can display the buildings within a jurisdiction that contain sprinklers. Blueprints can be linked to the building footprint with a complete diagram of the sprinkler system. The user can easily access this information.
- Locations of standpipe-equipped buildings – GIS can display all buildings within a jurisdiction that contain standpipes. The user can display relevant information concerning standpipes including blueprints.
- Local fire alarm buildings – GIS can display all the alarm boxes or buildings with alarms within a jurisdiction. GIS can be linked to alarm systems and display the location of incoming alarm activation. Information about each alarm, exact location, and so forth, can be contained in the underlying table.
- Display of risk occupancy – GIS can display the locations of all types of occupancy including worst or maximum, key or special, typical or routine, and remote or isolated along with other important information associated with each occupancy.
- Display of "hard to serve" areas – GIS can display all the areas that are difficult to serve due to one-way roads, long travel times, multiple addresses within a single building, or other complications. The underlying table can contain information describing why these areas are difficult to serve or actions to reduce service time delays.
- Hazardous materials point locations – GIS can display locations where hazardous materials are present. Each location can be color coded by degree of danger, and underlying tables can contain information about each hazardous material, safety precautions, and health hazards.
- Hazardous materials transportation corridors – GIS can display (on top of the road systems/railroads, etc., or other topographic features) where hazardous material transportation corridors exist. Pipelines that transport hazardous or toxic materials can also be identified along with valves, direction of flow, and so forth. Underlying tables can contain specific information

concerning when transportation corridor risk is highest, common types of materials transported in each corridor, and so forth.

– Modeling – GIS can display a model (plume, explosion, flood, earthquake, epidemic, etc.). The model can be used with other GIS data to analyze infrastructure damage, road closure requirements, casualties, and other issues important for planning and response to potential or unfolding events.

– High EMS demand area – GIS can conduct an analysis of historic EMS calls by geographic area. Those geographic areas with high call volume can be identified and compared to other important information – demographics, land use, and so forth – to determine possible relationships and mitigation strategies.

– Assessed valuation (by category) – GIS can display assessed valuation classified by geographic areas, color coded by valuation. Underlying tables can contain information concerning the assessment, values, ownership, land use, and so forth.

– Preplanned structure locations – GIS can identify structures by icon, color code, and so forth, where preplans exist. By clicking on the structure, all the preplan information for the desired structure can be displayed. Floor plans, specific preplan actions, contacts, shut-offs, hazardous materials, and so forth, can all be accessed.

– Fire prevention assignments – GIS can identify all the areas of specific fire prevention programs, compliance inspections, fire prevention inspectors assigned by areas, and the like. Underlying tables can identify specific program tasks, status of current program implementation, and so forth.

– Arson/Unknown fire locations – GIS can display known or historical arson areas, areas with criteria that meets arson potential, and so forth. Underlying tables can contain information concerning arson history, owners that have had multiple arson events, common arson devices, and the like. GIS can also identify known arsonist address locations, methods of operation, and arson history.

– Targeted occupancies for public education (by category) – GIS can display by color coding or placing icons on properties with occupancy classifications that require particular fire prevention education programs. Underlying tables can identify what programs have been completed, ownership of properties, and so forth.

– EMS call demand by type – GIS can analyze and identify by area EMS call type and response time performance averages or response times for each call. Underlying tables can contain information about each EMS call, victim, date, time, and so forth.

– Evacuation zone planning – GIS can analyze evacuation routes from specific areas, ideal shelter locations, and other geographic information about evacuation routes, shelters, and maximum amount of traffic flow and shelter capacity.

– Damage assessment modeling – GIS is ideal for conducting and displaying damage assessment related to disasters, fires, or complex emergencies. After assessments are conducted, GIS can determine the total damage or loss by value, property type, or other desired category.

– Emergency inventory resource location – GIS can identify and display emergency supply locations by supply needs, distance, travel times, airport access, and the like. Underlying tables can contain information about each emergency resource type, costs, handling procedures, and so forth.

– Support for communication/dispatch function – GIS can identify where communication/dispatch backup locations exist, where mobile dispatch centers can be deployed with maximum communication coverage, and so forth.

– Display of external service agreement coverage area – GIS can display external service agreement area locations, classify them by type of agreement, and display other agencies that respond. Underlying tables can contain specific information about each service agreement area.

– Underground tanks – GIS can display underground tank locations, tanks with known seepage problems, and tanks that are abandoned or scheduled for removal.

– Critical care facilities – GIS can display key community facilities such as hospitals, schools, and blood banks. Underlying tables can contain specific information concerning hospital trauma capabilities, areas suitable for staging area implementation or incident command posts, and so forth.

GIS provides fire personnel a tool that integrates isolated data systems that can be effectively analyzed and displayed for a variety of planning and preparedness functions.

5 RESPONSE

5.1 *First responders*

Lloyd Layman authored *Fire Fighting Tactics* in 1953 (first published in 1940 under the title *Fundamentals of Fire Fighting Tactics*) and developed the concept of "size-up." Size-up encompasses facts, probabilities, possibilities, plans of action, and so forth, for an incident. Layman writes, "If you are going to rush into an emergency, you better have your information together."

Firefighters and rescue workers, especially those involved with incident management, are well aware that the first few minutes of a call determine its outcome. A typical "room and contents fire" (the beginning of most house fires) reaches flashover within 7 to 10 minutes of ignition, and occupants who have not already escaped are not likely to survive. Likewise, a vehicle accident victim will begin to suffer brain damage if deprived of oxygen for more than six minutes.

The negative effect of lapsed time cannot be overestimated. As fire and EMS department services expand, the importance of GIS is becoming widely recognized. First responders must get to the emergency, size-up the emergency, and deploy. The first responder's mission is to save lives and protect property and natural resources. Information that GIS can provide to support the first responder mission includes

- Incident location
- Quickest route
- Hydrant locations
- Preplans
- Photographs
- Floor plans
- Hazardous material locations
- Utility control points

Touch screen technology coupled with mobile computers allows first responders to access information quickly to reduce size-up time and results in quicker, safer deployment.

5.2 *Chief officers*

When the chief officer or incident commander arrives at the emergency, he or she is responsible to manage the event rather than become part of the tactical deployment. The chief officer requires additional, different information to perform the command mission. Depending on the complexity and size of the incident, the information and data requirements may include the following:

- What other exposures or other facilities are threatened by this incident?
- Where should the incoming units be positioned to access hydrants and effectively support the units already on scene?
- If an equipment staging area on incident command post is required, where are parking lots, schools, churches, malls, or other suitable facilities located?
- If helicopter evacuation of victims is required, where are suitable landing sites?
- If medical triage or decontamination is required, where can it be implemented?
- If hazardous materials are involved or a chemical plume is being generated, where is it going, what does it threaten, and what actions are required to protect and evacuate the public?
- If an explosion is possible, who needs to be evacuated and where is an immediate evacuation facility?

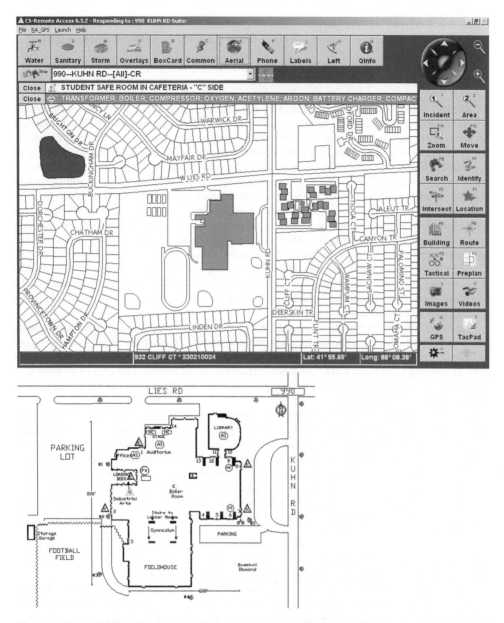

Figure 6. Tactical GIS applications provide first responders information and data for more efficient size-up.

All these decisions require maps, data, and a variety of information from different sources. Having access to GIS data, imagery, school locations, parking lots, adjacent exposures, and hydrant locations provides an accurate picture of the event and supports critical command decisions.

GIS can meet the challenge of providing first responders and incident commanders the right information, at the right time and place, that is easy to access and use. Much of the information first responders require has already been collected but resides in a variety of formats in and a variety of locations. GIS can integrate the information and provide it graphically to first responders through maps. GIS provides intelligent maps and allows the user to acquire other critical information. GIS can model explosions, plumes, and other potential emergencies. Instead of guessing or estimating evacuation requirements, transportation network problems, and other infrastructure threats, GIS will provide a more accurate prediction of the event and display the potential consequences.

6   MOBILE GIS

As wireless broadband networks continue to expand, GIS support for a variety of operations becomes possible. Mobile PCs, computer tablets, and handheld devices with GPS and wireless advances allow first responders to send and receive geographic information and incident updates. For example, the city of Los Angeles Fire Department deploys a mobile computer with GPS in its command helicopter. Upon arrival, the helicopter circles the incident (wildfires, floods, hazardous spills, explosions, etc.), and collects and records a GPS polygon of the primary and secondary damage perimeter on the mobile computer map in the helicopter. These GPS perimeters can be wirelessly transmitted to a server. The perimeter polygons are combined with other GIS data on the server including imagery, streets, and block group census data. This Web-based GIS application is accessible by authorized Los Angeles city personnel. The GIS application posts the perimeter data and automatically generates a population effects report. Personnel from various departments, such as Public Works, Parks and Recreation, and Emergency Operations, can observe where an incident is occurring; zoom in to the imagery to determine potential infrastructure damage and threats; understand how many people are affected, displaced, or injured; and locate possible evacuation requirements and shelter needs.

One of the most difficult tasks during the first hours of a large emergency is collecting accurate assessment information. Mobile devices now make it possible for emergency personnel to record and add damage information to their maps and send this data to computers at the incident command post, Emergency Operations Center, or other unified command centers. This creates an integrated view of the incident showing where the damage is most severe, affected critical infrastructure, and priorities for search and rescue.

In York County, South Carolina, the Fire Prevention Bureau created a database of routine flow and pressure tests conducted on fire hydrants. After soggy paperwork problems were experienced by field inspectors, handheld GPS-based units were programmed for point-and-click data collection.

The inspectors now perform more inspections each day because they no longer make frequent trips back to the office to hand-copy notes. They accurately locate each hydrant, which enables GIS to map them, something never previously done. In York County, the hydrant map data is now available to anyone by logging on to York County's Web-accessible GIS. Firefighters will be able to consult a hydrant map before they depart for a fire, see where the closest hydrants are, and know which hydrants have failed recent pressure tests.

E-911 dispatchers have direct access to the fire hydrant status layer of the GIS in their computer-aided dispatch systems. A dispatcher can warn responding emergency crews if a hydrant near an emergency scene is not in service and direct them to the closest alternate.

Local insurance agents now consult the county GIS via the Internet to measure the distance between structures and hydrants without having to call county staff. This information helps them calculate more equitable fire insurance rates for property owners and provides greater efficiency for York County.

6.1   *Public information*

Providing the public with emergency information is an important component of emergency management. Historically, maps have been used as one of the key products for answering questions such as Where is the emergency? What is the status of the emergency? Which roads are closed? Which business facilities are closed? and Where are the evacuation centers? The press (TV, radio, print) has to understand the story before they can communicate effectively to their customers. GIS can be invaluable, providing fire information officers' maps that inform the press and allow them to focus on important issues. GIS maps can be filmed by television news to provide viewers a better

Figure 7.   Incident perimeter mapping can be displayed using handheld mobile device and GIS.

understanding of where the dramatic fire footage is occurring. Maps can be printed in papers and other print media and posted on Web sites to provide accurate public information. Another benefit to anxious homeowners who have been displaced to an evacuation center is the ability to view damage maps to determine if their property survived the disaster.

7 EMERGENCY MANAGEMENT

During an emergency or crisis, maps play a critical role in responding to the event, search and rescue, mitigating further damage, and understanding the extent of the impacts. GIS is an appropriate platform to organize the extensive amount of spatial data both generated and utilized during an emergency event. A properly designed and implemented GIS will allow managers and responders to access critical location data in a timely manner so that lives and property can be protected and restored.

Fire departments are the front line of defense for emergencies of all types. Large-scale emergencies range from natural disasters (earthquakes, floods, hurricanes, ice storms, etc.) to industrial or technological emergencies (train car derailments, petroleum fires, hazardous material spills, etc.) to terrorist attacks. Large-scale emergencies often involve multiple casualties, critical infrastructure damage, and evacuations and can last for several days or months. Managing large-scale emergencies is complex. Complex emergencies include search and rescue operations; displaced citizens; loss of utility services; and coordination among many departments, agencies, levels of government, and the private sector. One of the most complex challenges of emergency management is determining where damage is most extensive, where lives are most threatened, and where to assign limited emergency response personnel and equipment. During a major search operation for a lost or overdue hiker, GIS can accurately determine which sectors have been searched adequately and which sectors need to be revisited.

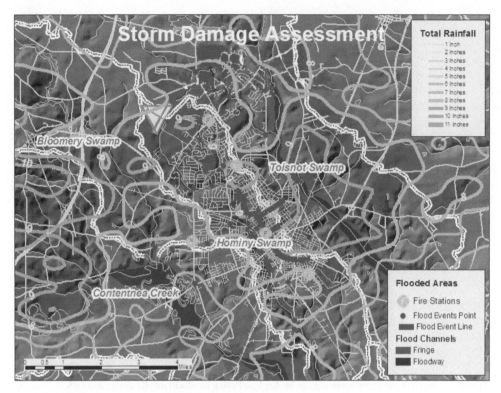

Figure 8.   Damage assessment displayed using GIS.

GIS provides a primary capability to organize, display, and analyze information for sound decision making. Using GIS data layers and imagery of the affected area(s), integrating damage and affected areas information, provides incident commanders a comprehensive view of the emergency. In effect, GIS is a primary capability for creating a "common operating picture" for the incident. Emergencies are very dynamic, and as circumstances change, GIS can reflect these changes. GIS can incorporate temporal information (weather, hazardous material locations, and emergency personnel locations) and model how an emergency might continue to evolve or what type of damage may be expected. These models could include

– Chemical plume dispersion
– Blast models
– Hurricane tracks and effects
– Flood damage
– Earthquake damage
– Wildfire spread

Models provide incident commanders a relative understanding of where evacuations may be necessary, potential damage to critical infrastructure, transportation network disruption, and so forth. Maps can be printed with the appropriate symbology to illustrate work assignments and incident facilities (incident command post helispots, staging areas, hot zones, etc.) and quickly dispersed to emergency personnel as part of the incident action plan. Areas with catastrophic damage are often confusing and disorienting to emergency personnel. GIS, with the appropriate data, provides a view of the area before the damage occurred, underground infrastructure, control points, potential hazardous material locations, and other information to support emergency response actions.

366

Table 1. Example of fire behavior vegetation models used by U.S. wildfire agencies.

Fuel model/Class	Model description/Typical complex
Grass and grass-dominated models	
1	Short grass (1 ft)
2	Timber (grass and understory)
3	Tall grass (2.5+ ft)
Chaparral and shrub fields	
4	High pocosin/chaparral (6+ ft)
5	Brush (2 ft)
6	Dormant brush, hardwood slash
7	Southern rough/low pocosin (2–6 ft)
Timber litter	
8	Closed timber litter
9	Hardwood litter
10	Heavy timber litter and understory
Slash	
11	Light logging slash
12	Medium logging slash
13	Heavy logging slash
Nonfuel	
14	Water
15	Bare/Nonflammable

8   WILDFIRE

Wildfire planning and analysis, suppression, fire prevention/education, and vegetation management techniques continue to evolve and change through information management technologies. GIS is one of the primary technologies influencing these changes.

8.1   *Analysis and planning*

8.1.1   *Analysis*
GIS allows fire personnel to better view and understand physical features and the relationships that influence fire behavior, the likelihood of a fire to occur, and the potential consequences of fire events.

8.1.1.1   Hazard identification
Hazard is the topography and vegetation on which a wildfire fire will burn. The identification and classification of hazard focus on understanding various topography types, vegetation types, and fire intensity expectations by individual landscapes.
   GIS data requirements for understanding hazard include

– Digital elevation models – Digital elevation models provide users the ability to extract information and data important for wildfire behavior such as
   • Slope – Steepness of slope
   • Aspect – Direction in which the slope faces (sun exposure and duration)
   • Contour lines – The ability to visualize terrain and elevation features.
– Vegetation
   • Vegetation polygons by species type.
   • Vegetation grouped or classified by fire behavior models or characteristics (potential fire spread and fire intensity characteristics groupings). An example of fire behavior vegetation models used by wildfire agencies in the United States is given in Table 1.

- For fire behavior modeling, additional datasets may be required such as
    (1) Canopy cover
    (2) Stand height
    (3) Duff loading
    (4) Coarse woody vegetation
    (5) Gridded fire weather indexes.

The result of hazard identification will be a GIS data layer that rates or classifies landscapes based on potential wildfire intensity under a given set of weather conditions. Landscapes can be assigned numerical values or given generalized descriptions such as

− Extreme
− High
− Medium
− Low

8.1.1.2   Risk identification

The identification and classification of risk focuses on understanding historical fire ignitions, land use, and natural ignited wildfires (lightning). The purpose of identifying risk is to classify landscapes based the likelihood of wildfire ignitions. Classification of the severity of risk is based on the amount and variety of risks within a geographic area. The types of GIS data include

− Historical incident occurrence by incident type and location
− Land use zoning
− Transportation corridors (roads, rail, off-road, etc.)
− Communities
− Recreation areas
− Industrial use

The result of the risk identification will be a GIS layer with risk polygons rated based on the density or accumulation of various risks within a concentrated area. Risk can be classified by numerical value or generalized descriptions such as

− Extreme
− High
− Moderate
− Low

8.1.1.3   Value identification

Value identification focuses on identifying values and their tolerance (or intolerance) to wildfire. The purpose of value identification is to understand what type of fire protection actions will be necessary to protect or preserve them. The types of GIS data required to identify values can include

− Housing developments
− Recreation developments
− Sensitive or protected plant species
− Sensitive or protected animal habitat
− Commercial timber landscapes
− Cultural resources
− Other natural resource values

The result of the value identification will be a GIS layer with all the important values identified and displayed.

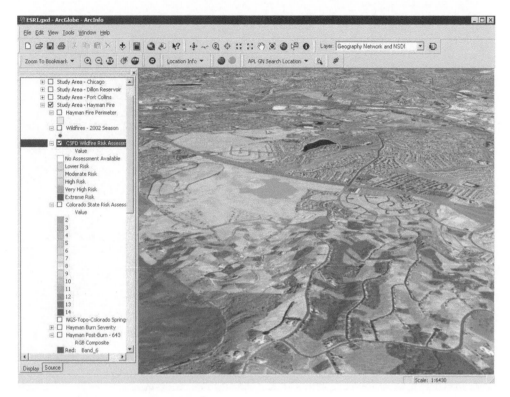

Figure 9.   GIS illustrates fire hazard areas.

8.1.1.4   Aggregation

The last step in the wildfire analysis is the aggregation of hazard, risk, and values. The result of aggregation will be a GIS layer of fire management landscapes with a fire behavior hazard rating (potential fire intensity), a risk rating (potential for fire ignition), and an inventory of values at risk. The aggregated fire management landscapes provide the foundation for a comprehensive wildfire protection plan.

8.1.2   *Planning*

Upon completion of the wildfire analysis, a comprehensive plan can be developed for each fire management landscape. The plan will identify priority fire prevention, vegetation management, wildfire detection, and wildfire suppression program needs based on the combination of hazard, risk, and values within each landscape. Wildfire programs are expensive, and it is critical they be implemented effectively. The appropriate mix of fire prevention, vegetation management, and suppression response capability can be prescribed based on the values at risk and the potential for wildfire ignition and intensity.

8.2   *GIS for fire prevention*

Wildfire prevention efforts can be focused where wildfires pose the greatest risk of resource loss. When intense fire areas (highly flammable landscapes) exist near high-risk areas (ignition sources) and high values, fire prevention becomes critical. Historical fire information can be viewed with all the other landscape information. Fire prevention officers can begin to determine an appropriate program strategy. This strategy may be one of fire prevention education, fire prevention enforcement, or fire prevention engineering, depending on the type of land use and historical fire causes. As housing development continues to expand and encroach into wooded and brush-covered

369

Figure 10.   GIS models potential fire protection strategies for the wildland urban interface.

areas, it becomes fire prone. The "urban interface" requires extensive fire prevention and fire protection measures. GIS can model and display potential fire prevention/protection strategies.

8.3   *GIS for vegetation management*

Landscapes that require vegetation management treatments (prescribed fire, mechanical treatments, etc.) will stand out when using GIS. Landscapes with high flammability characteristics (high hazard, high risk, and high value) become obvious candidates for vegetation or fuel treatment programs. GIS technology allows fire managers to identify prescribed fire and vegetation management projects with the highest benefit (meeting multiple goals for resource and fire management). Vegetation management tactics can include mechanical, chemical, or prescribed burning techniques. GIS can assist in modeling how a fire will behave and spread under a variety of conditions to assist in developing fire prescriptions.

8.4   *Fire suppression*

Maps are the foundation for fighting wildfires. They are used for communicating operational assignments and potential spread scenarios, providing a visual reference for incident team strategy discussion, and conveying incident assignments to line personnel. Maps answer questions such as What is the topography in and near the fire? What is the jurisdiction(s) of the fire and where is it likely to spread? Where can firefighters safely be deployed based on the topography, vegetation, and fire spread? Maps help managers deploy resources safely and assess the potential overall damage of the fire. It is difficult to obtain detailed and comprehensive information from a paper map. Information is often gathered from maps, documents, and technical personnel over several days. Using GIS, this information is immediately available and can be easily viewed and understood under extremely stressful conditions. It is now common to see a GIS team assigned within the planning section of an incident management team. GIS teams provide maps showing transportation routes, facilities, air hazards, spread prediction, operations, and other geographically related incident management information.

Incident commanders want accurate information and intelligence before making tactical decisions. Historically, information has often been gathered through the use of paper maps, on-scene resource advisors, and observations from the air and the ground. Maps are the foundation for fighting wildfires, particularly large fires. Maps are used for communicating operational assignments and potential spread scenarios and for providing a visual reference for incident team strategy discussion. Relevant information is often gathered from a variety of maps, documents, and technical personnel over several days. In GIS, this information can be immediately available, easy to access, view, and understand under extremely stressful conditions. Problems can be modeled and analyzed as required.

Laptop computers now have the storage, performance, screen brightness, and strength to be used in the field. It is possible to have a great deal of information in a laptop computer. It is now possible to have hundreds of maps, images, and detailed information in the fire manager's vehicle and at the incident command post in the initial stages of a wildfire. First response helicopters can capture the perimeter using GPS and wirelessly transmit it to operations personnel on the ground. Command decisions, such as placement of firefighters, suppression protection priorities, and potential evacuations, become more effective with accurate information. Incident Action Plan maps can be printed from the vehicle on a small printer or can be remotely transmitted to the dispatch center, multiple agency command centers, and national headquarters. Decisions concerning how to manage the wildfire (tactics and strategy), resources at risk, firefighter safety, resource protection priorities, and so forth, can be determined quickly and efficiently with the assistance of GIS tools.

The United States Department of Agriculture Forest Service commissioned a study to evaluate the use of GIS for large wildfires. This published document, "Study of Potential Benefits of Geographic Information Systems for Large Fire Incident Management" (Fox & Hardwick 1999), includes Project Findings as follows:

"The participants in this study (members of incident command teams based in California including individuals employed by federal and local government agencies) overwhelmingly agreed that GIS would be a useful tool for large fire management. It is a complement to existing tools, providing information that is not available now, and allowing certain information to be gathered in a more timely or cost effective manner than it can currently be collected. They believed GIS puts accurate information in the hands of those who need it, when and where they need it. Better information leads to better decision making, which in turn leads to fighting a fire more effectively, efficiently and safely. It will also facilitate the public information portion of fire management, freeing valuable resources for other efforts."

The publication identified the following tangible benefits of GIS use for wildfires:

– Saving money by not performing unnecessary tasks ($2 million during the Palm Fire, 1997)
– Improving safety by accurately identifying hazardous areas
– Improving crew confidence in management decisions by confirming human observations with hard data
– Saving time by reducing the extent of physical reconnaissance efforts.

8.5 Education and training

GIS is beneficial for education and training. Wildfire management requires several years of experience and training to become proficient. GIS provides access to detailed landscape information during wildfire events. Fire personnel begin to understand the complexities, fire effects, and fire behavior characteristics of various wildfires much sooner when using GIS. Modeling provides a better understanding of what a fire might do and what elements influence the wildfire most. Wildfire knowledge has traditionally been gained through years of experience, formal training, and

371

discussions with experienced fire personnel. GIS is now another resource fire personnel can use to expand their understanding of the variables and complexities that affect wildfires. As modeling becomes more precise, dispatchers will be able to determine the potential of new starts and possible dispatch requirements.

GIS is a vital tool for wildfire information management. GIS is a primary repository of information that can be quickly accessed and viewed when required. GIS is becoming more suitable for emergency field operation use and is integrating tools that allow real-time display of information. Rapid access to information, safety, efficiency, and better resource management decisions are being made with the use of GIS for wildfire management. Information is critical for wildland fire management. GIS is information, all in one place, easy to visualize and understand.

9   SUMMARY

The Fire Service Mission is to protect life and property from fire and other natural or manmade emergencies through planning and preparedness, incident response, public education, and code enforcement. In order to accomplish this mission, GIS is rapidly becoming an essential tool to analyze, define, clarify, and visualize community fire problems in the development and execution of fire protection policy. GIS can model a community or landscape; analyze and display features important to the fire service mission; and provide access to important documents, photographs, drawings, data tables, and so forth, associated with features on the GIS map display. GIS can analyze and measure response time capabilities; identify incident hot spots by time and day of week; and target hazards, hydrants, and other information important for deployment analysis. First responders can have immediate access to critical information for emergency incident locations, best route, and detailed information concerning the building or facility to which they are responding. Incident commanders can maintain better scene control with detailed maps and imagery of the emergency location as well as the exposures and features around the incident. GIS is essential for the management of large-scale emergencies or disasters where large numbers of public safety resources are deployed, with various resource assignments during a dynamic incident. Resource status, event prediction, incident facility identification, public information dissemination, and incident status are all more effectively and efficiently performed using GIS.

REFERENCES

Fox, B. & Hardwick, P. 1999. Study of Potential Benefits of Geographic Information Systems for Large Fire Incident Management. Salt Lake City: Pacific Meridian Resources.
Layman, L. 1940. Fundamentals of Fire Fighting Tactics (published as Fire Fighting Tactics in 1953). Quincy: National Fire Protection Association.
Shaw, E.M. 1876. A Complete Manual of the Organization, Machinery, Discipline, and General Working of the Fire Brigade of London. London: Charles and Edwin Layton.

Author index

Author index

Subject Index

377

ISPRS Book Series

1. Advances in Spatial Analysis and Decision Making (2004)
   Edited by Z. Li, Q. Zhou & W. Kainz
   ISBN: 978-90-5809-652-4 (HB)

2. Post-Launch Calibration of Satellite Sensors (2004)
   Stanley A. Morain & Amelia M. Budge
   ISBN: 978-90-5809-693-7 (HB)

3. Next Generation Geospatial Information: From Digital Image Analysis to Spatiotemporal
   Databases (2005)
   Peggy Agouris & Arie Croituru
   ISBN: 978-0-415-38049-2 (HB)

4. Advances in Mobile Mapping Technology (2007)
   Edited by C. Vincent Tao & Jonathan Li
   ISBN: 978-0-415-42723-4 (HB)
   ISBN: 978-0-203-96187-2 (E-book)

5. Advances in Spatio-Temporal Analysis (2007)
   Edited by Xinming Tang, Yaolin Liu, Jixian Zhang & Wolfgang Kainz
   ISBN: 978-0-415-40630-7 (HB)
   ISBN: 978-0-203-93755-6 (E-book)

6. Geospatial Information Technology for Emergency Response (2008)
   Edited by Sisi Zlatanova & Jonathan Li
   ISBN: 978-0-415-42247-5 (HB)
   ISBN: 978-0-203-92881-3 (E-book)

T - #0204 - 071024 - C0 - 246/174/22 - PB - 9780367387792 - Gloss Lamination